Point-of-Care Biosensors for Infectious Diseases

Point-of-Care Biosensors for Infectious Diseases

Edited by Sushma Dave and Jayashankar Das

WILEY-VCH

Editors

Prof. Dr. Sushma Dave
Jodhpur Institute of Engineering and Technology
Department of Applied Sciences
Jodhpur
India

Dr. Jayashankar Das
Director
Valnizen Healthcare
301 Prabhu Vijay Building
Vile Parle(W)
Mumbai
400056
India

Cover Images: © Elpisterra/Shutterstock;
© ImageFlow/Shutterstock

Library of Congress Card No.: applied for

British Library Cataloguing-in-Publication Data
A catalogue record for this book is available from the British Library.

Bibliographic information published by the Deutsche Nationalbibliothek
The Deutsche Nationalbibliothek lists this publication in the Deutsche Nationalbibliografie; detailed bibliographic data are available on the Internet at <http://dnb.d-nb.de>.

Print ISBN: 978-3-527-35045-2
ePDF ISBN: 978-3-527-83795-3
ePub ISBN: 978-3-527-83796-0
oBook ISBN: 978-3-527-83794-6

Typesetting Straive, Chennai, India
Printing and Binding CPI Group (UK) Ltd, Croydon, CR0 4YY

C120262_140623

Contents

Preface

The book "Point-of-Care Biosensors for Infectious Diseases" is full of information that is condensed into a succinct account of the key achievements to date in the area of biosensor-based diagnostics for infectious diseases. It includes the development done so far in the field and details the exciting emerging technologies. It also describes the importance of biosensors as a ubiquitous technology of the future for health and the maintenance of health. It is expected that the combined strategy of applying new approaches based on the knowledge gained by the different researchers from different areas of biosensor application will help in the development of critical diagnosis of infectious diseases, the development of new point-of-care (POC) techniques, and the rapid detection of diseases using more stable and new material application. Each year, infectious diseases are responsible for millions of deaths. Many of the infectious disease diagnostic tools used today require a great deal of time, a laboratory setting, and trained personnel. Due to this, the need for effective POC diagnostic tools is significantly increasing, emphasizing affordability, portability, sensitivity, specificity, timeliness, and ease of use. The book includes recent advancements in the novel POC technologies focusing on microfluidic and plasmonic-based approaches. Deadly infectious diseases are triggered by pathogenic microorganisms such as viruses, fungi, bacteria, and parasites.

Chapter 1 is on fundamentals, which describes various kinds of biosensor detection methods, are discussed in this chapter with a significant focus on optical, electrochemical, and mass-based techniques. Chapter 2, Nuts and Bolts of Modern Biosensing Technology: Smart Health Diagnostic Devices, introduces working elements and relevant principles when establishing innovative biosensors. In addition to traditional POC instrumentations, this chapter also draws attention to technological strategies for miniaturizing devices. The state-of-the-art integrated sensing platforms, particularly wearable bioelectronics, are discussed. It also includes a discussion of representative examples in their respective applications and their future directions to evolve conventional biosensing tools in biomedical analysis. Chapter 3 summarizes the use of biosensors and provides comprehensive insight into the detection of disease-associated biomarkers and the latest trends in electrochemical biosensors for disease-related detection. A detailed description of materials used for fabrication and the working principle of electrochemical biosensors is also given. Chapter 4 deals with POC-biosensing systems to detect, diagnose, and monitor cardiovascular disease, diabetes, cancer, and infectious diseases at their initial stages through their evolution as nanomaterial integration systems in point-of-care testing (POCT). Chapter 5 is based on organic- and inorganic-based nanomaterials for health care diagnostics and concentrates on the various hybrid nanomaterials and their applications. Chapter 6 summarizes the applications of CRISPR/Cas-based biosensors in the field of diagnostics. In Chapter 7, the authors have discussed the mechanism of biosensor as well as how piezoelectric materials are suitable for the construction of a piezoelectric biosensor. In addition to these, they have also illustrated the type of piezoelectric biosensor and their application for the early detection of infectious diseases. In Chapter 8, the authors have initially discussed different types of biosensors and then given special attention to gold, silver, platinum, copper, zinc oxides, and other metal oxide-based biosensors, followed by different biosensing techniques such as electrochemical, colorimetric, and fluorometric to detect infectious diseases. In Chapter 9, how biosensors for POC applications are developed and how they are replacing pathology labs by bedside devices are discussed. In Chapter 10, strategic synthesis of diagnostic novel materials and their role against infectious diseases are discussed. Similarly, Chapter 11 discusses the development of a diagnostic kit for POC biosensors, the fundamentals, and their applications in detail. The last chapter gives an overview of lab-on-a-chip devices for POC infectious disease diagnostics.

The book can be a referral book for a large number of researchers, faculties, and students in the areas of molecular diagnosis, infectious diseases, biosensors, and related fields.

Dr. Sushma Dave
Jodhpur
Rajasthan, India

Dr. Jayashankar Das
Bhubaneswar
Odisha, India

25 May 2023

1

Biosensors for Infectious Diseases-Fundamentals

Maheswata Moharana[1], Subrat K. Pattanayak[1], Fahmida Khan[1], and Sushma Dave[2]

[1]National Institute of Technology, Department of Chemistry, Raipur 492010, India
[2]Jodhpur Institute of Engineering and Technology, Department of Applied Sciences, Jodhpur 342802, India

1.1 Introduction

The rapid growth in the urbanization processes and its association with inadequate city planning, poor management of sanitary conditions as well as water supplies, high population density, and interference in previously unaffected ecosystem leads to the spread of infectious diseases [1]. The term "infectious diseases" refers to medical conditions caused by a variety of pathogenic microorganisms, which include bacteria, viruses, fungi, and parasites. The diseases can spread from one organism to another through direct or indirect contact that results in a number of ailments, some of which are fatal [2]. Infectious disease outbreaks continue to put a heavy burden on the world's population despite the surge in medical advancements. They consistently pose challenges to international healthcare systems, raising ongoing concern about the rising frequency of epidemics throughout the world. It was reported, in 2016, the infectious diseases were responsible for one-fifth of all deaths that were officially recorded worldwide. In addition, socioeconomic and environmental issues, such as climate change, migration, and population increase, will probably make this scenario worse, particularly in overpopulated places. The need for early detection systems has grown as the possibility of more frequent epidemics and disease outbreaks has increased. An illustration of the necessity for early detection techniques to track diseases outbreaks is the continuing COVID-19 pandemic brought on by the rapid transmission of the novel coronavirus [3]. The success of measures for disease zoning, control, or eradication is greatly influenced by the rapid detection of a virus or antigen due to the threat posed by infectious diseases. All public-health programs must include both disease surveillance and diagnosis as essential elements. Prior to a virus outbreak having disastrous effects on the economy, people, and the environment, it is crucial to stop it from spreading or minimizing its speed [4]. Infectious diseases are

Point-of-Care Biosensors for Infectious Diseases, First Edition.
Edited by Sushma Dave and Jayashankar Das.

categorized as (i) severe respiratory syndrome, which falls into the category of novel and previously unknown diseases, (ii) Foot and mouth diseases, on the other hand, fall into the category of recognized diseases that have risen in incidence, virulence, or in certain geographic range, and (iii) diseases such as avian influenza that are expected to become more prevalent in recent future [4]. In some cases, the diseases can spread through in a community in just a few hours, depending on the types of infectious diseases and the surrounding weather. The ongoing COVID-19 pandemic serves as a stark reminder: within two years of its emergence, more than 4.6 million lives have been lost, and the cost to the world economy is approaching US$7 trillion [5]. Finding the pathogens that cause infectious diseases is the first step in controlling them. By using an agar plate to grow bacteria on Petri's invention, Dr. Koch altered how we view diseases (i.e. the Petri dish). Laboratory culture-based detection of infectious agents has evolved into the "gold standard" in clinical microbiology. When combined with his novel microscopy, the technique enabled to link pathogens as the source of diseases commonly known as Koch's postulates [6]. Another significant development was introduced known as polymerase chain reaction (PCR), which includes the increase of detection limits of the infectious agents, even those that were slow-growing or uncultivable [7]. The diagnostic procedures are restricted only to the centralized medical laboratories due to the need for supporting infrastructure, such as highly skilled employees and capital equipment to perform PCR and culture assays. We need to make next technological leap to fast, economical, yet highly accurate diagnostic tests to better deal with infections that is emerging rapidly. The biosensor is one alluring device that can offer quick information on a disease outbreak [8]. It is commonly known that biosensors play a key role in environmental monitoring [9, 10], agriculture [11, 12], food and water analysis [13, 14], and medicine/clinical analysis [15, 16].

1.2 Biosensors Fundamental Aspects

Biosensors have a variety of definitions in the literature. However, according to 1999 IUPAC standards, they can be defined as a self-contained integrated receptor–transducer system that may provide specific quantitative or semiqualitative analytical information utilizing a biological recognition element [17]. Ideally, the biosensor should be a reagentless device that is typically employed in the detection process with the noble purpose of providing quick, accurate, and reliable information about the biochemical composition of its environment. It should also be able to respond continuously, reversibly, and without disrupting the sample. Different types of biosensors are available [18–21]. However, all of them essentially consist of a biological recognition component, or bioreceptor which interacts with the analytes to be detected and generates signals by the means of signal processing unit or transducer. A schematic representation of the typical components of biosensor is shown in Figure 1.1. An enzyme, antibody, nucleic acid (NA), cell/tissue, and hormones can be employed as bio-component. Its function is to selectively interact with the target analytes, and the outcome of the biochemical process is then turned into quantifiable signal through the transducer [22]. There are several types of transducing systems,

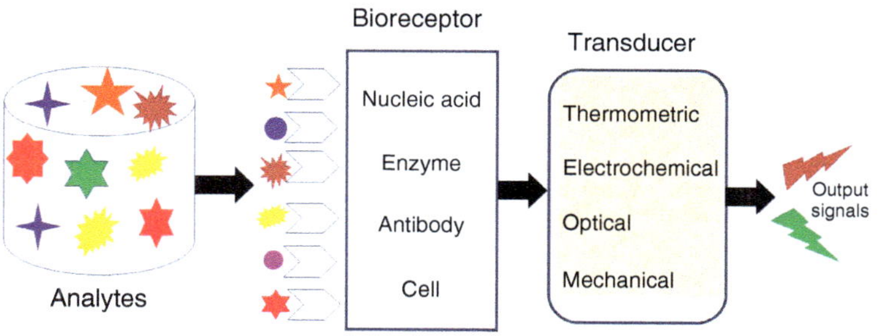

Figure 1.1 The schematic of biosensor concept representation.

including electrochemical, optical, piezoelectric, thermometric, and magnetic [23]. A "biosensor" is a device used to measure for analyte detection that combines a biological and physicochemical detector linked to a component. The design as well as function of biosensor determines the analyte detection. A noninvasive smartphone-based analyte biosensor can be tested on smartphones and other widely used devices. This enables rapid and cost-effective preliminary detection possible.

1.3 Classifications of Biosensor

In the late 1960s, Clarke and Lyons developed biosensors [24]. There are several perspectives to categorize biosensors, but the most frequently used two are biorecognition component and the signal transduction component. Based on the above two categories, biosensors classification is summarized in Table 1.1.

1.3.1 Biorecognition Perspective

Biosensors are categorized as nucleic acid, protein receptor-based immunosensors, enzymatic biosensors, and whole-cell biosensors based on the biological recognition component. The details of principles as well as examples are discussed in Sections 1.3.1.1–1.3.1.5.

1.3.1.1 Nucleic Acid Biosensors

A nucleic acid-based biosensor uses a complex DNA or RNA structure or an oligonucleotide with a known base sequence as detecting element. Nucleic acid biosensors can be used to find biological or chemical species, as well as DNA/RNA fragments. The analytes in the first application are DNA/RNA, and the hybridization reaction is used to detect it (a type of genosensor). In the second case, DNA/RNA acts as a receptor for particular biological/chemical species, such as drugs, contaminants, or target proteins [25]. A NA (natural and biomimetic forms of oligo- and polynucleotides) is integrated into nucleic acid (NA)-based biosensors as the biological recognition component. In DNA hybridization sensors, synthesized oligodeoxyribonucleotides are typically utilized as probes. The oligodeoxyribonucleotides are

Table 1.1 Classification of biosensors based on biorecognition elements and transduction perspective.

Biosensors category	Types	Examples
Bio-recognizing elements	Nucleic acid	Hybridization mechanism DNA-aptamers based
	Enzymes	Alcohol oxidase (Ethanol) Glucose oxidase (β-D-glucose)
	Protein receptors	Olfactory receptors Odorant-binding proteins
	Whole cells	β-galactosidase Green fluorescent protein
	Antibodies	Recombinant Polyclonal Monoclonal
Transduction through signals	Electrochemical	Potentiometric Impedimetric Voltametric Amperometric
	Optical	Surface plasmon resonance Absorbance-based Luminescence-based Reflectance-based
	Thermometric (Calorimetric)	Cholesterol oxidase Enzyme-glucose oxidase Enzyme-linked immune assay (ELISA)/thermometric ELISA (TELISA)
	Mass-sensitive	Electrochemical quartz crystal microbalance Piezoelectricity Chemical sensors
	Electrical	Dielectrophoresis Impedance based

immobilized to transducer surfaces using end-labels such as thiols, disulfides, amines, or biotin [26]. Especially, in the areas of clinical, environmental, and food analyses, DNA sensors have considerable potential for facilitating the accessibility of sequence-specific information [27]. The PCR and other amplification techniques, the effectiveness of the hybridization of the sequences, and the amount of background signal all play a role in determining the measurement sensitivity. Ionic strength, reaction temperature, and DNA computation circuit are some of the variables that affect the setting of specificity [28]. Biosensors based on DNA-aptamers have the ability to bind particular bacteria, viruses, proteins, and even small molecules and ions with exceptional specificity and affinity. As alternatives to antibodies, DNA-aptamers-based biosensors have been developed due to their low cost and great specificity [29].

1.3.1.2 Protein–Receptor Biosensor

Protein–receptor-based biosensors or non-catalytic proteins anticipate the protein's cell membranes to act as receptors are essential for biosensors. Multiple proteins work together and organize the sensing mechanism in the mammalian olfactory system. The olfactory receptor is a G-protein-coupled receptor (large protein family of receptors), and when a ligand molecule binds to the G-protein-coupled receptor, the second-messenger cascade of olfactory transduction is launched, which ultimately results in a cation influx through the ion channel associated with the system. Numerous studies have attempted to utilize this sensing capability of the membrane receptors for the development of biosensors since these receptors act as ligand-sensing elements [30]. There are 12 G-protein-coupled receptors that can sense and signal serotonin in humans [31].

1.3.1.3 Enzymatic Biosensor

The components of an enzymatic biosensor are an enzyme that detects and then reacts with the target analyte to produce a chemical signal, a transducer that converts the chemical signal into a physical signal, and an electronic amplifier that first prepares the signal before amplifying it [32]. Lactate, glucose, glutamate, and glutamine are few examples of the analytes that are essential to the metabolism of living beings. Glutamine and glucose help cells grow and function; lactate, which cells produce and use to measure how well their metabolism is working; and glutamate, an amino acid that is utilized by cells. For the detection of each of these analytes, a specific set of enzymes is required [33]. Interference is particularly difficult in biological samples since cells, proteins, small molecule metabolites and macromolecules, and electrochemical interferences are frequently present in the sample matrix. Hence, chemicals in the sample matrix have the potential to interfere with amperometric enzyme-based biosensors. According to the electron transfer mechanism used to measure the biochemical reaction or the degree of separation of the biosensor components (transducer, enzyme, mediators, and cofactors), amperometric enzyme biosensors are often categorized into three types. The existence of an enzyme is necessary in every step; thus, sensor performance depends on various factors, including working pH and temperature [34].

1.3.1.4 Whole-Cells Biosensors

A whole-cell biosensor is a kind of sensor that can find and recognize an element inside a cell or tissue. It is made up of several physical or chemical transducers and synthetic biomolecule recognition elements. Based on the differences in their molecular, cellular, and tissue sensing components, these biosensors can be divided into three groups. The reporting elements in the molecular-based biosensors are biologically active molecules such as enzymes, DNA, antigens, antibodies, and biofilms [35, 36]. The basic principle of a whole-cell biosensor is that it detects signals from its surroundings, such as small metabolites, chemicals, ions, temperature changes, or light, and then uses those signals to activate internal processing circuits [37, 38]. The applications of whole-cell biosensors include pharmacology, cell biology, environmental assessments, and toxicity. Drug delivery is one of the crucial applications of whole-cell biosensors [39].

1.3.1.5 Antibody-Based Biosensor

Antibodies can be utilized as bioreceptors in biosensors because of their selectivity. In vivo biosensor development has been aided by the incorporation of antibodies into biosensors. The antibody bioreceptors are traditionally mounted on the transducer surface in antibody biosensors. The analyte-containing solution is subsequently exposed to this [40]. All the molecules of antibodies follow the same structural principle, which is based on paired heavy and light polypeptide chains, and enable their integration into the immunoglobulin's common chemical class. Most of the time, immunoglobulin G (IgG) that predominates in serum uses biosensors [41]. The rapid and accurate identification of a variety of infections and related toxins is made possible by antibody-based sensors [42].

1.4 Transduction Through Signals

1.4.1 Electrochemical Biosensors

Electrochemical biosensors are one among the different types of biosensors, which have been utilized for various industrial applications since years [43]. According to the method of transduction, the electrochemical biosensors can be classified as amperometric, potentiometric, and impedimetric/voltametric. This type of biosensors analyzes interactions between the analyte and biorecognition element on the electrode surface to detect the changes in charge distribution and dielectric properties. Biological molecules, nucleic acids, proteins, disease biomarkers, and many more have been analyzed with the help of electrochemical biosensors [44].

1.4.2 Optical

Optical biosensors are one among the currently available various biosensing systems, which offer easy, portable, efficient, real-time, and cost-effective diagnostic tools with the advantages of sensitivity and specificity. Various innovative concepts like microelectronics, nanotechnologies, molecular biology, microelectrochemical systems, and biotechnology with chemistry are utilized to operate optical biosensors. It is highly essential for a simple, portable, and handheld optical biosensing instrument for the fast and accurate detection of harmful pathogens. Currently, the incorporation of intelligent nanomaterials in the form of gadgets offers significantly more sensitive and highly advanced sensors which may generate rapid results and help doctors and clinicians. Since years, optical biosensors have been developed for several applications. Over the past 10 years, a wide range of optical biosensing platforms, including surface plasmon resonance [45], interferometers [46], photonic crystals [47, 48], fiber-optics [49], and ring resonators [50], have studied for sensitive and label-free detection. The advantages of optical sensors are their sensitivity to electromagnetic interference, ability for remote sensing, capacity for minimization assays, inherent safety, and capability for multiplexed recognition within a single device [51]. One of the most important limitations of widely used optical sensing systems is the penetration depth of the evanescent field, which is frequently less essential than the average size of the optical field [52].

1.4.3 Thermometric (Calorimetric)

Biosensor systems that are capable of adapting new goals are in high demand nowadays. The enzymes used in enzyme-linked immunosorbent assays (ELISAs) produce heat as a result of an enzyme-catalyzed reaction, making it simple to customize and modify a calorimeter to detect enzymes as indicators of antigen [53]. Hydrogen peroxide is widely used as a substrate in ELISA tests given that it contains a variety of reaction enzymes and high reaction enthalpy (98 $kJ\,mol^{-1}$) (like catalase or horseradish peroxidase) [54]. Chemical reactions catalyzed by enzymes generate heat. ELISAs with optical-based detection have been developed for point-of-care application, although they lack quantitative data or require materials with specified optical properties [55]. An ELISA with a calorimetric readout of the heat produced by the enzyme reaction was initially developed by Mattiasson et al. and was called thermometry enzyme-linked immunosorbent assay (TELISA) [56]. The first TELISA calorimetric biosensor monitored with flow-through columns needs enormous sample amounts larger than finger prick, ambient temperature, and immobilized enzymes. Clinically relevant levels of phenylalanine and herceptin have been found in serum is one successful development of exceptionally sensitive nanocalorimeter TELISA systems [57].

1.4.4 Mass-Sensitive

A wide range of biosensors can be developed using mass-sensitive devices and the imprinting approach. They work as optical sensors for detecting numerous physiological activities [58]. The piezoelectric effect, which was discovered by Pierre and Jacques Curie in 1880, serves as the foundation for all mass-sensitive devices [59]. The function of a piezoelectric platform, also known as a piezoelectric crystal, or sensor component, is based on the theory that oscillations change when a mass is bonded to the surface of the piezoelectric crystal [60]. The sensing technique adopted everywhere is the change of mass and subsequent change in resonance frequency of the oscillating quartz plate. This is due to the analyte's morphological, optical, and functional characteristics not interfering with the resonating transducer's detection principle [61]. From the analytical chemistry standpoint, piezoelectricity is particularly suited for the development of physical sensors and biosensors. Piezoelectricity is particularly suited for the development of physical sensors and biosensors from an analytical chemistry standpoint. The principle of these assays can be explained by providing a simplified description. For example, two electrodes apply alternating voltage to the surface of the biosensor or sensor to excite it. When a crystal is placed in an oscillating circuit, alternating voltage causes it to oscillate mechanically and the frequency of the oscillations can be measured [62]. Analytes or other masses attached to the surface of crystal, or more precisely, the surface of electrodes poisoned on the crystal, create an oscillating frequency shift [63].

1.4.5 Electrical

Conventional methods for identifying medical complications are time consuming, cost effective, and required skilled personnel for analysis. Electrical biosensors are regarded as a clear choice in diagnostic applications due to their portability for screening, cost, usability, and online monitoring. In addition, electrical biosensors are utilized to detect targets in different matrices in real time, be selective, and without preparing samples. Over time, electrical biosensors have been developed employing various transducer technologies, such as field-effect transistors, interdigitated electrodes, and microelectrodes [64]. Due to advancements in the conversion of molecular analytical signals into electrical signals, significant efforts have been made to develop and enhance the sensitivity of electrical biosensors to detect dengue virus DNA. This type of electrical biosensors has been designed with the transducers of nanoscale structures, such as nanotubes, nanoparticles, and nanowires as the dimension is comparable to the feature sizes of chemical and biological species to be detected. Silicon nanowires are highly demonstrated due to their unique mechanical, optical, and electrical characteristics as well as high biocompatibility and high surface-to-volume ratio. These are also shown to have excellent electrical detecting capacities with good electron or hole transit in the detection. The synthesis of silicon nanowires involves either top-down or bottom-up approach [65]. Furthermore, compared to other devices, the electrical detection based on silicon nanowire has a higher influence on conductance, and faster response of detection. Due to the linear output, low power needs, good resolution with ultra-low-level sensitivity up to parts per trillion or sub-pico or femto molar range, electrochemical biosensors are recognized as being exceptionally sensitive forms of transducers. The repeatability and precision are also quite good. Point-of-care and point-of-need electrochemical sensors are also available for deployment [66]. Surface modification with nanomaterials helps electrochemical biosensing applications [67]. Nowadays wireless nanowire-based biosensors is helpful technology in diagnosing infectious diseases [68–72].

1.5 Conclusions

Due to benefits including high selectivity and sensitivity, the potential for downsizing, portability, low cost, and quick reaction, biosensors have increased their influence over the past 50 years in a variety of sectors, including therapeutic applications. The intricacies of numerous biological processes in health and disease are now being clarified by recent developments in biomarkers discovery and biotechnology, highlighting novel targets for diagnosis and treatments. This is crucial in the case of infectious diseases because of the continued high number of projected fatalities, the threat of pandemics and epidemics, the emergence and reemergence of diseases, and the drug resistance of pathogens. Therefore, having reliable diagnosis techniques available is essential. This study discusses current methods for diagnosing viral diseases, ideas about biomarkers, and ligand selection, in addition to emphasizing the prospects of biosensor technology. Additionally crucial to the democratization of diagnosis are biosensors. Due to their high cost, centralized nature, and need for trained specialists to operate, many present systems are inaccessible to a sizeable portion of the global population. Because of this, the potential for cost savings, mobility, and simplicity is greatly appreciable, particularly in the case of diseases that are often ignored. The future also seems optimistic. It is possible that developments in disciplines like genetics, epigenetics, chemistry, biochemistry, physiology, and bioinformatics will help to better understand the subtleties of biological processes in both health and sickness. Particularly in diagnosis and treatments, or possibly both, new research targets are becoming available and existing ones are becoming better understood (i.e. theragnostic). Therefore, combining these discoveries with innovative technologies like biosensors could alter the current scenario of medical diagnosis.

References

1 Neiderud, C.J. (2015). How urbanization affects the epidemiology of emerging infectious diseases. *Infection Ecology & Epidemiology* 5 (1): 27060.

2 Sarwar, M. (2015). Insect vectors involving in mechanical transmission of human pathogens for serious diseases. *International Journal of Bioinformatics and Biomedical Engineering* 1 (3): 300–306.

3 Jiménez-Rodríguez, M.G., Silva-Lance, F., Parra-Arroyo, L. et al. (2022). Biosensors for the detection of disease outbreaks through wastewater-based epidemiology. *TRAC Trends in Analytical Chemistry* 155: 116585. https://doi.org/10.1016/j.trac.2022.116585.

4 Pejcic, B., De Marco, R., and Parkinson, G. (2006). The role of biosensors in the detection of emerging infectious diseases. *Analyst* 131 (10): 1079–1090.

5 Cheon, J., Qin, J., Lee, L.P., and Lee, H. (2022). Advances in biosensor technologies for infection diagnostics. *Accounts of Chemical Research* 55 (2): 121–122.

6 (a) Weiss, R.A. (2005). Robert Koch: the grandfather of cloning? *Cell* 123 (4): 539–542;bFredericks, D.N. and Relman, D.A. (1996). Sequence-based identification of microbial pathogens: a reconsideration of Koch's postulates. *Clinical Microbiology Reviews* 9 (1): 18–33.

7 Siqueira, J.F. Jr. and Rôças, I.N. (2005). Exploiting molecular methods to explore endodontic infections: part 1 – current molecular technologies for microbiological diagnosis. *Journal of Endodontics* 31 (6): 411–423.

8 Chatterjee, N., Manna, K., Mukherjee, N., and Saha, K.D. (2022). Challenges and future prospects and commercial viability of biosensor-based devices for disease diagnosis. In: *Biosensor Based Advanced Cancer Diagnostics* (ed. R. Khan, A. Parihar, and S.K. Sanghi), 333–352. Academic Press https://doi.org/10.1016/B978-0-12-823424-2.00013-2.

9 Long, F., Zhu, A., and Shi, H. (2013). Recent advances in optical biosensors for environmental monitoring and early warning. *Sensors* 13 (10): 13928–13948.

10 Rodriguez-Mozaz, S., de Alda, M.J.L., Marco, M.P., and Barceló, D. (2005). Biosensors for environmental monitoring: a global perspective. *Talanta* 65 (2): 291–297.

11 Mufamadi, M.S. and Sekhejane, P.R. (2017). Nanomaterial-based biosensors in agriculture application and accessibility in rural smallholding farms: food security. In: *Nanotechnology* (ed. R. Prasad, M. Kumar, and V. Kumar), 263–278. Singapore. https://doi.org/10.1007/978-981-10-4573-8_12: Springer.

12 Smart, A., Crew, A., Pemberton, R. et al. (2020). Screen-printed carbon-based biosensors and their applications in agri-food safety. *TRAC Trends in Analytical Chemistry* 127: 115898.

13 Khansili, N., Rattu, G., and Krishna, P.M. (2018). Label-free optical biosensors for food and biological sensor applications. *Sensors and Actuators B: Chemical* 265: 35–49.

14 Leonard, P., Hearty, S., Brennan, J. et al. (2003). Advances in biosensors for detection of pathogens in food and water. *Enzyme and Microbial Technology* 32 (1): 3–13.

15 Masson, J.F. (2017). Surface plasmon resonance clinical biosensors for medical diagnostics. *ACS Sensors* 2 (1): 16–30.

16 Mascini, M. and Tombelli, S. (2008). Biosensors for biomarkers in medical diagnostics. *Biomarkers* 13 (7, 8): 637–657.

17 Thevenot, D.R., Toth, K., Durst, R.A., and Wilson, G.S. (1999). Electrochemical biosensors: recommended definitions and classification. *Pure and Applied Chemistry* 71 (12): 2333–2348.

18 Bai, Y., Xu, T., and Zhang, X. (2020). Graphene-based biosensors for detection of biomarkers. *Micromachines* 11 (1): 60.

19 Shah, J. and Wilkins, E. (2003). Electrochemical biosensors for detection of biological warfare agents. *Electroanalysis: An International Journal Devoted to Fundamental and Practical Aspects of Electroanalysis* 15 (3): 157–167.

20 Nigam, V.K. and Shukla, P. (2015). Enzyme based biosensors for detection of environmental pollutants – a review. *Journal of Microbiology and Biotechnology* 25 (11): 1773–1781.

21 García-Aljaro, C., Cella, L.N., Shirale, D.J. et al. (2010). Carbon nanotubes-based chemiresistive biosensors for detection of microorganisms. *Biosensors and Bioelectronics* 26 (4): 1437–1441.

22 Haupt, K. and Mosbach, K. (2000). Molecularly imprinted polymers and their use in biomimetic sensors. *Chemical Reviews* 100 (7): 2495–2504.

23 Kumar, S., Dilbaghi, N., Barnela, M. et al. (2012). Biosensors as novel platforms for detection of food pathogens and allergens. *BioNanoScience* 2 (4): 196–217.

24 Clarke, S.F. and Foster, J.R. (2012). A history of blood glucose meters and their role in self-monitoring of diabetes mellitus. *British Journal of Biomedical Science* 69 (2): 83–93.

25 Palchetti, I. and Mascini, M. (2008). Nucleic acid biosensors for environmental pollution monitoring. *Analyst* 133 (7): 846–854.

26 Thomas, S., Saji, K.J., and Jayaraj, M.K. (2022). An introduction to biosensors. In: *Nanomaterials for Sensing and Optoelectronic Applications* (ed. M.K. Jayaraj, P.P. Subha, and S. Thomas), 91–107. Elsevier.

27 Lucarelli, F., Tombelli, S., Minunni, M. et al. (2008). Electrochemical and piezoelectric DNA biosensors for hybridisation detection. *Analytica Chimica Acta* 609 (2): 139–159.

28 Stoecklein, N.H., Erbersdobler, A., Schmidt-Kittler, O. et al. (2002). SCOMP is superior to degenerated oligonucleotide primed-polymerase chain reaction for global amplification of minute amounts of DNA from microdissected archival tissue samples. *The American Journal of Pathology* 161 (1): 43–51.

29 Pfeiffer, F. and Mayer, G. (2016). Selection and biosensor application of aptamers for small molecules. *Frontiers in Chemistry* 4: 25.

30 Misawa, N., Osaki, T., and Takeuchi, S. (2018). Membrane protein-based biosensors. *Journal of the Royal Society Interface* 15 (141): 20170952.

31 Lengger, B., Hoch-Schneider, E.E., Jensen, C.N. et al. (2022). Serotonin G protein-coupled receptor-based biosensing modalities in yeast. *ACS Sensors* 7 (5): 1323–1335.

32 Ronkainen, N.J., Halsall, H.B., and Heineman, W.R. (2010). Electrochemical biosensors. *Chemical Society Reviews* 39 (5): 1747–1763.

33 Hertz, L. and Rothman, D.L. (2016). Glucose, lactate, β-hydroxybutyrate, acetate, GABA, and succinate as substrates for synthesis of glutamate and GABA in the glutamine–glutamate/GABA cycle. In: *The Glutamate/GABA-Glutamine Cycle* (ed. A. Schousboe and U. Sonnewald), 9–42. Cham: Springer.

34 Rocchitta, G., Spanu, A., Babudieri, S. et al. (2016). Enzyme biosensors for biomedical applications: strategies for safeguarding analytical performances in biological fluids. *Sensors* 16 (6): 780.

35 Bousse, L. (1996). Whole cell biosensors. *Sensors and Actuators B: Chemical* 34 (1–3): 270–275.

36 Gui, Q., Lawson, T., Shan, S. et al. (2017). The application of whole cell-based biosensors for use in environmental analysis and in medical diagnostics. *Sensors* 17 (7): 1623.

37 Liu, Q., Wu, C., Cai, H. et al. (2014). Cell-based biosensors and their application in biomedicine. *Chemical Reviews* 114 (12): 6423–6461.

38 Gu, M. B., Mitchell, R. J., & Kim, B. C. (2004). Whole-cell-based biosensors for environmental biomonitoring and application. Zhong, JJ. Biomanufacturing. Advances in Biochemical Engineering, 87. Springer, Berlin, Heidelberg. https://doi.org/10.1007/b13533 269-305.

39 Gupta, N., Renugopalakrishnan, V., Liepmann, D. et al. (2019). Cell-based biosensors: recent trends, challenges and future perspectives. *Biosensors and Bioelectronics* 141: 111435.

40 Zeng, X., Shen, Z., and Mernaugh, R. (2012). Recombinant antibodies and their use in biosensors. *Analytical and Bioanalytical Chemistry* 402 (10): 3027–3038.

41 Dzantiev, B.B. and Zherdev, A.V. (2013). Antibody-based biosensors. In: *Portable Biosensing of Food Toxicants and Environmental Pollutants* (ed. D.P. Nikolelis, T. Varzakas, A. Erdem, and G.-P. Nikole), 161–196. Boca Raton, FL: CRC Press Book.

42 Byrne, B., Stack, E., Gilmartin, N., and O'Kennedy, R. (2009). Antibody-based sensors: principles, problems and potential for detection of pathogens and associated toxins. *Sensors* 9 (6): 4407–4445.

43 Goud, K.Y., Kailasa, S.K., Kumar, V. et al. (2018). Progress on nanostructured electrochemical sensors and their recognition elements for detection of mycotoxins: a review. *Biosensors and Bioelectronics* 121: 205–222.

44 Goud, K.Y., Reddy, K.K., Khorshed, A. et al. (2021). Electrochemical diagnostics of infectious viral diseases: trends and challenges. *Biosensors and Bioelectronics* 180: 113112.

45 Kim, H.M., Uh, M., Jeong, D.H. et al. (2019). Localized surface plasmon resonance biosensor using nanopatterned gold particles on the surface of an optical fiber. *Sensors and Actuators B: Chemical* 280: 183–191.

46 Sharma, A., Mishra, R.K., Goud, K.Y. et al. (2021). Optical biosensors for diagnostics of infectious viral disease: a recent update. *Diagnostics* 11 (11): 2083.

47 Chen, Y.T., Liao, Y.Y., Chen, C.C. et al. (2019). Surface plasmons coupled two-dimensional photonic crystal biosensors for Epstein–Barr virus protein detection. *Sensors and Actuators B: Chemical* 291: 81–88.

48 Endo, T., Ozawa, S., Okuda, N. et al. (2010). Reflectometric detection of influenza virus in human saliva using nanoimprint lithography-based flexible two-dimensional photonic crystal biosensor. *Sensors and Actuators B: Chemical* 148 (1): 269–276.

49 Huang, J.C., Chang, Y.F., Chen, K.H. et al. (2009). Detection of severe acute respiratory syndrome (SARS) coronavirus nucleocapsid protein in human serum using a localized surface plasmon coupled fluorescence fiber-optic biosensor. *Biosensors and Bioelectronics* 25 (2): 320–325.

50 Guider, R., Gandolfi, D., Chalyan, T. et al. (2015). Sensitivity and limit of detection of biosensors based on ring resonators. *Sensing and Bio-sensing Research* 6: 99–102.

51 Luchansky, M.S. and Bailey, R.C. (2012). High-Q optical sensors for chemical and biological analysis. *Analytical Chemistry* 84 (2): 793–821.

52 Brecht, A. and Gauglitz, G. (1995). Optical probes and transducers. *Biosensors and Bioelectronics* 10 (9, 10): 923–936.

53 Nguyen, H.H., Lee, S.H., Lee, U.J. et al. (2019). Immobilized enzymes in biosensor applications. *Materials* 12 (1): 121.

54 Kazura, E., Mernaugh, R., and Baudenbacher, F. (2020). A capillary-perfused, nanocalorimeter platform for thermometric enzyme-linked immunosorbent assay with femtomole sensitivity. *Biosensors* 10 (6): 71.
55 Quesada-González, D. and Merkoçi, A. (2018). Nanomaterial-based devices for point-of-care diagnostic applications. *Chemical Society Reviews* 47 (13): 4697–4709.
56 Mattiasson, B., Borrebaeck, C., Sanfridson, B., and Mosbach, K. (1977). Thermometric enzyme linked immunosorbent assay: TELISA. *Biochimica et Biophysica Acta (BBA)-Enzymology* 483 (2): 221–227.
57 Lubbers, B.R. (2015). Nano-calorimetry for point of care diagnostics. Doctoral dissertation. Biomedical Engineering, Nashville, Tennessee.
58 Mujahid, A. and Dickert, F.L. (2012). Molecularly imprinted polymers for sensors: comparison of optical and mass-sensitive detection. In: *Molecularly Imprinted Sensors* (ed. S. Li, Y. Ge, S.A. Piletsky, and J. Lunec), 125–159. Elsevier.
59 Tombelli, S. (2012). Piezoelectric biosensors for medical applications. In: *Biosensors for Medical Applications* (ed. S. Higson), 41–64. Woodhead Publishing.
60 Pohanka, M. (2018). Overview of piezoelectric biosensors, immunosensors and DNA sensors and their applications. *Materials* 11 (3): 448.
61 Dickert, F.L. and Hayden, O. (2001). Non-covalent molecularly imprinted sensors for vapours, polyaromatic hydrocarbons and complex mixtures. In: *Techniques and Instrumentation in Analytical Chemistry*, vol. 23 (ed. A. Janshoff and C. Steinem), 503–525. Elsevier.
62 García-Martinez, G., Bustabad, E.A., Perrot, H. et al. (2011). Development of a mass sensitive quartz crystal microbalance (QCM)-based DNA biosensor using a 50 MHz electronic oscillator circuit. *Sensors* 11 (8): 7656–7664.
63 Cheng, C.I., Chang, Y.P., and Chu, Y.H. (2012). Biomolecular interactions and tools for their recognition: focus on the quartz crystal microbalance and its diverse surface chemistries and applications. *Chemical Society Reviews* 41 (5): 1947–1971.
64 Bacher, G., Bhand, S., and Deshpande, S. (2022). Electrical biosensors for virus detection. In: *Advanced Biosensors for Virus Detection* (ed. R. Khan, A. Parihar, A. Kaushik, and A. Kumar), 241–259. Academic Press.
65 Nuzaihan, M., Hashim, U., Arshad, M.M. et al. (2016). Electrical detection of dengue virus (DENV) DNA oligomer using silicon nanowire biosensor with novel molecular gate control. *Biosensors and Bioelectronics* 83: 106–114.
66 Kaushik, A., Yndart, A., Jayant, R.D. et al. (2015). Electrochemical sensing method for point-of-care cortisol detection in human immunodeficiency virus-infected patients. *International Journal of Nanomedicine* 10: 677.
67 Radhakrishnan, S., Kim, B.S., and Dave, S. (2022). Surface modification with nanomaterials for electrochemical biosensing application. In: *Advanced Nanomaterials for Point of Care Diagnosis and Therapy* (ed. S. Dave, J. Das, and S. Ghosh), 101–120. Elsevier.
68 Panwar, R., Churi, H., and Dave, S. (2022). Point-of-care electrochemical biosensors using CRISPR/Cas for RNA analysis. In: *Biosensors for Emerging and Re-Emerging Infectious Diseases* (ed. J. Das, S. Dave, Radhakrishnan, and P. Mohanty), 317–333. Academic Press.

69 Sahoo, M., Gadnayak, A., Nayak, A. et al. (2022). Advantages of silicon nanowire-based biosensors as wireless technology for infectious disease diagnosis. In: *Biosensors for Emerging and Re-Emerging Infectious Diseases* (ed. J. Das, S. Dave, S. Radhakrishnan, and P. Mohanty), 407–417. Academic Press.

70 Xue, Q., Kan, X., Pan, Z. et al. (2021). An intelligent face mask integrated with high density conductive nanowire array for directly exhaled coronavirus aerosols screening. *Biosensors and Bioelectronics* 186: 113286. https://doi.org/10.1016/j.bios.2021.113286.

71 Dave, S. and Kirubavathy, S.J. (2022). Biosensors based on metal-organic framework (MOF): paving the way to point-of-care diagnosis. In: *Electrochemical Applications of Metal-Organic Frameworks* (ed. S. Dave, R. Sahu, and B.C. Tripathy), 255–267. Elsevier.

72 Sahoo, S., Nayak, A., Gadnayak, A. et al. (2022). Quantum dots enabled point-of-care diagnostics: a new dimension to the nanodiagnosis. In: *Advanced Nanomaterials for Point of Care Diagnosis and Therapy* (ed. S. Dave, J. Das, and S. Ghosh), 43–52. Elsevier.

2

Nuts and Bolts of Modern Biosensing Technology: Smart Health Diagnostic Devices

Itthipon Jeerapan[1,2], Gabriela Valdés-Ramírez[3], and Barbara Brunetti[4]

[1]*Prince of Songkla University, Center of Excellence for Trace Analysis and Biosensor, 15 Karnjanavanit Road, 90110 Hat Yai, Songkhla, Thailand*
[2]*Prince of Songkla University, Division of Physical Science and Center of Excellence for Innovation in Chemistry, Faculty of Science, 15 Karnjanavanit Road, 90110 Hat Yai, Songkhla, Thailand*
[3]*Autonomous Metropolitan University (Iztapalapa), Chemistry Department, 186 San Rafael Atlixco Av., 093400 Mexico City, Mexico*
[4]*University of Milan, DeFENS, Via Celoria 2 I-20133, Milan, Italy*

2.1 Introduction

Modern healthcare sectors aim to improve individual wellness and minimize the risks of infectious diseases through speedy, sensitive, and reasonable diagnostics. The concept of modern diagnostics has led to the translation from centralized healthcare (i.e. relying on standard laboratories with large equipment) into decentralized formats [1, 2]. Existing conventional diagnostic approaches are limited by long analysis time, high costs, technical complications, and limited sensitivities. It is essential to combat the need for early detection by developing biosensing technology for convenient diagnosis among patients [3]. More clearly, the COVID-19 pandemic due to severe acute respiratory syndrome coronavirus 2 (SARS-CoV-2), which started in 2019, is rampaging through the healthcare of the globe. It continues to remind all of humanity how quickly and severely an infectious disease can affect the world. Clinical laboratories overrunning in many cases are struggling to keep pace with the demand of healthcare management. This accelerates the development of smart point-of-care (POC) instrumentations.

Various platforms of medical diagnostic testing are essential for the rapid determination of indicators and analytes linked to health status, such as pathogens. Smart POC platforms can be classified into organ-interfaced electronics, implantables, paper-based or microfluidic diagnostic systems, and wearables [4–6]. These smart and easy-to-use devices, particularly wearables and portable sensing devices, allow the nonexpert user and the healthcare personnel to identify the infection status or pathogens effectively. The main objective of developing new sensors is to

Point-of-Care Biosensors for Infectious Diseases, First Edition.
Edited by Sushma Dave and Jayashankar Das.

translate time-consuming conventional testing into rapid and efficient diagnosis. This aims to be independent of centralized laboratories, experienced personnel, and bulky or complex equipment. Traditionally, the detection of viruses and bacteria requires gold-standard molecular techniques, e.g. the reverse transcription-polymerase chain reaction (RT-PCR) [7, 8]. The classical protocols involve multistep isolation, culturing, and sophisticated biochemical tests. In addition, serological tests including the enzyme-linked immunosorbent assay (ELISA) are applied for analyzing antibodies and immunoglobulin [9]. Nevertheless, these laborious approaches usually take a long time to obtain results. Therefore, practical alternatives have emerged to provide a more convenient and rapid way of testing.

Several platforms are applied to realize convenient biosensors and diagnostic devices. These include surface-enhanced Raman scattering (SERS)-based biosensors [10], electrochemical sensors [11], fluorescence biosensors [12], colorimetric biosensors [13], chemiluminescence biosensors [14], and surface plasmon resonance (SPR)-based biosensors [15]. These advanced techniques aim to cover multiple biomarkers (such as metabolites) and plenty of biological species causing the illness in human. According to a report, 1415 species of viruses, prions, bacteria, rickettsia, fungi, protozoa, and helminths can cause infection in humans [16].

Different targets require specific strategies for engineering detection mechanisms. The key is the recognition element (Figure 2.1) which is further discussed in Section 2.2 of this chapter. The understanding of principles is critical for

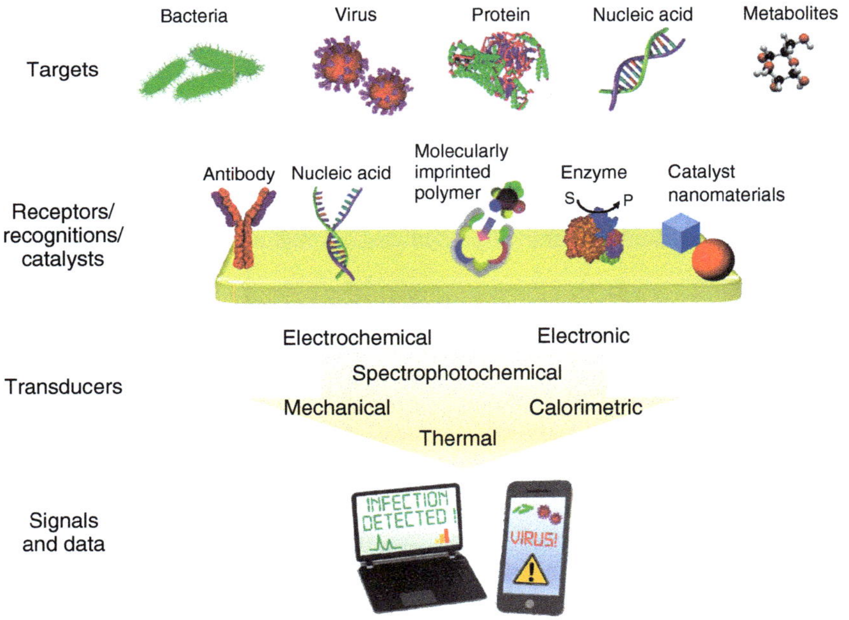

Figure 2.1 Schematic illustration showing operation principles of biosensors. The targets, such as pathogens and biomarkers, are detected with the assistance of corresponding receptors, catalysts, or nanomaterials, followed by signal transduction and output.

engineering efficient diagnostic devices. Sensors relying on electrochemistry are among the most attractive platforms for practical sensors as electrochemical sensors can analyze the content of biological samples (such as blood and other biofluids) by converting an event to an electronic signal. This electrical signal is then further integrated with smart digital devices. Therefore, easy-to-use sensing devices with minimal user involvement can be realized. Similarly, sensing platforms relying on simple spectroscopic methods are also developed. Importantly, conventional diagnostic tools can be connected to an internet network or modern communication technology to allow remote and rapid monitoring. In addition to chemical parameters, vital signs, such as body temperature and blood pressure, are also important. The routine and convenient monitoring of such vital signs by the individual and medical professionals would need a new platform to ensure that the user and professionals (particularly, in the remote area) can assess to diagnose the general physical health or the infection status. The integration of using modern chemical and physical sensors will show rapid diagnostic results, which can be taken to help to evaluate the problem, manage the problem, and show progress toward recovery.

This chapter gives insights into the fundamental mechanisms used for advancing biosensors for reading vital information and biomarkers, helping to warn the sickness or disorder caused by pathogens. We focus on POC devices along with state-of-the-art wearable platforms that can monitor significant bio-/and chemical parameters available in biofluids (such as a small volume of blood, sweat, urine, interstitial fluids, saliva, and tears) that can be obtained in noninvasive or minimally invasive ways by patients themselves. We highlight advanced sensors exploiting various analytical principles, such as amperometric, potentiometric, voltammetric, impedance spectroscopy, and colorimetric principles. With some examples of smart sensors, we can access interstitial fluids using subcutaneous needles, collect and check sweat using epidermal devices, analyze saliva with mouthguards, screen blood using small implants, and analyze saliva. The example list of analytes includes pyocyanin (PyoC), pyoverdine (PyoV), cortisol, cytokines, and metabolites (such as glucose, phosphate, and glycerol). In addition, this chapter discusses the remaining challenges in this research area, as well as the prospects for future developments.

2.2 Nuts and Bolts for Point-of-Care (POC) Biosensor-Based Testing

Due to the continuous evolution of diseases and a variety of pathological problems, it is essential to design new POC or personalized sensors to diagnose various problems of pathological states. POC biosensors give the user a unique advantage because the analysis can be done in the direct proximity of the user. More clearly, state-of-the-art wearable biosensors are applicable platforms that can closely match daily life. The key goal of engineering modern biosensors is to facilitate speedy and easily accessible testing at the POC.

As shown in Figure 2.1, a biosensor is an analytical device mainly integrating **(i) responsive elements or sensing receptors** and **(ii) a physicochemical detector** that helps to convert clinical parameters (such as bio/chemical concentration of the target) into a measurable signal (such as current, potential, and spectrophotochemical indication). Eventually, the system yields a digital electronic signal (which can be connected to smartphones or modern electronic devices) or simply observable colors by the naked eye that is proportional to the concentration of a specific target. The details of components in typical biosensors are described in the following categories.

2.2.1 Analytes

An **analyte** is a target or a bio/chemical constituent that will be identified and measured. Fundamentally, biosensors include qualitative and quantitative analysis of different targets by converting their biological, chemical, or physical actions into detectable signals. For example, an analyte in a biosensor designed to identify malaria is histidine-rich protein [17]. For detecting SARS-CoV-2, an analyte can be immunoglobulin M (IgM), or immunoglobulin G (IgG) [18–20].

2.2.2 Receptors and Sensing Elements

A sensing **receptor** is a unit that specifically recognizes the analyte. Basically, a bioreceptor is a unit that precisely recognizes the analyte. A variety of bioreceptors used in developing biosensors include enzymes, cells, proteins, aptamers, deoxyribonucleic acid (DNA), and antibodies. For example, multiple bioreceptors used in the developed biosensors for SARS-CoV-2 diagnosis include DNA aptamer [21], G-quadruplex nucleic acid (NA) structure [22], and specific antibodies [23]. Biorecognition event due to the interaction of the bioreceptor with the analyte yields the signal generation in the charging form of electrochemical signal, light, pH, mass, or color. The recognition of a specific analyte with the bioreceptor should be designed to yield proportional expressions, hence indicating the presence and the level of the analyte. The signal due to the biorecognition expression can be used to assess disease progression and therapeutic efficacy. In general, detection mechanisms can be divided into label-free and labeled assays.

Label-free assays determine the existence of an analyte via the direct reaction or the direct interaction between the analyte and the receptor layer on a transducer surface without the use of complicated tags (such as radioisotopes, quantum dots, or fluorescent dyes). Examples of the interaction of molecules are ligand–receptors, antibodies–antigens, and proteins–proteins. The analyte binding with receptor elements immobilized on solid support (such as the electrode) can cause measurable changes (e.g. interfacial capacitance or resistance). This straightforward principle provides an advantage because the label-free strategy needs only a single recognition element. This simplicity is essential for designing conventional sensing devices. Importantly, the detection can be realized in a short time with reasonable reagent costs. This label-free strategy is usually suitable for small molecular analytes that can be bound within the binding space of the receptor. Another benefit of the label-free mode is the high potential to realize real-time quantitative measurements. This enables the device to monitor the analyte continuously. We will elaborate on label-free applications for detecting various analytes, such as the detection of bacteria by using the interaction between the bacterial cell wall and ampicillin (Amp) (in Section 2.3.1.1), the detection of viruses according to the interactions between antibodies and antigens (in Section 2.3.1.2), and the detection of cortisol by using the interaction between cortisol monoclonal antibody (McAb) and cortisol molecule (in Section 2.3.2).

Labeled assays are common for classical and robust sensing designs. For instance, ELISA is the standard sandwich immunoassay for diagnosis in clinical laboratories and sensing applications in home test kits. For labeled assays, the analyte is sandwiched between capture and detector elements. Both capture and detector elements can recognize the analyte specifically. Note that the recognition events to form standard sandwich labeled assays require many steps. Therefore, labeled assay systems are complicated, time-consuming, and costly. However, capture and detector elements have different binding sites. Advantageously, these unique binding sites enhance the specificity and minimize the background. Importantly, the detection unit in labeled assay design needs to be tagged with a signaling component (such as enzymes, radioisotopes, fluorophores, nanoparticles, or quantum dots). Depending on sensing designs, this signaling part can be integrated with electrical, mechanical, or optical transducers. In general, the capture element is attached to a solid surface, e.g. glass chips, conductive electrodes, or nano/microbeads. Examples of signaling integrations are electrochemical sensors to detect redox reactions from enzyme tags and optical sensors to detect fluorescentor luminescent tags, or the simple observation by naked eyes to detect colorimetric tags. For instance, ELISA generally relies on a capture antibody and a detector antibody conjugated with an enzyme tag for catalyzing the conversion of chromogenic substrate into colored molecules. A typical example of a sensitive chromogenic substrate is 3,3′ 5,5′-tetramethylbenzidine (TMB). This chromogenic substrate is applied when using horseradish peroxidase (HRP) as a tagged enzyme. During the enzymatic reaction to decompose hydrogen peroxide by HRP (which is tagged on the detector element), TMB can display a blue color that can be quantified spectrophotometrically at a wavelength of 650 nm.

It is important to ensure that the design of the sensor can provide high selectivity to the sought analyte. Appropriate materials to form the sensing component should be carefully selected or engineered. The selection of the element depends on the application and the relevant transductor mechanism. The following elements are typical choices.

One of the most common elements acting as bioreceptors is an **enzyme**. Enzymes provide effective binding and catalytic activities. Therefore, we can integrate an enzyme with a transducer to build a biocatalytic biosensor that produces an analytical signal, proportional to the concentration of the analyte. The reaction with the assistance of enzymes can cause a range of changes, such as the measurable change of the proton concentration, generation, or depletion of chemicals (such as oxygen, ammonia, and peroxide), light emission, absorption, reflectance, or heat emission. The corresponding transducer includes electrical parameters such as current and potential via electrochemical means, absorption of light via spectrophotochemical means, or temperature change via thermal measurements. Another category uses a specific **antibody**. This is important for designing immunosensors that apply to bind the **antigen** with high affinity and specificity. One of the most robust and specific recognition strategies is using the **DNA** concept. This strategy relies on the natural recognition that single-stranded NA can bind to the complementary strands through stable hydrogen bonds. Therefore, DNA-based biosensors represent promising opportunities for specific, sensitive, and rapid detection of infectious disease-associated target molecules.

Moreover, direct detection, e.g. nonenzymatic sensing strategy, is also desirable because researchers worldwide are working to overcome the stability challenge of biological elements. In this approach, inorganic catalysts or nanomaterials are engineered to mimic biorecognition or provide the catalytic activity to increase the rate of a reaction. For example, metal oxide-based materials have been widely applied in the sensor due to their high catalytic capability for glucose oxidation [24]. Copper oxide/polypyrrole/reduced graphene oxide (GO) composite is an example to use as a reactive surface for designing a glucose sensor [25]. This composite material on the electrode allows the detection of glucose via oxidation at a low potential of 0.2 V vs SCE (saturated KCl). However, biological elements typically provide a highly specific affinity between the target and the sensing layer, thereby increasing the biosensor's specificity.

To mimic the function and structure of antibodies and biological receptors, customized polymers are developed to recognize the target with high sensitivity and selectivity. This category can be called "plastic antibodies" or **molecularly imprinted polymer** (MIP). An MIP is a polymeric network with specific molecular recognition pockets that mimics the affinity function and morphology of antibodies, which are biological receptors that target specific analytes. During the polymerization of monomers with cross-linkers in the presence of the analyte, specific molecular recognition cavities can be engineered (acting as a template molecule). When we perform this polymerization, the added analyte is confined in the polymeric matrix. Afterward, the analyte (i.e. the template) is removed. Therefore, we can build specific cavities for enabling molecular recognition. The created cavities can match the analyte in size, shape, and chemical functional group orientation. This biomimicking MIP can be assembled into sensors by combining it with various signal amplification or transduction platforms. These polymeric sensors have been studied and have shown promising potential in the detection of a wide range of target molecules [26]. For example, MIP-based electrochemical detection of COVID-19 is demonstrated [27].

2.2.3 Transducer

The **transducer** is another significant unit of the biosensor. The function of the transducer is to convert the biorecognition event into a detectable signal. A variety of techniques to read the detectable signal can be electrochemical techniques (amperometry, voltammetry, potentiometry, conductometry, and impedimetry), optical modes (colorimetric, fluorescence, luminescence, interferometry), calorimetry, mass-monitoring modes (using piezoelectric or acoustic principles). The biosensors used to detect and monitor pathogen infections can be classified into several groups, based on the detectable signal. The main classifications are (i) optical, (ii) electrochemical, and (iii) mass-based biosensors, which will be discussed in this chapter. In general, the obtained signal from transducers displays the relationship between the signal and the quantity of analyte–receptor recognition (determining the concentration of the analyte).

2.2.4 Signal Processing Unit

Once the transduced signal is generated, we require a **signal processing unit** to display the relevant data to the user. This unit uses electronics that can have additional functions, e.g. amplification and transformation of analytical signals from analog into a digital system. The signal processing unit can integrate hardware and software to produce the easy-to-understand output of the biosensor. The processed signals can eventually be sent to the traditional display or the receiver (including computers or smartphones) for the end user.

2.3 Advances in Biosensing Technology

Vital signs are indicators of how well our body is functioning. These signs include body temperature [28], respiratory rate [29], and blood oxygen level [30]. These signs are generally checked at healthcare offices. In addition to these vital signs, modern sensing platforms should be able to detect bio/chemical parameters (such as the presence or even the concentrations of pathogens and relevant metabolites). In Sections 2.3.1 and 2.3.2, we describe details of such biosensing technologies.

2.3.1 Advanced Sensors for Detecting Pathogens

The term "pathogen" is used to refer to all kinds of microorganisms that can cause infection of organisms' host, initiating illness in the host. The meaning can be extended to an infectious agent or a germ (such as a small collection of genetic code) that triggers diseases. The main types of pathogens are fungi, protists, parasitic worms, bacteria, and viruses. In Section 2.3.1, we focus on the significant classes: bacteria and viruses.

Bacteria and viruses are responsible for many mild, moderate, and severe infections and diseases. Bacteria are free-living small single cells with a membrane that encapsulates their genetic material and can reproduce on their own. Viruses are a small collection of genetic material, either DNA or RNA which is enclosed by a protein coat. In contrast to bacteria, viruses require a host to replicate. During the replication process, the host cells are killed producing damage to the host organism [31].

Human skin, as well as body fluids, are environments where bacteria can be found. Although most bacteria are harmless to the human body, some varieties can cause infections and diseases. As soon as harmful bacteria and viruses infect the host, they can start replication/reproduction processes, and new pathogens can easily spread to a new host. When the human body is the host, the infections can spread through the respiratory and gastrointestinal tract through coughing or sneezing, by direct contact with body fluids from an infected host, or by direct contact with contaminated surfaces, food, or water. Some infections can produce symptoms almost immediately. However, others need an incubation period. The incubation time depends on the infecting microorganism as well as the health of the host. Diseases, such as pneumonia, tuberculosis, cholera, anthrax, leprosy, diarrhea, influenza, or acquired immunodeficiency syndrome (AIDS), are caused either by bacteria or viruses. Several methodologies have been developed to diagnose and control infections. These methodologies include polymerase chain reaction (PCR), ELISA, or nucleic acid testing (NAT). These approaches are performed in clinical and medical laboratories, requiring expensive equipment, sample preparation, long-time analysis, and qualified trained personnel. To facilitate early pathogen detection, several biosensing methods are under development. The aim of the biosensing methodologies is the development of rapid and highly sensitive detection of infectious diseases for POC devices [32, 33].

In Section 2.3.1.1, we explain ideas of optical and electrochemical sensing systems for bacteria (such as *Staphylococcus aureus, Salmonella,* and *Pseudomonas aeruginosa*) detection in Section 2.3.1.1. Additionally, Section 2.3.1.2 describe sensors for viruses (such as Zika virus, hepatitis B virus, or Ebola virus) and biosensors for COVID-19 detection.

2.3.1.1 Biosensors for Bacteria Detection

Some bacteria can change the pH of the media; this capability of bacteria is used for bacteria detection through a colorimetric sensor. For instance, Figure 2.2A describes the biosensing response to the microorganism in the wounded site [34]. Wounds are sites where pathogens can enter the host and grow, causing infections that can lead to significant injuries and fatalities worldwide. Therefore, wounds are considered one of the major causes of morbidity and mortality. The pathogens detected in this example are five bacteria, commonly found on the human skin, including three gram-positive (***S. aureus, Micrococcus luteus***, and ***Corynebacterium amycolatum***), and two gram-negative (***Escherichia coli*** *and* ***P. aeruginosa***). The sensor for early bacteria detection shown in Figure 2.2A relies on the development of a chromatic flexible biosensor. The sensing platform uses polydiacetylene (PDA) as the sensitive indicator, which possesses optical and fluorescence properties. PDA is polymerized together with polyvinyl butyral (PVB) by photopolymerization by ultraviolet (UV) exposure. The PVB–PDA polymer forms flexible recognition membrane fibers blue color. When the prepared polymer membranes are exposed to the pathogen agent and tested at room temperature or 37 °C, a color change from no color to pink–red color is observed due to pH changes. For the exposure to *E. coli* and *S. aureus*, there is a color change to pink–red color even at room temperature in the first hour. However, for the other three pathogens, the color changes are observed only at 37 °C. The chromic behavior of PDA depends on the polymer side chains, solvent, and temperature. In the sensing PVB–PDA fibers, the color change is due to

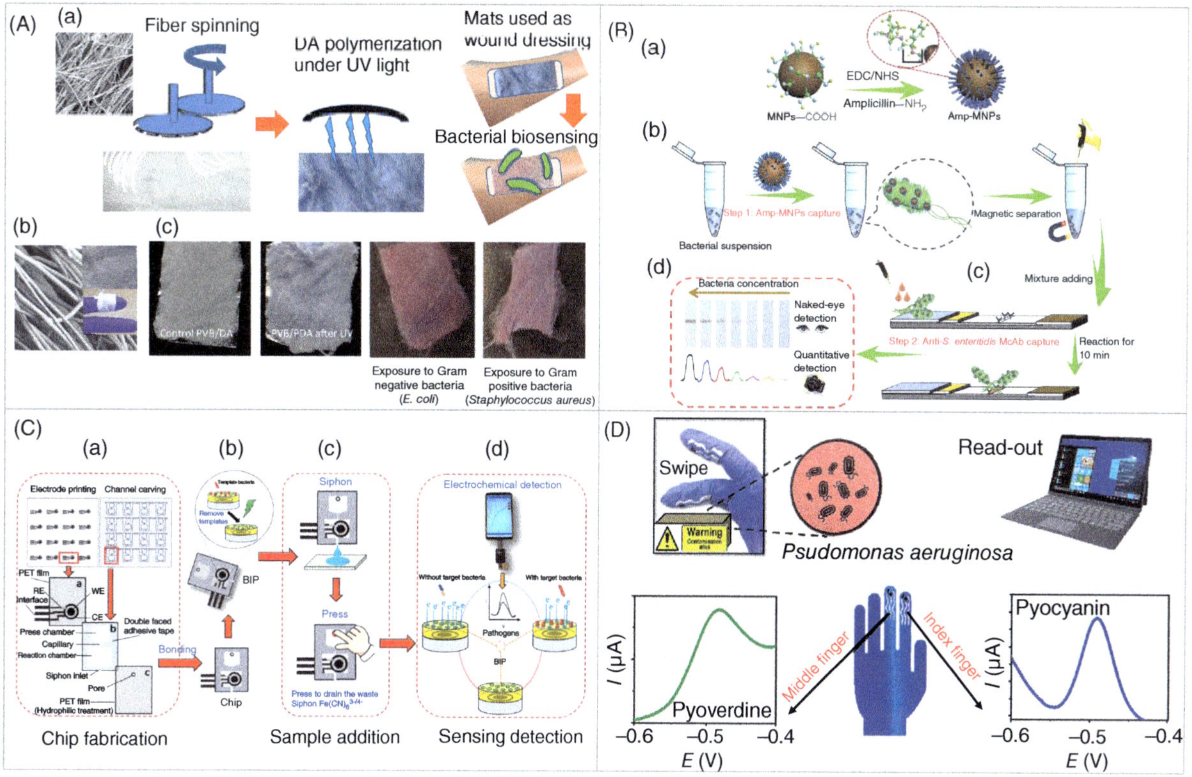

Figure 2.2 Biosensors for bacteria detection. (A) Schematic illustration of colorimetric bacteria flexible biosensor (a) PVB/DA fibers fabrication by force spinning technique followed by DA polymerization under UV light, and biosensor detection principle by color change after bacteria exposure (blue to pink/red color). (b) Images for flexible polymer fibers, (c) PVB–PDA membrane exposed to bacteria. Source: Reproduced with permission from Vidal et al. [34]; © 2019, Elsevier B.V. (B) Graphic illustration of an optical strip sensor for salmonella detection. (a) MNPs modification with Amp. (b) Bacteria separation through ampicillin (Amp)-MNPs/Bacteria interaction. (c) LFIA biosensor. (d) Sensor strip color line by the presence or absence of bacteria concentrations detected by the naked eye. Source: Reproduced with permission from Bu et al. [35]; © 2020, Elsevier B.V. (C) Schematic illustration of the self-driven PET chip-based electrochemical imprinting sensor. (a) screen-printed electrode (SPE) fabrication coupled to PET film and reaction chamber. (b) MIP film through bacteria template. (c) Chip/sample interaction (d) chip working principle by electrochemical detection. Source: Reproduced with permission from Jiang et al. [36]; ©2021 Elsevier B.V. (D) The illustration of bacteria interaction to SPEs printed in the index and middle fingers for its electrochemical detection through SWV. Source: Reproduced with permission from Ciui et al. [37]; © 2018, American Chemical Society.

the interaction of the bacteria surface and PDA. Figure 2.2A(a) illustrates the fabrication and detection processes, and Figure 2.2A(b) shows the sensing polymer fibers. Figure 2.2A(c) exemplifies the fibers before and after UV exposure and color change after bacteria exposure. The flexible membranes show promising results to be used as scaffolds/bandages to alert patients of potential infections by the microbial attack before the individual appears symptomatic. The sensing membrane developed in this study has the potential to detect a wide range of pathogens that interact with the PDA, making it a promising material for wearable sensors used to detect the presence or absence of bacteria.

Another simple tool for sensing bacteria is a colorimetric sensor strip. An example is described for ***Salmonella*** detection. *Salmonella* is a common pathogen that can be found in food processing and food circulation. Once these bacteria penetrate the gastric system, they can cause abdominal pain, diarrhea, vomiting, dehydration, and even death in severe cases. For rapid *Salmonella* detection, different optical and electrochemical sensor systems have been developed [38, 39]. Sensor strips, on the other hand, are among the most well-known sensors that are also used for bacteria detection, including *Salmonella*. Figure 2.2B illustrates a lateral flow immunoassay (LFIA) method for ultrasensitive bacteria detection [35]. The system uses *Salmonella enteritidis* as a bacteria model, the antibiotic Amp is selected as the capture reagent and magnetic nanoparticles (MNPs) are employed to bind and concentrate the bacteria from a sample. The LFIA biosensor consists of four pads (i.e. sample pad, conjugate pad, reaction membrane, and absorbent pad). Into the LFIA system, a T line is prepared with a McAb dispersed onto a nitrocellulose (NC) membrane. For bacteria detection, first, MNPs with carboxyl groups are modified with Amp by cross-linking (Figure 2.2B(a)). The Amp-MNPs are employed as a capture probe instead of antibody–MNPs as is performed in a conventional system. The Amp-MNPs are incubated into the sample containing *S. enteritidis*. The bacteria are captured by Amp-MNPs through the proteins of the bacteria's surface. The complex is separated by a magnet (Figure 2.2B(b)), and later, the concentrated portion is dropped on the sample pad onto de LFIA (Figure 2.2B(c)); the solution migrates toward the absorption pad toward the antibody line. The color on the test line will gradually change from no color to yellow–brown color, which can be seen by the naked eye, depending on the amount of complexation (Figure 2.2B(d)).On the LFIA biosensor, if no bacteria were present in the sample, there would be no band. The Amp-MNPs allow to concentrate and separate bacteria samples without the need for long centrifugation and separation steps. The developed LFIA system can detect from 10^2 to 10^7 CFU ml^{-1}. A challenge of the described system is the low selectivity because Amp can strongly bind other bacteria (not only *S. enteritidis*); however, the platform detection method can be used to detect other bacteria by employing the proper antibiotics.

Salmonella can also be detected by electrochemical systems as illustrated in Figure 2.2C. Imprinted electrochemical electrodes (modified or bare electrodes) onto a variety of substrates are currently used for the development of portable electrochemical sensors. In a chip-based imprinted electrochemical sensor is presented [36]. The system is based on screen-printed technology. The SPEs are assembled onto a polyethylene terephthalate (PET) substrate. A polydimethylsiloxane (PDMS) microfluidic reaction channel is integrated over the SPEs. The working electrode is modified by a polymeric imprinting layer. The recognition element on the working electrode is fabricated by electropolymerization of pyrrole-*Salmonella* solution. The bacteria (template) are removed from the polymeric recognition layer with lysozymes and dodecyl sulfate, providing the imprinting recognition cavity for *Salmonella*. The detection is via the electrochemical signal of a given concentration of $Fe(CN)_6^{3-/4-}$ (a negatively charged redox probe). The testing sample is placed at the reaction channel and incubated for 15 minutes, if *Salmonella* exists in the sample, the bacteria coupes the free space in the recognition template, thus decreasing the signal of $Fe(CN)_6^{3-/4-}$. The fabrication, modification, and testing processes are illustrated in Figure 2.2C. The data can be output via a mobile phone connected to an electrochemical analyzer. This example of the sensing platform has potential uses in the POC for other pathogens and offers rapidity, low-cost, simple operation, and sensitive detection.

Some pathogens can produce an electroactive metabolite; therefore, the pathogen can be easily detected by electrochemical methods without the need of recognition elements or sensor modification. This is the case of *P. aeruginosa*. This electrochemical detection is illustrated in the following example. A glove-based electrochemical sensor for the detection of the bacteria *P. aeruginosa* is described in Figure 2.2D. *P. aeruginosa* is responsible for infections in chronic wound burns, bacteremia, burns, pneumonia, cystic fibrosis, and urinary tract infections [40]. The bacteria *P. aeruginosa* can be detected through two of its metabolites (PyoC and PyoV). Both are electrochemically active spices [37]. The used sensing platform is fabricated by screen printing technology, utilizing conductive inks for SPEs assembly on the index and middle finger of a glove. The electrochemical detection is performed by the oxidation signal of PyoC and PyoV, utilizing square wave voltammetry (SWV). For PyoC an oxidation peak at −0.5 V vs Ag/AgCl is observed, whereas the oxidation peak for PyoV is observed at 0.4 V vs Ag/AgCl. The oxidation of PyoC is caused by the reaction involving two electrons and two protons, whereas the oxidation of PyoV is caused by the one electron and proton interchange. The sensing platform can be used for liquid samples (such as covering the three electrodes system in each finger with a drop of the tested sample) as well as solid samples (such as bacteria found in surgical instrumentation, operation tables, doors, or sinks). For solid samples, a hydrogel film is used to maintain electrical contact between the three electrodes. Once the sample is incubated for a few seconds in the hydrogel, the SWV can be recorded. The presence of oxidation peaks indicates that the samples have been contaminated with *P. aeruginosa*. Sensors fabricated in this method are easy to prepare, low cost, and reusable, thus providing potential applications in clinical areas.

The bacteria sensing platforms have demonstrated effective and useful tools in the biomedical field for fast in situ bacteria detection and/or quantification. However, modifications and sensor improvements are still required for real-world applications and potential commercialization. Furthermore, portable electrochemical instrumentation systems can be useful in facilitating their commercialization.

2.3.1.2 Biosensors for Detecting Viruses

Designing Biosensors for Detecting Zika, Hepatitis B, and Ebolaviruses

Viruses are another kind of microorganism responsible for a variety of diseases around the globe. The detection of these microorganisms typically is through their NA. Hence, biosensors based on a single strand of DNA or RNA are typically used for the detection of viruses, in addition to microorganisms, or genetic traits [41, 42]. This kind of biosensors has potential applications for daily diagnosis detection at home and in developing countries where diseases caused by viruses such as influenza, human immunodeficiency virus (HIV), Ebola, Zika, chikungunya, yellow fever, or dengue are a health and even a social problem. The last four mentioned viruses are primarily transmitted by a bite of the *Aedes genus* mosquito. Influenza A and B viruses are the major viruses highly contagious to humans that cause seasonal respiratory diseases [43].

Recombinase polymerase amplification (RPA) has been combined with NA sensors to improve virus detection. The RPA can use primers that replicate rapidly at human body temperature. Hence, when NA-RPA sensors are combined with wearable technology, body temperature can be used for triggering fast amplification. An example of the flexible optical biosensor system is described in Figure 2.3A [44]. A wearable RPA sensor composed of inlets, outlets, and a reservoir for a visual readout is used for NA fragments of virus detection (such as **Zika** virus). At this POC system, human heat is used as the RPA reaction temperature, and the recombinase protein seeks the homologous sequence in the double-strand deoxyribonucleic acid template (dsDNA), which is the biorecognition element. The DNA strand is assembled with the single-stranded binding protein after the primers are carried out by the *Bsu* polymerase in presence of dNTPs. The flexible sensor is mounted into a flexible bandage. This sensor is coupled with a microfluid channel and reservoir for the reaction. DNA primers are placed at the side of the microchannel and incubated for 10 minutes; heat temperature starts the strand replication. After 10 minutes, SYBR Green (a dye used as a NA stain in molecular biology) [47] is added to the reaction chamber. When the virus is present in the sample, the fluorescence probe is intercalated into the DNA; consequently, fluorescence can be detected after its exposure to UV–vis light. Figure 2.3A(a) illustrates the described detection system colorimetric wearable POC for Zika detection, (b) fluorescence signal for wearable RPA sensor system at different fragment NA-Zika concentration. Although the detection system is qualitative, the fluorescence intensity provides an insight of the virus concentration in a sample. The detection principle for the developed system may be used for other viruses utilizing the appropriate DNA primers. The system is a promising tool for rapid tropical virus detection.

Biosensors also can be used for quantification by integration of portable devices. One case is by the integration of biosensors with smartphones, which are potential alternative platforms for portable, simple sensitive, and selective tools for the POC systems. Figure 2.3B exemplifies the integration of an amperometric immunosensor with a smartphone-controlled via near field communication (NFC) system [45]. The biosensor system is fabricated by modified screen-printed graphene electrodes

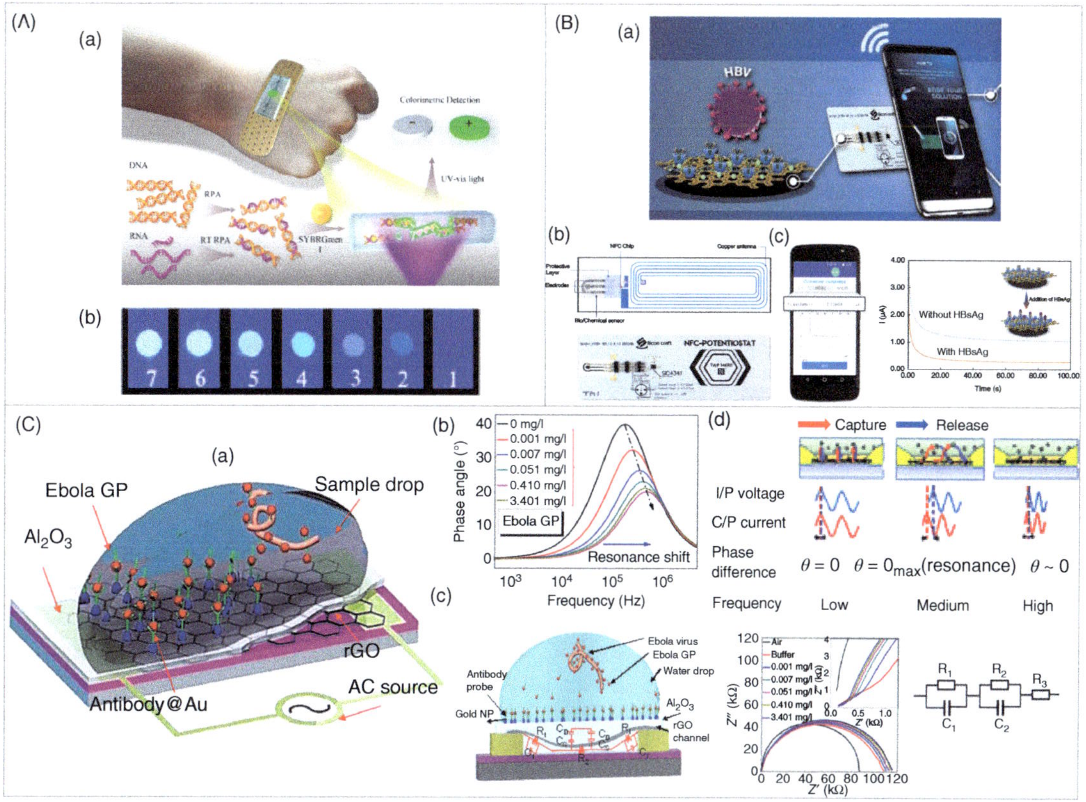

Figure 2.3 Biosensors for detecting Zika, Hepatitis B, and Ebola viruses. (A) Optical, flexible bandage biosensor for NA from Zika virus detection (a) general working principle, (b) fluorescence intensity at different NA concentrations. Source: Reproduced with permission from Yang et al. [44]; © 2019, Elsevier B.V. (B) NFC smartphone-based system for hepatitis B electrochemical detection. (a) sensor system coupled to a smartphone, (b) circuit board and portable NFC potentiostat, (c) real-time amperometric response measurement on a smartphone screen, and the amperometric detection principle. Source: Reproduced with permission from Teengam et al. [45]; © 2021, Elsevier B.V. (C) Schematic diagram for Ebola electrochemical detection, (a) biosensing layer, (b) impedimetric face angle at different GP concentrations, (c and d) Nyquist plot and equivalent circuit for different GP-Ebola concentrations. Source: Reproduced with permission from Maity et al. [46]; © 2018, American Chemical Society.

(SPGEs). The working electrode is modified with gold nanoparticles (AuNPs). β-cyclodextrin (β-CD) is electropolymerized on the electrode surface to capture specific antibodies for detecting **Hepatitis B** antigen (HBsAg); such antibodies work as the biorecognition agents in this sensing platform. Figure 2.3B(a) illustrates the integration of the electrochemical sensor and a smartphone of the detection system for hepatitis B. Figure 2.3B(b) shows the NFC-enabled smartphone for virus detection. The used NFC sensor is a card-sized type-2 Tag, which can connect with a potentiostat-type sensor. The operating circuits contain a SIC4341 NFC chip, antenna, and electrode printed circuit board (PCB) ($4.5 \times 12.5 \times 0.1\,cm^3$). The sensor system is incubated in HBsAg solution for one hour followed by 30 minutes of incubation on bovine serum albumin (BSA) for surface blocking. The quantification of HBsAg is performed through the redox signal of $(Fe(CN)_6)^{3-/4-}$ before and after antigen addition. The diminution on the $(Fe(CN)_6)^{3-/4-}$ signal is interpreted as the presence of HBsAg. According to this detection principle, Figure 2.3B(c) illustrates different amperometric signals before and after HBsAg sensor exposure. Such amperometric measurements are carried out applying 0.3 V vs Ag/AgCl. The calibration curve indicates a linear response in the range of 10–200 $\mu g\,ml^{-1}$ and a detection limit of 0.17 $\mu g\,ml^{-1}$. The described system can provide results in real-time helping the diagnosis by the physicians. The system may be extended to other specific NAs (antibodies) for the detection of other specific viruses.

Uncontrolled virus spread has resulted in pandemics, which have affected many countries in the last century, such as those caused by Ebola and HIV viruses. An effective, rapid, reliable, and user-friendly sensing platform is required to stop the spread of viruses. Hence, there is a demand for a rapid POC detection system. The Figure 2.3C illustrates an electronic-resonance-frequency modulation to detect **Ebola** through a glycoprotein (GP) [46]. The electronic resonance-based system can perform measurements in one to two– minutes, providing stable detection and avoiding external noise. The developed system uses a standard lithographic method to fabricate gold electrodes onto SiO_2-layer, GO that is attached to gold (Au) surface through electrostatic interactions, an annealing process for 10 minutes at 400 °C is used to reduce the GO to improve the contact with the Au electrodes. Onto GO, a layer of Al_2O_3 passivation layer is deposited by atomic layer deposition (ALD). This is followed by AuNP sputtered layer. Cystamine (AET) is used to surface modification and a glutaraldehyde solution is added onto the surface. Afterward, the electrodes are incubated into the biorecognition agent solution (antibody probe solution KZ 52), which detects recombinant EBOV GP. The sensors are incubated in a tween 20 solution to prevent nonspecific bindings of analytes (as shown in Figure 2.3C(a)). When the sensors are exposed to GP at various concentrations, impedance measurements are taken to detect Ebola. The measurements of resonance frequency show that the system can detect GP from 0.001 to 3.401 $mg\,ml^{-1}$ (Figure 2.3C(b)). The proposed equivalent circuit for the GP detection system is shown in Figure 2.3C(c), where a second capacitance C_2 is added when the sensor is exposed to the solution. C_2 is coupled with gate-voltage-induced modified channel trans resistance R_2 forming a different time constant. Figure 2.3C(d) shows the channel-oxide trapping mechanism at different frequencies and the effect on phase shift modulation. In AC measurements, the resonance-frequency shift occurs when the top-gate field is generated by the antigen–antibody interactions, allowing the Ebola detection. The developed system has the potential to serve as an alternative measurement platform for field effect transistor (FET) biosensors and bioelectronics in healthcare.

Designing Biosensors for the Detection of SARS-CoV-2

SARS-CoV-2 has spread over the world since 2019. The rapid spread of the virus has caused a pandemic situation. The fast transmission of the virus indicates that a rapid diagnosis can help to stop the virus spread. Tracking symptoms such as headache, fever, fatigue, difficulty breathing, and cough was one of the first methods used to detect SARS-CoV-2; nevertheless, the symptoms can be common to other infections [48]. As a result, more accurate detection methods are required. The RT-PCR is one of the most accurate laboratory techniques for SARS-CoV-2 detection. However, the technique requires a sample collection, long processing time, expensive instrumentation, and skilled trained personnel. ELISA is another technique used for SARS-CoV-2 detection; this is less expensive and requires less time for analysis when compared to RT-PCR. Nonetheless, the simple technique is critical for early-stage diagnosis [49–55]. The demand for cheaper and portable sensing systems is increasing. In this regard, optical and electrochemical biosensors are promising alternatives that are currently under development; some of the strategies are described in the examples below.

Several strategies have been leveraged for developing POC electrodes systems for SARS-CoV-2 detection. One strategy is detection via IgM and IgG. The IgM can be found in the human body during viral infections, and the IgG is present in the human body as a long-term immunity biomarker. IgM can be detected in blood three or six days after infection begins, while IgG can be detected eight days after infection begins. Therefore, the detection of IgM and IgG provides information about recent SARS-CoV-2 exposure. Figure 2.4A illustrates a colorimetric POC-LFIA for the simultaneous detection of IgG and IgM in human blood within 15 minutes [18]. The POC strip consists of five parts: plastic backing, sample pad, conjugate pad, absorbent pad, and membrane (Figure 2.4A(a)). The sensor strips are prepared by immobilization of three biorecognition agents: antihuman-IgM, antihuman-IgG, and anti-rabbit-IgG, which are immobilized at the M, G, and C lines in the trip. The conjugated pad is sprayed with a mixture of AuNP-COVID-19 recombinant antigen conjugate and AuNP-rabbit-IgG. The sample pad has a pretreatment with BSA and tween solution before the sensor strip is used. The sample, either blood or serum mixed with buffer solution, is placed into the sample port, and incubated for around 15 minutes. In the sensor strip, three lines of red/pink color can appear, the presence of lines (M and G) indicates the presence of anti-SARS-CoV-2-IgM and or anti-SARS-CoV-2-IgG or both, if only a pink/red line at C appears, the sample is negative to SARS-CoV-2 as shown in Figure 2.4A(b).

Other strategies for virus detection including the SARS-CoV-2 is to employ electrochemical biosensors. As in optical sensors, the recognition layer can be fabricated employing different techniques including MIP [26]. Microorganisms and viruses can be detected using MIP-based sensors. MIP monolayers can detect protein fractions

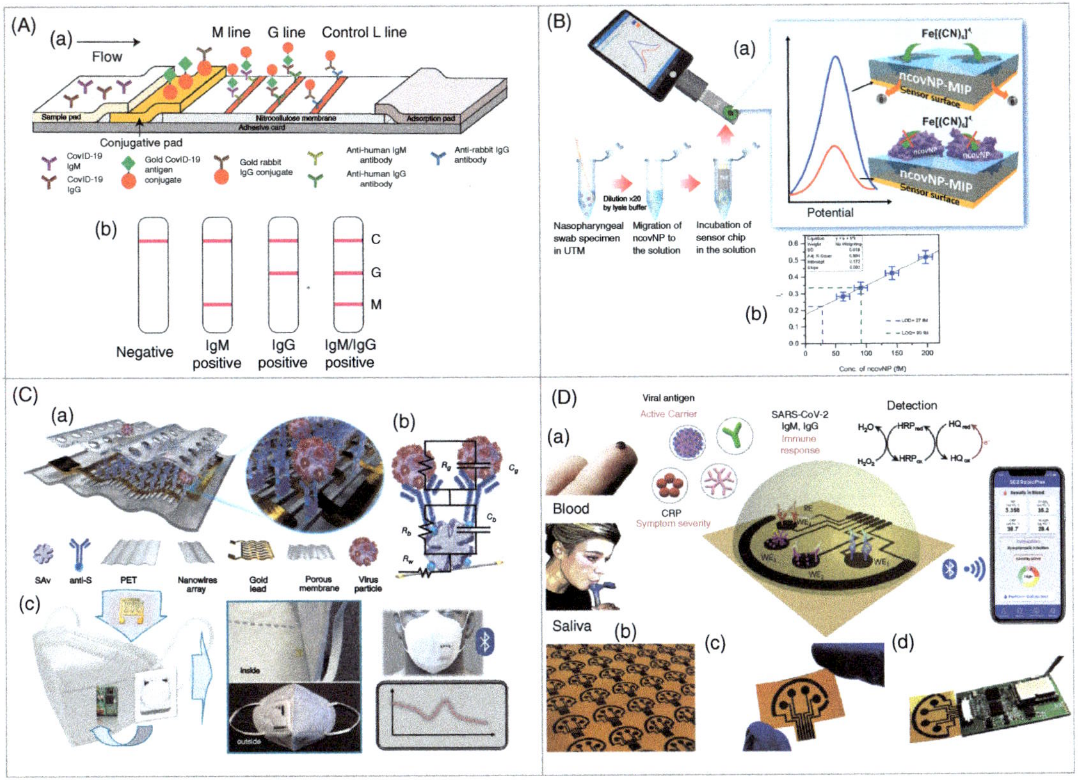

Figure 2.4 Biosensors for the detection of SARS-CoV-2. (A) Optical strip biosensor for SARS-CoV-2 through detection of IgG and IgM in blood or serum samples. A pink/red line means a negative result, two or three lines indicate the presence of SARS-CoV-2. Source: Reproduced with permission from Li et al. [18]. (B) Electrochemical MIP sensor based on nucleoprotein SARS-CoV-2 detection. (a) Schematic illustration of sample preparation and sensor detection principle. (b) Calibration plot for ncovNP detection. Source: Reproduced with permission from Xue et al. [56]; © 2021 Elsevier B.V. (C) Schematic illustration for a SARS-CoV-2 sensor face mask. (a) sensor layers, (b) equivalent circuit for (S-protein) detection, and (c) sensor integration in the intelligent mask face. Source: Reproduced with permission from Xue et al. [56]; © 2021 Elsevier B.V. (D) Rapid multiplex system for electrochemical detection of SARS-CoV-2 in blood and saliva samples, (a) integrated detection system and detection principle, (b) mass production sensor, (c) flexible biosensors with four working electrodes, a common reference electrode, and a common auxiliary electrode, (d) sensor array connected to a printed board circuit. Source: Reproduced with permission from Torrente-Rodríguez et al. [19]; © 2020 Elsevier Inc.

on pathogen surfaces selectively, resulting in low-cost, sensitive, fast, and portable sensors. In the case of SARS-CoV-2, the MIP layer can be fabricated by introducing in the template different fractions of a characteristic protein on the surface of the virus as is mentioned [57]. Figure 2.4B shows an example of a MIP electrochemical biosensor for SARS-CoV-2 detection [57]. The MIP electrochemical sensor is based on the detection of the nucleoprotein (ncovNP) through differential pulse voltammetry (DPV) in the presence of a redox probe. The sensors are prepared using gold-based thin-film electrodes (Au-TFEs). The MIP biorecognition film is prepared by ncovNP immobilization onto Au surface previously modified with 3,3′-dithiobis[sul fosuccinimidyl propionate] (DTSSP). After immobilization of the protein ncov-NP-MIP, a film is formed by electropolymerization of poly-*m*-phenylenediamine (PmPD), which is then treated with an ethanolic solution and acetic acid to remove the protein. Each preparation step is characterized by cyclic voltammetry of the redox probe (that is, $K_3[Fe(CN)_6]/K_4[Fe(CN)_6]$). The modified sensor is used in conjunction with a portable potentiostat that is controlled by a tablet or smartphone. The SARS-CoV-2 virus can be detected through the measurement of the reduction signal from the redox probe. When SARS-CoV-2 is present in the sample, the protein ncov-NP will attach to the MIP film and block the surface to access the redox probe, thus decreasing the DPV signal. The sensor is used with nasopharyngeal samples. Figure 2.4B(a) illustrates the MIP layer and the detection sensor principle; this figure also shows the sample preparation and sensor incubation in the test sample before ncov-NP detection. The sensor shows a linear response from 2.22 to 111 fM with a limit of detection (LOD) of 15 fM and a limit of quantitation (LOQ) of 50 fM as can be seen in the calibration plot in Figure 2.4B(b). The developed sensor is a potential POC express test that can distinguish some interfering proteins. However, proteins found in coronaviruses, such as MERS, must still be validated.

The efforts to develop a POC system for the detection of SARS-CoV-2 are not limited to strip development. Figure 2.4C illustrates an intelligent face mask on which the sensing system is based on a flexible immunosensor based on high-density conductive nanowire array [56]. In traditional sensors, rigid electrodes (metals or semiconductors) are inflexible and difficult to wear. A biocompatible and portable feature of flexible sensors makes them an ideal component of wearable devices [58]. A face mask can collect 95% of the aerosol particles produced by breathing, coughing, talking, or sneezing, allowing sampling to be performed without the need for additional infectious waste and without exposing other people. The disposable nanosensor can be embedded in a respiratory protective device (N95 mask) as shown in Figure 2.4C. The spacing wires in the nanosensor match the size of the virus enabling its capture. The nanosensor is made of three layers: (i) the external layer is a PC porous membrane that serves as a sensor protector and a droplet collector; (ii) the middle layer is a nanowire array that serves as a sensitive layer; and (iii) the internal layer is a flexible PET substrate that serves as a bottom supporting layer that facilitates adhesion to the mask. The nanowires at the sensing layer are modified with a bio-ink based on PEDOT:PSS. This bio-ink is water-soluble; thus, it can be directly patterned by soft printing. The surface of the bio-ink electrodes is modified with anti-spike proteins from SARS-CoV-2, which is the biorecognition agent. The anti-spike protein is immobilized through specific streptavidin (SAv)–biotin interactions. Sensor layers are shown in Figure 2.4C(a). The detection principle relies on the impedance model shown in Figure 2.4C(b). When the target particle (spike proteins [S-protein]) is captured, the surface is equivalent to the impedance of the antigens. When the virus concentration in the samples increases, the impedance of the captured antigens also increases. The generated digital data is transmitted via wireless to the smartphone as shown in Figure 2.4C(c). The sensor integration in the mask is a wearable device for the rapid on-site screening of SARS-CoV-2 infections. The LOQ of this system is 7 pfu ml^{-1}. Pfu stands for a plaque-forming unit, which is a measure of how many viruses can develop plaques per unit volume.

When false positive or false negative results are obtained, the results must be confirmed with an additional test, which requires additional time. To avoid this problem, a multiplexed, portable, and wireless electrochemical platform for rapid detection of COVID-19: SARS-CoV-2 is developed. The integrated system is illustrated in Figure 2.4D [19]. This platform detects and quantifies some specific biomarkers such as nucleocapsid protein (NP), specific immunoglobulins (Igs) against SARS-CoV-2 S1 (S1-IgM and S1-IgG) as well as CRP all at physiological relevant ranges in saliva and blood. The platform is based on the immobilization of antigens or antibodies (four working electrodes Figure 2.4D(a)) immobilized onto laser-engraved graphene (LEG) onto a polymide (PI) substrate (Figure 2.4D(b), (c)). The recognition agents are immobilized through 1-Pyrenebutyric acid (PBA), the unreacted sites are blocked with BSA to avoid nonspecific adsorption of other molecules in the interest sample. The electrochemical detection is performed by DPV and open-circuit potential-electrochemical impedance spectroscopy (OCP-EIS). The samples are incubated for 10 minutes prior to electrochemical measurements to ensure the highest sensitivity for each sensor; however, only one-minute incubation provides qualitative information about the absence and the presence of a target molecule, providing a rapid POC device. The amperometric measurements for NP viral antigen and CRP proteins are based on a double-sandwich and a sandwich configuration respectively. Direct immunoassays are used to detect S1-IgG and S1-IgM. The amperometric readings are sent via bluetooth to a user device. The electronic system includes a PCB and a lithium battery as can be observed in Figure 2.4D(d). This described POC system could be used at home for telemedicine care and remote monitoring.

The quick diagnosis of pathogens demonstrates promising sensing devices for accurate analysis in various biofluid samples; however, selectivity for some of the systems needs to be improved, and the incubation time for real samples should ideally be short. Researchers still work on adapting different technologies, such as wearable substrates, microfluidic chips, and screen printing, and combining them with mobile phones and miniaturized instruments. The challenges remain in mass-production fabrication, low-cost, high sensitivity and specificity, rapid response, and easy operation to deliver a POC system that can be used at home or at physician offices without any need for complex laboratory analysis samples. The development of smart sensing technologies will aid in early detection, reducing the scale of pathogen outbreaks and improving infection management.

2.3.2 Advanced Biosensors for Monitoring Metabolites

Metabolites are the intermediate products of metabolic reactions that occur within tissues and organs. Metabolite levels have been extensively studied in clinical diagnostics because they may reflect the functional status of the human body. A variety of metabolites have been specifically identified as targets for disease diagnosis, state, and progression. The body employs several defense systems to control or eliminate invading microorganisms, resulting in fluctuations in the levels of metabolites. Many metabolites are particularly useful as biomarkers because of their stability in media and ease of measurement. These compounds are a chemically heterogeneous group with distinct properties (such as reactivity, polarity, and solubility). The levels of metabolites are continuously changing because they are extensively exchanged with the environment (food and medications intake, pathological stimuli, and excretion). A relatively new research area, called metabolomics, has emerged. Metabolomics is "the quantitative measurement of the dynamic multiparametric metabolic response of living systems to pathophysiological stimuli or genetic modification" [59]. Metabolomics can assist in identifying metabolite patterns associated with diseases as well as tracking the efficacy and side effects of a treatment. Common techniques to trace metabolites in biological specimens are mass spectrometry, nuclear magnetic resonance (NMR) spectroscopy, and matrix-assisted laser desorption/ionization-time of flight (MALDI-TOF) coupled with complex multivariate statistical methods. All these techniques suffer from several limitations, such as bulky and expensive instrumentation and long analysis time. Moreover, they are not suitable for mobile health care systems, real-time detection, or untrained operators.

In recent years, numerous biosensor systems for POC applications have been developed. These strategies have demonstrated high effectiveness in detecting a wide range of metabolites in decentralized settings. The required characteristics of good stability, high specificity and sensitivity, ease of use, rapidity, low cost, and biocompatibility (for in vivo analyses) were also satisfied. This aspect makes biosensors important as sidekicks for diagnosis and other health-related fields.

Metabolites are present in a variety of bodily fluids, cells, and tissues. Their determination could be performed not only in blood, plasma, and serum, but also in other fluids such as tears, sweat, saliva, and urine for a noninvasive way. As it was frequently pointed out [60, 61], the quantity of many health-related species in these fluids relates to that circulating in blood.

The best-known example of a metabolite connected to a disease is **glucose**, whose continuous monitoring is essential in diabetes managing. Glucose levels in various bodily fluids are investigated. Several noninvasive systems have been developed. The levels of glucose in sweat, as well as other metabolites (such as lactic acid and uric acid) are related to that in blood. Metabolites such as glucose are also present in human tears along with electrolytes, proteins, and lipids [61].

An example of a noninvasive colorimetric system for the determination of glucose in tears is shown in Figure 2.5A [62]. The sensing system is integrated in contact lenses. The receptor is enzyme-based and comprises cerium nanoparticles (CeNPs) conjugated with glucose oxidase (GOx) using poly(ethylene glycol) (PEG). The CeNPs-PEG-GOx-laden lens is obtained using a common contact lens mold and photopolymerization, as reported in Figure 2.5A(a). In detail, a (hydroxyethyl)-methacrylate-based contact lens solution is sonicated with a photoinitiator and the CeNPs-PEG-GOx complex, the solution is poured into contact lens molds, then the molds are exposed to UV light. The sensing mechanism is the following: the enzyme oxidizes glucose to H_2O_2, which rapidly reduces colorless Ce^{3+} to yellow Ce^{4+}. The yellow CeNPs-PEG-GOx complex can be analyzed to quantify glucose using a smartphone equipped with a newly developed image-processing algorithm. The correction algorithm allows the measurement of color change with the same accuracy of a standard spectrometer. The wearability and cytotoxicity of the modified lens were evaluated, demonstrating that the entrapped complex is stable and does

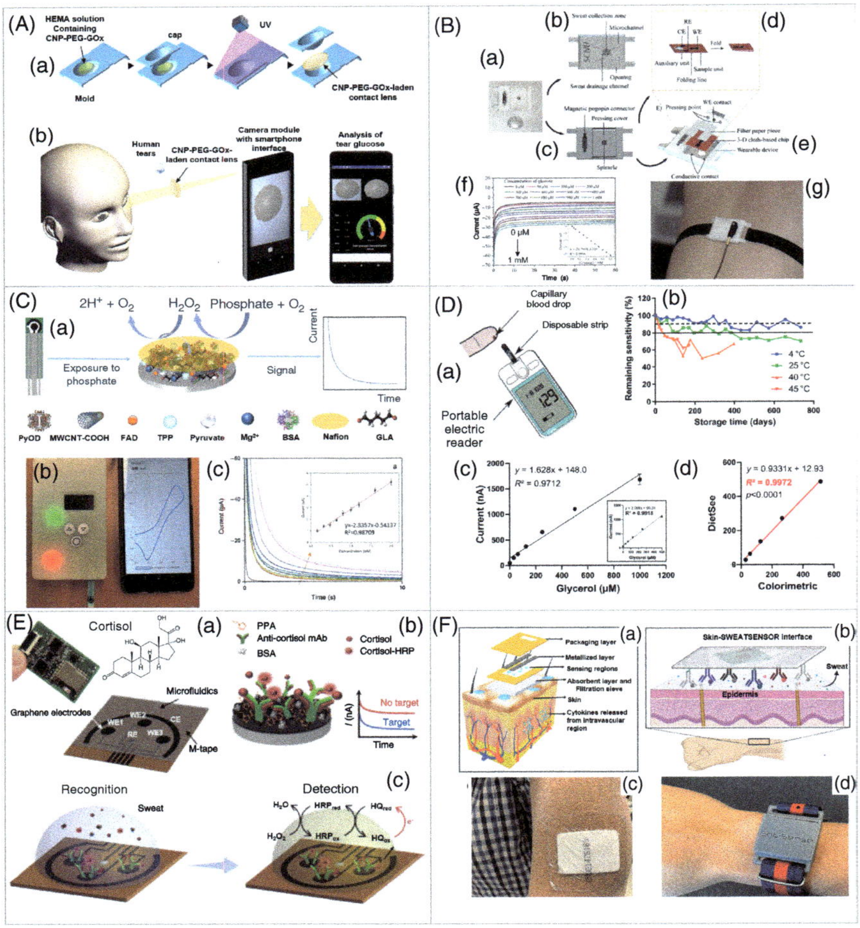

not leak from the lens. The lens was tested on diabetic rabbits showing a proportional correlation between the detected tear glucose and blood glucose measured by a conventional method. Moreover, the biosensor was also evaluated using tear samples from diabetic and healthy volunteers. The biosensor's performance was compared with the standard finger-prick method, and the results showed good agreement between the two methods.

A wearable cloth-based electrochemical biosensor for the determination of glucose is shown in Figure 2.5B [63]. This device comprises a network of microchannels with micromachined openings for sweat collection and a disposable three-dimensional cloth-based chip as detector (Figure 2.5B(a)–(e)). The chip includes two units (the unit containing the working electrode and an auxiliary unit with reference and counter electrodes). The working electrode is obtained by screen

Figure 2.5 Examples of advanced sensors for detecting biomarkers such as glucose, phosphate, cortisol, and cytokines. (A) Schematic illustration of the colorimetric contact lens-based platform for glucose monitoring. (a) Fabrication process for a CeNPs-PEG-GOx-laden contact lens from a general contact lens mold using photopolymerization. (b) RGB color analysis using the smartphone-based image-processing algorithm. Source: Reproduced with permission from Park et al. [62]; © 2021, American Chemical Society. (B) A wearable electrochemical biosensor for detection of glucose in sweat (WCECS). (a) Photo of the WCECS. (b) and (c) Structural schematics of the bottom and top of the wearable device. (d) Schematic of the 3-D cloth-based chip. (e) WCECS assembled from the wearable device and cloth-based chip. (f) Responses of the WCECS to different glucose concentrations from 0 to 1 mM (the inset shows the relevant calibration curve). (g) WCECS tied onto the back of a human subject. Source: Reproduced with permission from Zheng et al. [63]; © 2021, Elsevier B.V. (C) Biosensor system for saliva phosphate detection. (a) Illustration of the detection mechanism. (b) Electrochemical measurement using wireless portable potentiostat and smartphone. (c) Amperometric responses of the sensor to different concentrations of phosphate from 0 to 2 mM (the inset shows the relevant calibration curve). Source: Reproduced with permission from Bai et al. [64]; © 2021, Elsevier B.V. (D) Dietsee device for glycerol monitoring. (a) Schematic illustration of the device. (b) Stability of sealed DietSee test strips stored at different temperatures. (c) Calibration curve for glycerol as determined by the DietSee device. The inset shows the linear range. (d) Correlation between measurements of glycerol concentrations in PBS obtained using DietSee and standard colorimetry. Each measurement was performed in triplicate. Mean values are represented. Source: Reproduced with permission from Degrelle et al. [65]; © 2021, Elsevier B.V. (E) A wireless device for monitoring sweat cortisol. (a) Design of the microfluidic three-working-electrode sensor array for cortisol detection and photograph of the PCB with the graphene sensor patch for signal processing and wireless communication. WE, working electrode; CE, counter electrode; RE, reference electrode. (b) Representation of the affinity-based electrochemical cortisol sensor construction and sensing strategy: HRP, horseradish peroxidase; HQ, hydroquinone; PPA, pyrrole propionic acid; BSA, bovine serum albumin; McAb, monoclonal antibody. (c) Schematic of the electrochemical detection of cortisol in human sweat. Source: Reproduced with permission from Torrente-Rodríguez et al. [66]; © 2020, Elsevier B.V. (F) Wearable sweat sensor for detecting cytokines as an infection biomarker. (a) Layers of the strip on contact with skin. (b) Skin-SWEATSENSER device interface, where SWEATSENSER is functionalized with specific antibodies to capture the study biomarkers. (c) PharmChek sweat patch worn by subject on arm. (d) SWEATSENSER device worn on hand by a subject recruited for the study. Source: Reproduced with permission from Jagannath et al. [67]; © 2022, John Wiley & Sons. CC BY 4.0.

printing carbon ink, then modifying it by drop deposition of multiwalled carbon nanotubes with Prussian Blue (MWCNTs-PB) solution, then a GOx/chitosan mixture in phosphate buffer solution (PBS). GOx is used as a biorecognition element to selectively sense glucose as the analyte. A hand-held bipotentiostat is used for the read-out of the amperometric signal. Glucose is oxidized by means of the oxidase enzyme to produce gluconic acid and H_2O_2. MWCNTs promote the electron transfer among H_2O_2 and PB. PB is an electrochemical mediator that is frequently used in biosensing applications [68]. PB is used to decrease the potential required in the process of oxidizing H_2O_2, minimizing the interference of other electroactive compounds usually present in physiological samples. The detection of H_2O_2 is of great importance for biosensors that utilize oxidase enzymes as recognition elements since the concentration of H_2O_2 is proportional to its substrate. Without PB, direct H_2O_2 oxidation occurs at a higher potential where other electroactive compounds can be oxidized (i.e. ascorbate, urate, etc.). An example of the sensor responses to different glucose concentrations from 0 to 1 mM glucose is reported in Figure 2.5B(f) along with the relevant calibration curve. Selectivity, reproducibility (intra-assay and inter-assay), and storage stability were also tested in a model sweat solution. For real-time monitoring of glucose levels, the sensor could be tied onto the back of the human subject, as reported in Figure 2.5B(g).

Several other biosensors for remote glucose monitoring with noninvasive features were developed in recent years. As an example, a self-powered device for urine samples was proposed [69]. The electrochemical sensor is directly mounted on a diaper for monitoring elderly patients' conditions, and it is provided with a wireless transmitter. Some remarkable examples comprise a sensing system integrated in a pacifier for monitoring not only glucose but also other saliva biomarkers [70], and a nanoelectronic system able to work with a single drop of raw sweat, tears, or saliva [71]. Many microfluidic colorimetric systems with related smartphone readouts were also developed. Among them, researchers also developed a sensor integrated into a contact lens for multiplexed detection of some metabolites in tears [72] and one integrated into a mouthguard for monitoring glucose and nitrite in saliva [73].

Phosphate is a metabolite frequently used for disease diagnosis and monitoring of physiological conditions [64]. Hyperphosphatemia (high concentration of phosphate) can be found in hemodialysis patients and is linked to chronic kidney disease and uremia.

For this metabolite determination, an inkjet-printed electrochemical enzyme-based biosensor was developed [64], as shown in Figure 2.5C. Because the system can be easily managed using a smartphone, it is an excellent choice for POC phosphate detection. The immobilization of the enzyme (pyruvate oxidase) is realized through layer-by-layer printing strategy on a commercial SPE. Functional layers are printed in sequence. In the bottom layer, functionalized MWCNTs are introduced in the printing ink with the enzyme as carrier to enhance its loading. Cofactors are also added to support the enzyme reaction. In the middle layer, glutaraldehyde as crosslinking agent is deposited together with a nonionic surfactant containing a hydrophilic polyethylene oxide chain and an aromatic hydrocarbon hydrophobic side (Triton-X) as a viscosity modifier. The top protective layer is obtained by depositing a sulfonated tetrafluoroethylene-based fluoropolymer-copolymer (Nafion). The role of the MWCNTs was demonstrated by comparing voltammograms recorded at electrodes modified with and without MWCNTs in a $K_3Fe(CN)_6$ solution. The addition of the MWCNTs allows the favorable decrease of peak-to-peak separation, and a significant increase of peak current, suggesting an improved electron transfer kinetic.

The detection mechanism is detailed in Figure 2.5C(a). An Android app was developed for processing the signal and outputting the information from a portable potentiostat to a smartphone through bluetooth transmission, as shown in Figure 2.5C(b). The reported linear range is 160–2000 μM. The biosensor was tested in artificial saliva with a good recovery. Selectivity, stability, and reproducibility were also tested showing satisfying results. The lower measuring time and linear range extension show a clear improvement with respect to previously reported electrochemical sensors for saliva phosphate detection. The ease of operation, miniature size of the sensor, and portable instrumentation demonstrate the possibility of POC application.

Glycerol is a known biomarker of lipolysis. The rise of metabolites levels in blood are not only in response to physical exercise but also in relation to pathological conditions such as metabolic and cardiovascular diseases or cancer cachexia, making its monitoring essential in many healthcare settings [65].

A POC glycerometer device, called DietSee, was reported (Figure 2.5D) [65]. It is based on a strip-type biosensor that enables the quantification of glycerol directly from whole blood. This is particularly interesting because previously reported methods require the extraction of plasma from blood. The entire system is already patented, and it resembles well-known commercial glucometers (Figure 2.5D(a)). The strip-type biosensor consists of a SPE modified with a mixture of biochemical components (ATP, glycerol kinase, glycerol phosphate oxidase, peroxidase). The detection mechanism is reported in Scheme 2.1: glycerol is first phosphorylated by ATP to glycerol-3-phosphate (G3P) and ADP in a reaction catalyzed by glycerol kinase. Then, G3P is converted to dihydroxyacetone phosphate and H_2O_2 by glycerol phosphate oxidase. Finally, in the presence of a peroxidase, H_2O_2 is reduced to water and, simultaneously, a redox mediator is oxidized. The resulting electrical current, which is directly proportional to the glycerol concentration, is measured in six seconds by the electronic reader and is presented on the display.

$$\text{Glycerol} + \text{ATP} \xrightarrow{\text{Glycerol kinase (GK)}} \text{Glycerol-3-phosphate} + \text{ADP}$$

$$\text{Glycerol-3-phosphate} + O_2 \xrightarrow{\text{GPO}} \text{Dihydroxyacetone phosphate} + H_2O_2$$

$$H_2O_2 + \text{M(red)} \xrightarrow{\text{Peroxidase}} H_2O + \text{M(ox)}$$

$$\text{M(ox)} + ne^- \xrightarrow{\text{Electrode}} \text{M(red)}$$

Scheme 2.1 The detection mechanism of the glycerol sensor.

The biosensor was calibrated in several settings, by analyzing spiked samples of human and mouse blood and comparing the results to those obtained from standards in PBS (Figure 2.5D(c)). Results showed that the device is highly suitable for the detection of glycerol concentration in whole blood under most circumstances. Storage stability was analyzed through the course of regular use over a long period at different storage temperatures (Figure 2.5D(b)). Accuracy tests were made by comparing DietSee results with those obtained by a commercial colorimetric assay, showing a strong correlation between the two methods, as reported in Figure 2.5D(d). The effect of potential interferents was also evaluated, considering both endogenous (bilirubin, cholesterol, creatinine, etc.) and exogenous substances (therapeutic agents and vitamins). The device demonstrates a detection range suitable for humans, high selectivity, rapid response, and long storage stability. Moreover, its low price, ease of operation, and portability facilitates real-time analysis in healthcare settings.

Cortisol is one of the best-known steroid hormones and it is secreted by the adrenal gland in case of psychological or physical stress. The abnormal increase in cortisol could be related to severe infections or pathologies, such as autoimmune disease, cardiovascular complications, type-2 diabetes, and neurological disorders [74]. In contrast, abnormally low levels can lead to Addison's disease, which results in hypercholesterolemia, abnormal weight loss, and chronic fatigue [75]. Cortisol can be detected in a variety of fluids, such as blood, urine, saliva, and sweat.

An example of an integrated wireless device for monitoring sweat cortisol was designed [66]. It is an affinity-based electrochemical system, and the key component is a flexible five-electrode graphene sensor patch fabricated on a polyimide substrate via laser engraving. It consists of a microfluidic three-working-electrode sensor array, one Ag/AgCl reference electrode, and one graphene counter electrode as depicted in Figure 2.5E(a). The device also includes a microcontroller unit, a low-pass filter, and digital-to-analog/analog-to-digital converters for signal processing and wireless transmission. The detection in sweat and the sequential surface modification of the working electrodes for cortisol determination are shown in Figure 2.5E(b, c). Sweat cortisol and HRP-labeled cortisol compete for binding onto the antibody-modified electrode and the enzymatic reduction of H_2O_2 mediated by hydroquinone generates a cathodic current that is inversely proportional to the concentration of the target (Figure 2.5E(b)). Covalent attachment of the antibody on the working electrodes is achieved initially by activating the surface of the electrodes with carboxylate moieties and then by drop-casting the anti-cortisol antibody solution. Deactivation of unreacted sites with BSA is then performed. After a short incubation with sweat containing the enzymatic tracer (HRP-labeled cortisol), amperometric response in the presence of detection substrate (hydroquinone/H_2O_2) is recorded (Figure 2.5E(c)). The sensor was tested in buffer, sweat, and saliva samples from healthy subjects. Analytical parameters, such as LOD and concentration range, are suitable for physiological samples. Accuracy was established comparing the obtained data with those of a standard ELISA test. Since graphene has a large surface area and fast electron mobility, it increases sensitivity, while an immunosensing design offers high selectivity. Other remarkable features are the fast, noninvasive, monitoring ability and the opportunity for rapid and low-cost production of the sensor patch. All these reasons make this representative device suitable for becoming a wearable and portable platform, and it offers the possibility of personalized healthcare.

Cytokines are proteins produced by the immune system in response to pathogens or inflammatory events. Their level is, therefore, related to the progression of the infection. As a result, real-time monitoring of these species is critical for prognosis and treatment management. For their detection, current methods rely primarily on blood and derivatives (plasma, serum) or saliva as biofluid.

As shown in Figure 2.5F, an innovative approach was proposed [67], where determination of cytokines is carried out in passively expressed eccrine sweat. This example presents a newly developed system. The device consists of a replaceable sweat-sensing strip mounted onto a wearable electronic reader for real-time, continuous reporting of the signal. The output results can be sent via bluetooth to a smartphone. The strip is an electrochemical biosensor functionalized with receptors specific to the metabolites, i.e. the corresponding antibodies since the system works on an affinity-based mechanism. As detailed in Figure 2.5F(a, b), the strip also includes multiple fluid transport sites for effective capture of sweat. An absorbent layer of a patch interfaces with the skin to capture the sweat that diffuses through the porous sieve. The next layer, with sensing regions, functionalized with specific antibodies via a cross-linker on semiconducting ZnO nanofilm, entraps the target analytes (pro- and anti-inflammatory cytokines). The electrochemical binding interactions between the cytokines and their antibodies are transduced through the metallized layer. The sweat then diffuses into the next layer to be released out. The top-most layer is a packaging layer that allows for the used sweat to release out and prevent any external moisture from entering. EIS is used as the detection mode to determine the response of the binding interactions between the capture probes and target cytokines. An input sinusoidal voltage is applied to the electrode and the change in impedance due to the binding interaction resulting in charge modulation is recorded at a fixed frequency. Calibration curves for each biomarker were developed by measuring the impedance response for varying concentrations over the physiological range of 0.2–200 pg ml^{-1}. This concentration range covers the levels for determining acute infections including influenza, SARS-CoV-2, and sepsis. Accuracy was checked by comparing the proposed method with a standard ELISA test. For this purpose, sweat was passively collected using an FDA-approved sweat collection PharmChek patch as reported in Figure 2.5F(c). Selectivity and specificity were also verified. In addition, many clinical evaluations were made. The results of these tests demonstrate the actual applicability of the proposed system for on-field testing. It is worth mentioning that the device also monitors other vital signs such as skin temperature and perspiration. Figure 2.5F(d) shows the device worn on a subject wrist.

2.4 Conclusion and Prospects

The biosensing systems described in this chapter have great potential for the development of wearable/portable healthcare devices. They were proven able to provide rapid, sensitive, and selective detection of a broad range of clinically relevant target analytes, from common metabolites to those produced as response to invading microorganisms, to pathogens. The ease of use and the noninvasive features could expand their use to an ample variety of individuals, from healthcare personnel to common people. These devices could then facilitate the early diagnosis of many diseases associated with changes in body fluids and provide guidance during treatment monitoring. With the ongoing digital health trend, sensor technology has become a common part of people's daily routines. As more and more people adopt this technology, there is a growing number of individuals who are self-tracking their health. This has resulted in a significant amount of patient-generated health data that is progressively catching the attention of health care professionals and patients themselves.

Biosensor systems for POC analysis can be designed with different features to suit different applications. Many systems are meant to be self-administered. Self-assessment enable patients to take charge of their own health management and reduce the need for frequent medical visits and professional tests, resulting in lower travel expenses and time spent. Both routine tests for chronic disease patients and single use tests are needed. In the first case, it is preferable to have a portable reader equipped with single-use strips, chips, or similar components, made of low-cost and, preferably, environmentally friendly materials for better disposal. In contrast, single-use affordable devices such as commercially available LFIA devices are sought for one-shot tests such as infection assessment. In both cases, the readout should be easy to understand and unmistakable since incorrect interpretation of results could lead to several inconveniences. Moreover, painless, and comfortable or noninvasive body fluids collection is sought, especially for fragile patients. In this perspective, when other body fluids are not available, microneedle systems for interstitial fluids could be a viable alternative to standard finger prick methods.

For testing made by healthcare personnel, small dimension and portability of the device is still a preferable characteristic, but not always mandatory. In this case, the most sought aspect is the speed, along with the obvious low cost. As the global epidemic showed us, processing as many samples as possible in a short time is extremely important. Moreover, providing analytical testing results faster can make a significant difference in treating rapidly progressing illnesses and can even be a matter of life and death.

The main challenges are related to the complexity of some working mechanisms and the difficulty in translating manufacturing into large-scale production. Many aspects should be considered. Regarding the assembly of the sensor, a crucial step is the immobilization of the bioreceptor. This is because it needs to be stable, functional, and, in some cases, also in the appropriate configuration. It is also necessary to gain a deeper understanding of the kinetics and reactions between sensing components (enzymes, antibodies, mediators, and nanoparticles). The use of novel recognition elements and various transducing elements has contributed to solving the associated weaknesses.

Specificity and selectivity issues must be carefully evaluated to distinguish true signals from false positives. Reliable calibration methods should be evaluated since biological fluids have a very complex composition. To minimize differences and measurement uncertainty, it is important to use similar conditions for both standard calibration (in vitro conditions) as well as for real samples under physiological conditions. Notably, the materials used in the described novel systems have demonstrated good compatibility with biological samples and should not cause any additional issues. It is very important to consider that the composition of bodily fluids can vary, depending on a variety of factors such as health, age, drugs taken, or diet. The number of possible interferents and their combined actions is endless. A minority of devices developed for POC testing in academic laboratories have been used to detect analytes in untreated biological fluids. For these reasons, newly developed sensors should undergo extensive testing on a diverse range of individuals.

High sensitivity and low detection/quantification limits are clearly appealing features, but analytes levels in specific biological fluids should always been considered. As an example, additional research is needed to determine the relationship of biomarkers in sweat, saliva, tears, and breath to specific conditions and diseases. There is still room for improvement such as providing more reliable tests for the early detection of deadly diseases like cancer. Multidisciplinary collaboration, including industry, scientists, researchers, managers, and clinicians, can accelerate developments in this field and bridge the gap between laboratory proof-of-concept testing and full-scale clinical applications.

References

1 Sempionatto, J.R., Jeerapan, I., Krishnan, S., and Wang, J. (2020). Wearable chemical sensors: emerging systems for on-body analytical chemistry. *Analytical Chemistry* 92 (1): 378–396. https://doi.org/10.1021/acs.analchem.9b04668.

2 Parihar, A., Ranjan, P., Sanghi, S.K. et al. (2020). Point-of-care biosensor-based diagnosis of COVID-19 holds promise to combat current and future pandemics. *ACS Applied Bio Materials* 3 (11): 7326–7343. https://doi.org/10.1021/acsabm.0c01083.

3 Dwivedi, P. and Singha, M.K. (2021). IoT based wearable healthcare system: post COVID-19. In: *The Impact of the COVID-19 Pandemic on Green Societies: Environmental Sustainability* (ed. C. Chakraborty, S. Roy, S. Sharma, and T.A. Tran), 305–321. Cham: Springer International Publishing https://doi.org/10.1007/978-3-030-66490-9_13.

4 Jeerapan, I., Moonla, C., Thavarungkul, P., and Kanatharana, P. (2022). Lab on a body for biomedical electrochemical sensing applications: the next generation of microfluidic devices. *Progress in Molecular Biology and Translational Science* 187 (1): 249–279. https://doi.org/10.1016/bs.pmbts.2021.07.025.

5 Farokhzad, N. and Tao, W. (2021). Materials chemistry-enabled platforms in detecting sexually transmitted infections: progress towards point-of-care tests. *Trends in Chemistry* 3 (7): 589–602. https://doi.org/10.1016/j.trechm.2021.03.009.

6 Li, P., Lee, G.-H., Kim, S.Y. et al. (2021). From diagnosis to treatment: recent advances in patient-friendly biosensors and implantable devices. *ACS Nano* 15 (2): 1960–2004. https://doi.org/10.1021/acsnano.0c06688.

7 Fukushima, T., Kabata, H., Yamamoto, R. et al. (2021). The real-time reverse transcription-polymerase chain reaction threshold cycle values for severe acute respiratory syndrome coronavirus 2 predict the prognosis of coronavirus disease 2019 pneumonia. *Respiratory Investigation* 59 (3): 360–363. https://doi.org/10.1016/j.resinv.2020.12.011.

8 Hirotsu, Y., Mochizuki, H., and Omata, M. (2020). Double-quencher probes improve detection sensitivity toward severe acute respiratory syndrome coronavirus 2 (SARS-CoV-2) in a reverse-transcription polymerase chain reaction (RT-PCR) assay. *Journal of Virological Methods* 284: 113926. https://doi.org/10.1016/j.jviromet.2020.113926.

9 Gao, L., Yang, Q., Wu, P., and Li, F. (2020). Recent advances in nanomaterial-enhanced enzyme-linked immunosorbent assays. *Analyst* 145 (12): 4069–4078. https://doi.org/10.1039/D0AN00597E.

10 Kim, K., Kashefi-Kheyrabadi, L., Joung, Y. et al. (2021). Recent advances in sensitive surface-enhanced Raman scattering-based lateral flow assay platforms for point-of-care diagnostics of infectious diseases. *Sensors and Actuators B: Chemical* 329: 129214. https://doi.org/10.1016/j.snb.2020.129214.

11 Castle, L.M., Schuh, D.A., Reynolds, E.E., and Furst, A.L. (2021). Electrochemical sensors to detect bacterial foodborne pathogens. *ACS Sensors* 6 (5): 1717–1730. https://doi.org/10.1021/acssensors.1c00481.

12 Huang, Y., Chen, W., Chung, J. et al. (2021). Recent progress in fluorescent probes for bacteria. *Chemical Society Reviews* 50 (13): 7725–7744. https://doi.org/10.1039/D0CS01340D.

13 Nguyen, Q.H. and Kim, M.I. (2020). Nanomaterial-mediated paper-based biosensors for colorimetric pathogen detection. *TRAC Trends in Analytical Chemistry* 132: 116038. https://doi.org/10.1016/j.trac.2020.116038.

14 Sharma, A., Mishra, R.K., Goud, K.Y. et al. (2021). Optical biosensors for diagnostics of infectious viral disease: a recent update. *Diagnostics* 11 (11): https://doi.org/10.3390/diagnostics11112083.

15 Nor, S.N.S., Rasanang, N.S., Karman, S. et al. (2022). A review: surface plasmon resonance-based biosensor for early screening of SARS-CoV-2 infection. *IEEE Access* 10: 1228–1244. https://doi.org/10.1109/ACCESS.2021.3138981.

16 Woolhouse, M.E.J., Dye, C., Taylor, L.H. et al. (2001). Risk factors for human disease emergence. 356 (1411): 983–989. https://doi.org/10.1098/rstb.2001.0888.

17 Lo, Y., Cheung, Y.-W., Wang, L. et al. (2021). An electrochemical aptamer-based biosensor targeting *Plasmodium falciparum* histidine-rich protein II for malaria diagnosis. *Biosensors and Bioelectronics* 192: 113472. https://doi.org/10.1016/j.bios.2021.113472.

18 Li, Z., Yi, Y., Luo, X. et al. (2020). Development and clinical application of a rapid IgM-IgG combined antibody test for SARS-CoV-2 infection diagnosis. 92 (9): 1518–1524. https://doi.org/10.1002/jmv.25727.

19 Torrente-Rodríguez, R.M., Lukas, H., Tu, J. et al. (2020). SARS-CoV-2 rapidplex: a graphene-based multiplexed telemedicine platform for rapid and low-cost COVID-19 diagnosis and monitoring. *Matter* 3 (6): 1981–1998. https://doi.org/10.1016/j.matt.2020.09.027.

20 Limsakul, P., Charupanit, K., Moonla, C., and Jeerapan, I. (2021). Advances in emergent biological recognition elements and bioelectronics for diagnosing COVID-19. *Emergent Materials* 4 (1): 231–247. https://doi.org/10.1007/s42247-021-00175-9.

21 Abrego-Martinez, J.C., Jafari, M., Chergui, S. et al. (2022). Aptamer-based electrochemical biosensor for rapid detection of SARS-CoV-2: nanoscale electrode-aptamer-SARS-CoV-2 imaging by photo-induced force microscopy. *Biosensors and Bioelectronics* 195: 113595. https://doi.org/10.1016/j.bios.2021.113595.

22 Xi, H., Juhas, M., and Zhang, Y. (2020). G-quadruplex based biosensor: a potential tool for SARS-CoV-2 detection. *Biosensors and Bioelectronics* 167: 112494. https://doi.org/10.1016/j.bios.2020.112494.

23 Yakoh, A., Pimpitak, U., Rengpipat, S. et al. (2021). Paper-based electrochemical biosensor for diagnosing COVID-19: detection of SARS-CoV-2 antibodies and antigen. *Biosensors and Bioelectronics* 176: 112912. https://doi.org/10.1016/j.bios.2020.112912.

24 Dong, Q., Ryu, H., and Lei, Y. (2021). Metal oxide based non-enzymatic electrochemical sensors for glucose detection. *Electrochimica Acta* 370: 137744. https://doi.org/10.1016/j.electacta.2021.137744.

25 Moozarm Nia, P., Meng, W.P., Lorestani, F. et al. (2015). Electrodeposition of copper oxide/polypyrrole/reduced graphene oxide as a nonenzymatic glucose biosensor. *Sensors and Actuators B: Chemical* 209: 100–108. https://doi.org/10.1016/j.snb.2014.11.072.

26 Jamalipour Soufi, G., Iravani, S., and Varma, R.S. (2021). Molecularly imprinted polymers for the detection of viruses: challenges and opportunities. *Analyst* 146 (10): 3087–3100. https://doi.org/10.1039/D1AN00149C.

27 Amouzadeh Tabrizi, M., Fernández-Blázquez, J.P., Medina, D.M., and Acedo, P. (2022). An ultrasensitive molecularly imprinted polymer-based electrochemical sensor for the determination of SARS-CoV-2-RBD by using macroporous gold screen-printed electrode. *Biosensors and Bioelectronics* 196: 113729. https://doi.org/10.1016/j.bios.2021.113729.

28 Song, E., Chen, M., Chen, Z. et al. (2022). Mn^{2+}-activated dual-wavelength emitting materials toward wearable optical fibre temperature sensor. *Nature Communications* 13 (1): 2166. https://doi.org/10.1038/s41467-022-29881-6.

29 Sharma, P., Hui, X., Zhou, J. et al. (2020). Wearable radio-frequency sensing of respiratory rate, respiratory volume, and heart rate. *NPJ Digital Medicine* 3 (1): 98. https://doi.org/10.1038/s41746-020-0307-6.

30 Davies, H.J., Williams, I., Peters, N.S., and Mandic, D.P. (2020). In-ear SpO_2: a tool for wearable, unobtrusive monitoring of core blood oxygen saturation. *Sensors* 20 (17): https://doi.org/10.3390/s20174879.

31 Riglar, D.T. and Silver, P.A. (2018). Engineering bacteria for diagnostic and therapeutic applications. *Nature Reviews Microbiology* 16 (4): 214–225. https://doi.org/10.1038/nrmicro.2017.172.

32 Ali, A.A., Altemimi, A.B., Alhelfi, N., and Ibrahim, S.A. (2020). Application of biosensors for detection of pathogenic food bacteria: a review. *Biosensors* 10 (6): https://doi.org/10.3390/bios10060058.

33 Riu, J. and Giussani, B. (2020). Electrochemical biosensors for the detection of pathogenic bacteria in food. *TRAC Trends in Analytical Chemistry* 126: 115863. https://doi.org/10.1016/j.trac.2020.115863.

34 Vidal, P., Martinez, M., Hernandez, C. et al. (2019). Development of chromatic biosensor for quick bacterial detection based on polyvinyl butyrate-polydiacetylene nonwoven fiber composites. *European Polymer Journal* 121: 109284. https://doi.org/10.1016/j.eurpolymj.2019.109284.

35 Bu, T., Yao, X., Huang, L. et al. (2020). Dual recognition strategy and magnetic enrichment based lateral flow assay toward *Salmonella enteritidis* detection. *Talanta* 206: 120204. https://doi.org/10.1016/j.talanta.2019.120204.

36 Jiang, H., Jiang, D., Liu, X., and Yang, J. (2021). A self-driven PET chip-based imprinted electrochemical sensor for the fast detection of *Salmonella*. *Sensors and Actuators B: Chemical* 349: 130785. https://doi.org/10.1016/j.snb.2021.130785.

37 Ciui, B., Tertiş, M., Cernat, A. et al. (2018). Finger-based printed sensors integrated on a glove for on-site screening of *Pseudomonas aeruginosa* virulence factors. *Analytical Chemistry* 90 (12): 7761–7768. https://doi.org/10.1021/acs.analchem.8b01915.

38 Cinti, S., Volpe, G., Piermarini, S. et al. (2017). Electrochemical biosensors for rapid detection of foodborne *Salmonella*: a critical overview. *Sensors* 17 (8): https://doi.org/10.3390/s17081910.

39 Paniel, N. and Noguer, T. (2019). Detection of *Salmonella* in food matrices, from conventional methods to recent aptamer-sensing technologies. *Food* 8 (9): https://doi.org/10.3390/foods8090371.

40 Li, Y., Hu, Y., Chen, T. et al. (2022). Advanced detection and sensing strategies of *Pseudomonas aeruginosa* and quorum sensing biomarkers: a review. *Talanta* 240: 123210. https://doi.org/10.1016/j.talanta.2022.123210.

41 Du, Y. and Dong, S. (2017). Nucleic acid biosensors: recent advances and perspectives. *Analytical Chemistry* 89 (1): 189–215. https://doi.org/10.1021/acs.analchem.6b04190.

42 Chen, Y., Qian, C., Liu, C. et al. (2020). Nucleic acid amplification free biosensors for pathogen detection. *Biosensors and Bioelectronics* 153: 112049. https://doi.org/10.1016/j.bios.2020.112049.

43 Davis, L.E. (2014). Influenza virus. In: *Encyclopedia of the Neurological Sciences*, 2e (ed. M.J. Aminoff and R.B. Daroff), 695–697. Oxford: Academic Press https://doi.org/10.1016/B978-0-12-385157-4.00381-X.

44 Yang, B., Kong, J., and Fang, X. (2019). Bandage-like wearable flexible microfluidic recombinase polymerase amplification sensor for the rapid visual detection of nucleic acids. *Talanta* 204: 685–692. https://doi.org/10.1016/j.talanta.2019.06.031.

45 Teengam, P., Siangproh, W., Tontisirin, S. et al. (2021). NFC-enabling smartphone-based portable amperometric immunosensor for hepatitis B virus detection. *Sensors and Actuators B: Chemical* 326: 128825. https://doi.org/10.1016/j.snb.2020.128825.

46 Maity, A., Sui, X., Jin, B. et al. (2018). Resonance-frequency modulation for rapid, point-of-care ebola-glycoprotein diagnosis with a graphene-based field-effect biotransistor. *Analytical Chemistry* 90 (24): 14230–14238. https://doi.org/10.1021/acs.analchem.8b03226.

47 Dragan, A.I., Pavlovic, R., McGivney, J.B. et al. (2012). SYBR Green I: fluorescence properties and interaction with DNA. *Journal of Fluorescence* 22 (4): 1189–1199. https://doi.org/10.1007/s10895-012-1059-8.

48 Quer, G., Radin, J.M., Gadaleta, M. et al. (2021). Wearable sensor data and self-reported symptoms for COVID-19 detection. *Nature Medicine* 27 (1): 73–77. https://doi.org/10.1038/s41591-020-1123-x.

49 Santiago, I. (2020). Trends and innovations in biosensors for COVID-19 mass testing. 21 (20): 2880–2889. https://doi.org/10.1002/cbic.202000250.

50 Abid, S.A., Ahmed Muneer, A., Al-Kadmy, I.M.S. et al. (2021). Biosensors as a future diagnostic approach for COVID-19. *Life Sciences* 273: 119117. https://doi.org/10.1016/j.lfs.2021.119117.

51 Behera, S., Rana, G., Satapathy, S. et al. (2020). Biosensors in diagnosing COVID-19 and recent development. *Sensors International* 1: 100054. https://doi.org/10.1016/j.sintl.2020.100054.

52 Srivastava, M., Srivastava, N., Mishra, P.K., and Malhotra, B.D. (2021). Prospects of nanomaterials-enabled biosensors for COVID-19 detection. *Science of the Total Environment* 754: 142363. https://doi.org/10.1016/j.scitotenv.2020.142363.

53 Mahshid, S.S., Flynn, S.E., and Mahshid, S. (2021). The potential application of electrochemical biosensors in the COVID-19 pandemic: a perspective on the rapid diagnostics of SARS-CoV-2. *Biosensors and Bioelectronics* 176: 112905. https://doi.org/10.1016/j.bios.2020.112905.

54 Rasmi, Y., Li, X., Khan, J. et al. (2021). Emerging point-of-care biosensors for rapid diagnosis of COVID-19: current progress, challenges, and future prospects. *Analytical and Bioanalytical Chemistry* 413 (16): 4137–4159. https://doi.org/10.1007/s00216-021-03377-6.

55 Ranjan, P., Singhal, A., Yadav, S. et al. (2021). Rapid diagnosis of SARS-CoV-2 using potential point-of-care electrochemical immunosensor: toward the future prospects. *International Reviews of Immunology* 40 (1, 2): 126–142. https://doi.org/10.1080/08830185.2021.1872566.

56 Xue, Q., Kan, X., Pan, Z. et al. (2021). An intelligent face mask integrated with high density conductive nanowire array for directly exhaled coronavirus aerosols screening. *Biosensors and Bioelectronics* 186: 113286. https://doi.org/10.1016/j.bios.2021.113286.

57 Raziq, A., Kidakova, A., Boroznjak, R. et al. (2021). Development of a portable MIP-based electrochemical sensor for detection of SARS-CoV-2 antigen. *Biosensors and Bioelectronics* 178: 113029. https://doi.org/10.1016/j.bios.2021.113029.

58 Jeerapan, I. and Poorahong, S. (2020). Review – flexible and stretchable electrochemical sensing systems: materials, energy sources, and integrations.

Journal of the Electrochemical Society 167 (3): 037573. https://doi.org/10.1149/1945-7111/ab7117.

59 Tounta, V., Liu, Y., Cheyne, A., and Larrouy-Maumus, G. (2021). Metabolomics in infectious diseases and drug discovery. *Molecular Omics* 17 (3): 376–393. https://doi.org/10.1039/D1MO00017A.

60 Sonner, Z., Wilder, E., Heikenfeld, J. et al. (2015). The microfluidics of the eccrine sweat gland, including biomarker partitioning, transport, and biosensing implications. *Biomicrofluidics* 9 (3): 031301. https://doi.org/10.1063/1.4921039.

61 Farandos, N.M., Yetisen, A.K., Monteiro, M.J. et al. (2015). Contact lens sensors in ocular diagnostics. 4 (6): 792–810. https://doi.org/10.1002/adhm.201400504.

62 Park, S., Hwang, J., Jeon, H.-J. et al. (2021). Cerium oxide nanoparticle-containing colorimetric contact lenses for noninvasively monitoring human tear glucose. *ACS Applied Nano Materials* 4 (5): 5198–5210. https://doi.org/10.1021/acsanm.1c00603.

63 Zheng, L., Liu, Y., and Zhang, C. (2021). A sample-to-answer, wearable cloth-based electrochemical sensor (WCECS) for point-of-care detection of glucose in sweat. *Sensors and Actuators B: Chemical* 343: 130131. https://doi.org/10.1016/j.snb.2021.130131.

64 Bai, Y., Guo, Q., Xiao, J. et al. (2021). An inkjet-printed smartphone-supported electrochemical biosensor system for reagentless point-of-care analyte detection. *Sensors and Actuators B: Chemical* 346: 130447. https://doi.org/10.1016/j.snb.2021.130447.

65 Degrelle, S.A., Delile, S., Moog, S. et al. (2021). DietSee: an on-hand, portable, strip-type biosensor for lipolysis monitoring via real-time amperometric determination of glycerol in blood. *Analytica Chimica Acta* 1155: 338358. https://doi.org/10.1016/j.aca.2021.338358.

66 Torrente-Rodríguez, R.M., Tu, J., Yang, Y. et al. (2020). Investigation of cortisol dynamics in human sweat using a graphene-based wireless mHealth system. *Matter* 2 (4): 921–937. https://doi.org/10.1016/j.matt.2020.01.021.

67 Jagannath, B., Lin, K.-C., Pali, M. et al. (2021). Temporal profiling of cytokines in passively expressed sweat for detection of infection using wearable device. 6 (3): e10220. https://doi.org/10.1002/btm2.10220.

68 Ricci, F. and Palleschi, G. (2005). Sensor and biosensor preparation, optimisation and applications of Prussian Blue modified electrodes. *Biosensors and Bioelectronics* 21 (3): 389–407. https://doi.org/10.1016/j.bios.2004.12.001.

69 Shitanda, I., Fujimura, Y., Takarada, T. et al. (2021). Self-powered diaper sensor with wireless transmitter powered by paper-based biofuel cell with urine glucose as fuel. *ACS Sensors* 6 (9): 3409–3415. https://doi.org/10.1021/acssensors.1c01266.

70 García-Carmona, L., Martín, A., Sempionatto, J.R. et al. (2019). Pacifier biosensor: toward noninvasive saliva biomarker monitoring. *Analytical Chemistry* 91 (21): 13883–13891. https://doi.org/10.1021/acs.analchem.9b03379.

71 Sun, M., Pei, X., Xin, T. et al. (2022). A flexible microfluidic chip-based universal fully integrated nanoelectronic system with point-of-care raw sweat, tears, or saliva glucose monitoring for potential noninvasive glucose management. *Analytical Chemistry* 94 (3): 1890–1900. https://doi.org/10.1021/acs.analchem.1c05174.

72 Moreddu, R., Elsherif, M., Adams, H. et al. (2020). Integration of paper microfluidic sensors into contact lenses for tear fluid analysis. *Lab on a Chip* 20 (21): 3970–3979. https://doi.org/10.1039/D0LC00438C.

73 de Castro, L.F., de Freitas, S.V., Duarte, L.C. et al. (2019). Salivary diagnostics on paper microfluidic devices and their use as wearable sensors for glucose monitoring. *Analytical and Bioanalytical Chemistry* 411 (19): 4919–4928. https://doi.org/10.1007/s00216-019-01788-0.

74 Torpy, D.J. and Ho, J.T. (2007). Value of free cortisol measurement in systemic infection. *Hormone and Metabolic Research* 39 (06): 439–444. https://doi.org/10.1055/s-2007-980200.

75 McEwen, B.S. (1998). Protective and damaging effects of stress mediators. *New England Journal of Medicine* 338 (3): 171–179. https://doi.org/10.1056/NEJM199801153380307.

3

Disease Related Detection with Electrochemical Biosensors

Anulipsa Priyadarshini[1], Niharika Das[1], Saraswati Soren[1], Jashobanta Sahoo[2], Raghabendra Samantray[3], and Rojalin Sahu[1]

[1]*Future Materials Laboratory, School of Applied Sciences, Kalinga Institute of Industrial Technology (KIIT) Deemed to be University, Patia, Bhubaneswar, Odisha 751024, India*
[2]*Hindol College, Utkal University affiliated, Higher Education Department, Government of Odisha, Khajuriakata, Dhenkanal, India*
[3]*School of Chemical Technology, KIIT Deemed to be University, Bhubaneswar 751024, India*

3.1 Introduction

Human life has been enhanced by advances in science, technology, and industry. Health issues and diseases related to a longer lifespan, have become major concerns in the world today [1]. To decrease treatment costs and improve patient health, detection of diseases at the earliest stages is important. It is essential to employ highly sensitive and versatile techniques to detect diseases as soon as possible [2].

Many infectious diseases that cause outbreaks and have become a major concern in the modern world include dengue virus (DENV), Ebola virus (EBV), human immunodeficiency virus (HIV), Zika virus (ZIKV), influenza virus, hepatitis virus, and most recently, COVID-19. The rapid spread of these viruses makes them genetically stable and a constant threat to public health. Pathogenic viruses are imperative to detect rapidly and accurately due to several factors such as slow diagnostics, the rise of false-negatives and false-positives, and the requirement of extra kits. Researchers have worked to develop alternative approaches and devices to overcome these challenges. The use of electrochemical biosensors in environmental and agricultural applications, along with clinical laboratory testing, has become widely accepted. Electrochemical glucose biosensors are commonly used to monitor glucose levels of blood, diabetes, and to detect celiac disease, breast cancer, and prostate cancer, etc. [3].

A sensor gathers information about biological, chemical, or physical changes that converts the information into measurable information in the form of a signal [4] and the device that make a response in a selective way toward a specific sample and transforms chemical information, ranging from the concentration to total

Point-of-Care Biosensors for Infectious Diseases, First Edition.
Edited by Sushma Dave and Jayashankar Das.

composition of that particular analyte to an analytically helpful signal are termed as chemical sensors. Generally, chemical sensors have two components, i.e. a chemical recognition system and a physicochemical transducer, which are connected in series [5, 6]. Biosensors are the types of chemical sensors that are used for sensing a specific analyte by detecting biochemical molecular recognition [2, 7].

A biosensor is a simple and highly accurate analytical device that can detect a wide variety of biological and chemical analytes that converts chemical and biological reaction into a measurable signal. Due to low cost, fast response, portability, accuracy, as well as suitability for point-of-care diagnosis, these analyzers have steadily gained popularity in medical, environmental, industrial, food, and pharmaceutical analysis over the years [5, 6]. Combination of physicochemical transducers, i.e. electrical/optical/piezoelectric and biological recognition elements, i.e. biomacromolecules or whole cell/microorganisms senses the biological interaction between living environment and receptor of biosensors [8]. Depending on mode of physicochemical transduction, biosensors can be divided into optical, electrochemical, thermal, and piezoelectric biosensors [9].

Research and development of electrochemical biosensors are becoming comprehensively studied and have many potential applications contributing to upcoming generation medicines and point-of-care detection of biomarkers for diseases [10]. Electrochemical biosensors have gained popularity in recent years due to their high sensitivity, ease of use, low cost, ability to measure turbid samples, and potential compatibility, unlike other biosensor types [10, 11]. Furthermore, electrochemical biosensors have a biological recognition element, i.e. enzymes/antibodies/proteins or receptors that react with target sample and provide an electrical signal resembling to concentration of the given sample and a high specificity electrochemical transducer [12], which is briefly explained in Section 3.2. This chapter gives an overview of the various methods for making electrodes, immobilizing them, measuring them, and determining the relative materials associated with electrochemical biosensors. Additionally, some limitations of some recent works are also discussed.

3.2 Electrochemical Biosensors

For nearly 50 years, basic and applied research on electrochemical biosensors has been conducted. In 1975, the first commercial biosensor was used to perform a fast glucose assay on diabetic blood samples. There are currently many proposed and commercialized devices based on the biosensor principle, including those for pathogens and toxins, some of which are multichannel in nature [13].

Electrochemical biosensors with high sensitivity and specificity are used in biological recognition processes. The most common feature of electrochemical biosensors is the presence of a suitable element (enzyme, proteins, antibodies, nucleic acids, cells, tissues, or receptors) in the biorecognition layer that with the target analyte providing electro amplification, i.e. measurable signal selectively reacts by transducer [12, 14].

There are two major categories of electrochemical biosensors according to the nature of biological recognition: biocatalytic devices and affinity sensors whose detailed explanation will be provided in this section [4, 5, 11].

Electrochemical biosensors are classified according to their electrical input: direct current (DC) or alternating current (AC). Biosensors based on DC measure current variations as a readout for microbial interactions with sensor platforms. Depending on the waveforms of the input potential, the sensors can be categorized as voltametric or amperometric, where impedance is the frequency-dependent effective resistance of an electric circuit in response to an AC [15].

There are many types of biosensors, but electrochemical biosensors are used most often. An electrochemical detector measures an electrochemical signal by consuming or generating electrochemical species as a result of a bio-interaction process. In this, electrochemical detectors are used to measure electrochemical signals produced as a result of bio-interactions in which electrons are consumed or generated, producing an electrochemical signal. An electrochemical detector measures electrochemical signals produced by an electrochemical reaction as a result of the consumption or generation of electrochemical species during a bio-interaction [16].

According to biological recognition process, electrochemical biosensors are majorly classified into biocatalytic devices and affinity sensors. Enzymes, microorganisms, antibodies, and tissue slices are used in biocatalytic devices to recognize the target analyte and produce electroactive species. The analyte and a biological component such as an antibody, or receptor form a selective binding interaction in affinity sensors [4, 16].

But, due to a lack of surface architectures, electrochemical biosensors have been limited in their ability to identify biochemical events with high sensitivity. As a result, nanotechnology is increasingly being used to reduce dimensions of electrochemical sensors to increase their signal-to-noise ratios. Taking advantage of advances in electro and biochemistry, solid-state physics, and data analytics, highly sensitive and reliable micro (bio-)chemical sensors and sensor arrays can be generated [17]. We are therefore presenting a summary of recent progress in this field as well as discussing its future prospects. These days, numerous electrochemical biosensors are available for different health application needs. Some examples include glucose sensors, lactate sensors, catecholamine sensors, nucleic acid sensors, and uric acid sensors [7].

3.2.1 Materials

To begin with, it should be understood that electrode material plays an important role in the development of electrochemical sensors for detection of target molecules. As a result, the material selection criteria must include conductivity, cost, availability, workability, and biocompatibility. In this section, we discuss a variety of materials that have been used in the fabrication of sensors [18], including ceramics, semiconductors, organic materials, metals, and metamaterials [4].

For this purpose, a large number of nanomaterials, including nanoparticles, nanotubes, and nanowires have been proposed. These materials are widely available and have gradually gained popularity for use in different aspects of electrochemical biosensor systems, such as electrode fabrication, reporting molecules, capture probes, and coatings [4]. Nanomaterials based on metals, semiconductors, and organic compounds can improve magnetic electrical, optical, and chemical properties that are utilized for sensing devices. Therefore, they are frequently employed in research [19]. Some functional nanomaterials, such as carbon nanotubes, graphene, and silica nanoparticles, are commonly used in the development of highly effective electrode-supporting matrices because of their large surface area, high electrically conductive property, etc. [20]. An electrode surface with metal nanoparticles facilitates enzyme contact with the active site. The performance of a biosensor can be enhanced by using gold nanoparticles as electron relays. Because of their inherent anisotropies and effective electron transportation and excitons within the small dimension, 1D nanomaterials such as nanotubes and nanowires play crucial role in electrochemical biosensors. For instance, carbon nanotubes are commonly used as enzyme electrodes in electrochemical biosensors [4, 20].

Calcium carbonate ($CaCO_3$) has been shown to improve enzyme efficiency [20, 21]. Silver nanoparticles (AgNPs) have several advantages over other noble metal nanoparticles, by facilitating electron flow and accommodating more active sites on their surfaces [22]. Cao et al. developed an electrochemical immunosensor based on silver nanowires, even though the current assay system targets antigens, it can easily be extended to detect other antigens or biomaterials [23].

Carbon nanomaterials are introduced here as current electrode materials for electrochemical biosensors. Graphene is widely used as a replacement for traditional electrodes in electrochemical biosensors. Because of its excellent electronic and mechanical properties, graphene has been thought to be the best support material for label-free biosensors [19, 24].

In an investigation of α-fetoprotein quantitation using an electrochemical biosensor, Wang et al. investigated a label-free method [20, 25]. Multi-functionalized graphene (TB-Au-Fe_3O_4-rGO) was used for changing the electrode surface. Tuberculosis (TB) produces the electrochemical signal here as a type of redox probe. An electrochemical sensor reported by Trindade et al., for Cystatin C with intrinsic redox behavior mediated by ferrocene on a functionalized graphene base which was a failed biomarker [20, 26].

In one study, Patolsky et al. demonstrated that single-wall CNTs can be used to produce aligned reconstituted glucose oxidase that is readily attached to electrode surfaces and using single-walled carbon nanotubes (SWCNTs), we demonstrate that the enzyme's active site and electrode are electrically connected by the nano connector, while transport of electrons is controlled by the length of SWCNTs and distances greater than 150 nm [27]. CNTs possess numerous desirable mechanical and electronic properties dependent on their structure, so they can be used to make superior electrochemical biosensors [4]. SWCNTs have excellent mechanical and electronic properties and received a marginal attention in electrochemical biosensors due to their physicochemical properties [28]. Shirsat et al. [20, 29] created a SWCNT–polypyrrole multilayer biosensor that is coated with a polyvinylidene fluoride membrane, and used for glucose monitoring. Nonetheless, MWCNTs with excellent conduction and electro-catalytic properties were used as a modified scaffold on the electrode. An electrochemical biosensor based on polyethyleneimine-wrapped MWCNT electrodes was reported by Viswanathan et al. [30]. Mingdang Li et al. used AuNPs hybrid MWCNTs-SO_3H, which was primarily coupled by physisorption, to develop an ultrasensitive electrochemical biosensor [31].

MAX phases are a novel material class that combines metal and ceramic characteristics. They are $M_{n+1}AX_n$ materials, where A is an A-group element, M is an early transition metal, X is either a nitrogen or carbon group, and n ranges from 1 to 6, for example Ti_3AlC_2, NB_4AlC_3, and Cr_2AlC. MXenes, a 2D derivative of MAX phases, were first synthesized in 2011 and have already received significant attention because of their highly conductive nature, which can yield $1.5\times10^6\,S\,m^{-1}$ under optimal synthesis conditions. As of today, they have been functionalized with enzymes, DNA, and antibodies to enhance sensitivity and selectivity. The amorphous materials are nontoxic and have a wide range of functionalizations that are good for surface modification [18].

3.2.2 Working Principle

Generally, in (bio-)electrochemistry, a sensing device is utilized various types of recognition elements, but electrochemical detection mainly employs enzymes, primarily to measure currents (amperometry), potentials (potentiometry), and conductivities (conductometry) in between those electrodes. Additionally, field-effect and impedimetric methods utilize transistor technology to detect current through a potentiometric effect at a gate electrode to measure both resistance and reactance (impedance) [14].

In electrochemical biosensors, when electrochemical reactions are being monitored, an electrical current is generated, a charge accumulation is observed, or the medium between electrodes is altered in its conductive properties [10]. Using amplification and separation, useful information can be obtained. One of the most popular prototypes used to measure blood glucose levels in diabetic patients is an electrochemical glucose biosensor and it is also used to detect celiac disease [4], breast cancer [32], prostate cancer [33], hepatitis B virus [34], and other diseases [35].

The working electrode acts as a transducer, while the counter electrode provides a connection to an electrolytic solution, which enables a current to flow through the working electrode. During this process, chemical stability and conductive properties are both important for these electrodes. To maintain a known and stable potential, a reference electrode, usually made from Ag/AgCl. The working electrode, also known as the redox or sensing electrode, is usually referred to a reference electrode, a counter electrode, or auxiliary electrode is usually involved in the process of electrochemical sensing [14].

Biosensors are based on a different method that is directly related to protein electron transfer, acting as a transduction element. One field with a lot of potential for transferring electrons is proteins or redox enzymes [4]. Protein-film voltammetry (PFV) was a popular method in early years, to achieve direct electron transfer [36].

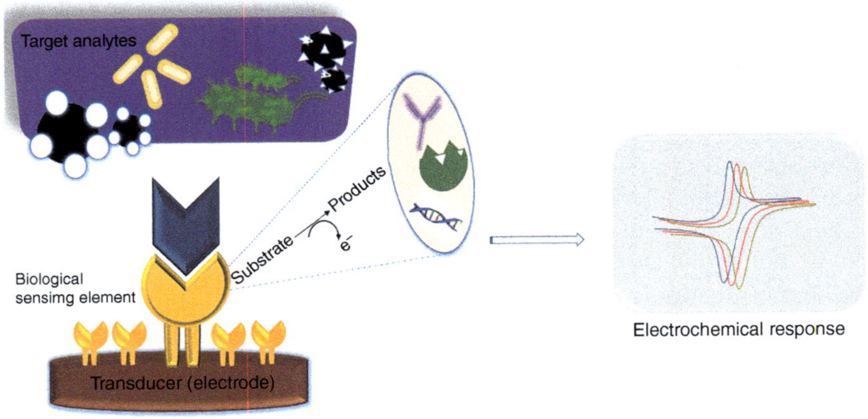

Figure 3.1 Schematic illustration of working principle of electrochemical biosensors.

Protein films absorbed on electrode surfaces are scanned by PFV. Signals are obtained from small sample quantities after the protein is immobilized on the electrode surface. In electrochemical biosensors, direct electron transfer increases sensitivity by integrating the recognition element with the transducer [4]. General working mechanism of electrochemical biosensors is illustrated in Figure 3.1.

3.3 Immobilization of Different Biomolecules

Immobilization of biomolecules on the transducer surface is an important procedure in the design of biosensors. The use of various immobilization procedures may contribute to differences in biosensor stability and sensitivity. For better immobilization, it is important that the antibody retains its functional conformation [4].

Shen et al. [37] investigated the impact of antibody immobilization on immunoassay performance parameters. The immobilization of enzymes is a crucial step for developing high-performance biosensors because it affects both the loading and the bioactivity of the enzymes. Enzyme immobilization has been investigated in various ways, including covalently attaching enzymes to the surface of substrates or incorporating enzymes into different matrixes. Many nanostructured materials have been synthesized and used to immobilize enzymes in various sizes, shapes, and compositions [38]. Several biosensors immobilize enzymes in three-dimensional matrixes, such as polydimethylsiloxane (PDMS), electropolymerized films (electrochemical polymerization), photopolymers, carbon pastes, and polysaccharides. The most commonly used covalent attachment techniques are glutaraldehyde and carbodiimide. A well-known approach used in glucose biosensors [39] is cross-linking, where enzymes are immobilized with glutaraldehyde. These techniques have been shown to be simple to implement, allowing for surface optimization. A novel H_2O_2 and glucose sensing platform based on the immobilization of Pd helical carbon nanofiber (Pd-HCNF) hybrid nanostructures was reported by Wagberg and coworkers [40]. Malhotra et al. fabricated a number of nanomaterials, which were used as electrode materials for immobilizing bienzymes (cholesterol oxidase [ChOx], cholesterol esterase [ChEt]). Amperometric biosensors using these bioelectrodes have demonstrated significant improvements in cholesterol biosensing [38].

Based on functional nanomaterials, Du and coworkers created a series of robust organophosphorus pesticides (OPs) biosensors. For example, to detect methyl parathion sensitively and selectively, a novel hydrolase biosensor based on self-assembly of methyl parathion hydrolase (MPH) on the Fe_3O_4@Au nanocomposite was developed. Fe_3O_4 nanocomposite makes it simple to build an enzyme biosensor. In addition, Adding AuNPs to the Fe_3O_4 core, blocks enzyme aggregates from forming and increases stabilization of enzyme immobilization [41].

Lipid membranes have a moderately biocompatible surface, and sensors built with lipid membranes have fast response rates and high sensitivities (BLMs). The incorporation of biological recognition molecules into lipid layers and their immobilization on BLMs are both influenced by the degree of access to reactive sites on the molecules [42]. Nikolelis et al. developed a method for building an immunosensor based on a self-assembled BLM [43].

According to Zhang and coworkers [44], a Pt@BSA nanocomposite and covalent adsorption of GOx enable sensitive glucose detection. Because of its simplicity and versatility, immobilization of enzymes with preserved activity with LBL assembly has been proven to be a reliable technique.

3.4 Different Types of Techniques Used in EC Biosensors for Detection of Various Diseases

Electrochemical biosensors typically rely on an enzymatic catalysis reaction between immobilized biomolecules and the targeted analyte, which generates electrons and influences the electrical properties of the solution. As discussed in Section 3.2, electrochemical biosensor devices can be classified according to their detection principle and application, as piezoelectric, potentiometric, conductometric, polarographic, voltametric, capacitive, impedimetric, and amperometry [3].

By reviewing representative devices and techniques from the aforementioned categories, a detailed overview of electrochemical sensing in conjunction with other well-established sensing methodologies is presented in this section [17]. In the next part, we will discuss each type of electrochemical biosensor based on a variety of techniques.

3.4.1 Voltametric Biosensor

Voltametric biosensors use a potential to a working electrode in comparison to a reference electrode and measure current via electrochemical reduction or oxidation at the working electrode shown in Figure 3.2 [4]. This is an electroanalytic method that involves the generation of current between a counter electrode and the working

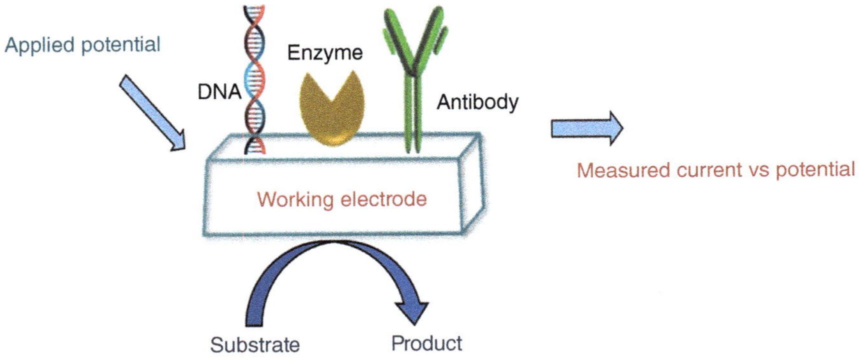

Figure 3.2 Schematic representation of function of voltametric biosensors.

electrode is monitored while the potential between the working electrode and a reference electrode is swept [8, 45]. In voltametric biosensors, biorecognition occurs between the analyte and the recognition layer, resulting in the current response through redox processes. Usually, current responses are observed as peaks that are correlated with electroactive species concentrations [17]. Voltametric sensors are noise-free and produce reliable results with high sensitivity [46]. In addition, research has been conducted to develop voltametric biosensors capable of simultaneous detection, as this can be applied to a smaller sample size, reducing analysis time, and reducing costs of detection.

A significant amount of attention has been paid to the toxins and pathogens of infectious diseases due to their importance for human health. The use of voltametric biosensing has enabled the detection of a number of toxins and pathogens. Voltametric techniques are classified as differential pulse voltammetry (DPV), linear sweep voltammetry (LSV), square wave voltammetry (SWV), and cyclic voltammetry (CV) based on the applied potential waveform [47].

A worldwide pandemic of COVID-19 has recently appeared, posing a serious public health threat. Researchers fabricated a biosensor for detection of COVID-19 in spiked saliva with an immobilized screen-printed electrode (eCovSens) and compared it to a potentiostat-based sensor that used a gold nanoparticle-immobilized FTO electrode. Using both sensing strategies, the eCovSens device provides high stability and rapid diagnosis [48].

In another study, quantum dots (QDs) based on bismuth citrate-modified graphite were used to create a novel and simple immunosensors, screen printed electrodes (SPE) that was originated for determination of CRP (C-reactive protein) in human serum by voltammetry. Under ideal conditions, a detection limit of 0.05 ng ml^{-1} was obtained in this study [49].

3.4.2 Electrochemical DNA Biosensors

Electrochemical DNA biosensors commonly use nanoscale interactions between the recognition layer, the target in solution [50], and a solid electrode surface.

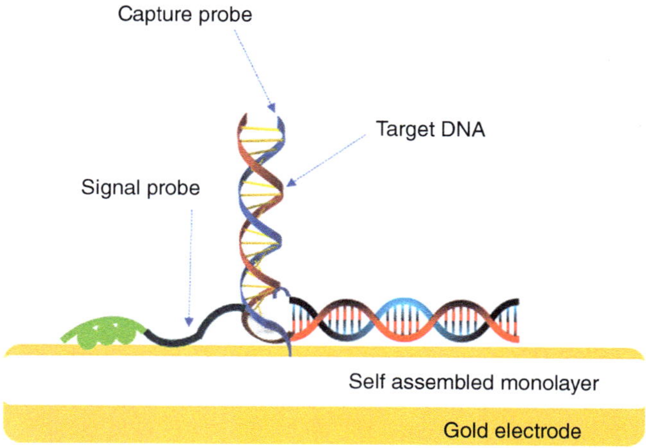

Figure 3.3 Detection principle of electrochemical DNA sensors.

The number of DNA biosensors has increased in recent years for the detection of oral cancer, human immunodeficiency virus type 1 (HIV-1), influenza viruses (e.g. H1N2, H3N2), Hepatitis B virus, etc. and the detection process is shown in Figure 3.3 [4]. Using a disposable electrochemical DNA biosensor, Azek et al. [51] were able to detect human cytomegalovirus (HCMV)-related DNA sequences. The detection limit for the HCMV amplified DNA fragment was 0.6 mol ml^{-1}. Using the method, the author demonstrated a 23 000-fold increase in sensitivity over electrophoresis and an 83-fold increase over the microtiter plate colorimetric hybridization assay.

Kara et al. [52] used the DPV method to identify HSV-related DNA sequences from PCR-amplified real samples and to separate Type 1 from Type 2 HSVs, as a result of DNA microarray technology, it can be used to detect the specific gene sequences associated with various viruses, bacteria, or even inherited diseases. Yan et al. developed a novel electrochemical assay that could simultaneously detect two DNA targets related to HIV and TB by using different QDs as well as sandwich hybridization [53]. An electrochemical ssDNA biosensor for sensing breast cancer was developed by Benvidi et al. In this electrochemical impedance spectroscopic (EIS) and CV techniques were used to investigate biosensor fabrication using the electrochemical redox pair of $[Fe(CN)_6]^{-3/-4}$ [32].

Detection of avian influenza A (H7N9) virus with the help of a novel tetrahedral DNA nanostructure has been reported in 2015. It exhibited that DNA tetrahedra look to have high potential as a probe for electrochemical biosensors, which have the potential to detect and control avian influenza A (H7N9) viruses and other pathogens at the gene level, which may help to prevent and control disease caused by them [54]. Ma et al. [55] modified a GCE of electrochemical biosensor with GO and AuNPs to detect Salmonella ssDNA. The immobilization of the thiolate Salmonella aptamer ssDNA sequence might decrease current, thus increasing resistance. As a result, the limit was calculated at 3 cfu ml^{-1}, which is extremely low. Aside from

DNA, microRNA detection is important in a variety of diseases, particularly in early clinical diagnosis.

3.4.3 Impedance Biosensors

The EIS technique has been used by chemists for more than a century. It is a method used for investigating the electrical properties of nanomaterials. An advantage of impedimetric methods is that samples can be detected without the use of enzyme labels. EIS techniques use the changes in charge transfer resistance (Rct) or electrode–electrolyte capacitance to detect biochemical changes at the surface of sensor [45]. Due to its ability to separate surface binding from solution resistance, EIS has also become a powerful tool to quantify signals. In contrast, Bogomolova et al. found that perturbations with small amplitudes did not have any negative effects on biological interactions [69]. Using the EIS method, George et al. invented CHIKV-nsP3, which causes replication of alphavirus.

Mishra and coworkers reported a higher impedance when measuring CD^{4+} (human cells) with a microelectrode. This study investigates the using battery-powered electronics construction of a small microelectrode which permits rapid detection and quantification of human $CD4^+$ cells within a minimum volume of blood through impedance measurements. Furthermore, when the number of captured cells increased, impedance increased. Others have investigated the use of EIS for the detection and quantification of virus- or drug-induced death in cell cultures [70]. The detection of COVID-19 by impedimetric means is in the early stages of development. COVID-19 could be detected using the EIS method due to its low LOD, rapid response, real-time monitoring of samples, cost-effective, and label-free detection [3].

To detect synthetic antigen sequences of ZIKV, Antonio et al. developed a label-free, impedance DNA biosensor with a detection limit of 25 nM, the sensor was capable of directly detecting the ZIKV sequences. In this study, a label-free impedimetric electrochemical DNA biosensor is described here for sensing of ZIKV. An electrode was manufactured by thermal evaporation on PET substrates in which nanometric gold layer was covered in a three-contact arrangement [64]. Moreover, Chowdhury and coworkers recently reported another promising study [71]. Their aim was to develop an ultrasensitive impervious impedance immunosensor capable of detecting the hepatitis E virus (HEV). A remarkable feature of the sensor is that it achieved a sensitivity comparable to real-time quantitative reverse transcription. Types of electrochemical biosensors used for detecting various disease is shown in Table 3.1.

3.4.4 Amperometric Biosensors

According to the literature, amperometric and potentiometric transducer platforms were the foundation of electrochemical biosensors. The first step towards the development of biomedical field was an amperometric glucose biosensor which is of low-cost, dependable, quick, and portable point-of-care diagnostic devices. At optimum

Table 3.1 Various diseases detected by different types electrochemical biosensors.

Target analyte	Biological sensing element	Types of electrochemical biosensors	Disease	References
Adiponectin	Antibody	Electrochemical immunosensor	Diabetes mellitus	[56]
Glucose	Enzyme	Glucose electrochemical biosensor	Diabetes mellitus	[57]
Anti-MBP	Antibody	Impedimetric immunosensors	Multiple sclerosis	[58]
β-Amyloid	Protein	—	Alzheimer's disease	[59]
α-Fetoprotein	Antibody	Electrochemical immunosensor	Liver cancer	[60]
p53 Protein	—	Amperometric biosensor	Tumour	[61]
miRNA-196b	DNA	Electrochemical biosensor	Pancreatic cancer	[62]
α-Fetoprotein	Antibody	Label-free immunosensor	Liver cancer	[63]
ZKV	Primer DNA	Electrochemical DNA biosensor	Dengue and yellow fever	[64]
Norovirus	DNA	Electrochemical DNA biosensors	Influenza virus A (H1N1)	[65]
CHIKV-nsP3	Antibody	Immunosensors	Chikungunya	[66]
miRNA-141	DNA-RNA hybrid duplexes	Electrochemical biosensor	Prostate cancer	[66]
L-Aβ40O	Antibody	Electrochemical biosensor	Alzheimer's disease	[67]
L-Fucose	—	Amperometric L-fucose biosensor	Liver cancer	[68]

potentials, amperometry measures the time-dependent electrooxidation/reduction of electroactive species. Due to its excellent selectivity, the potentiostatic technique is widely used in the development of chemical sensors. Furthermore, the amperometric technique's wide detection range and low-cost instrumentation have made it a favored among sensor developers. The advantage of amperometry is that it minimizes the charging current, which impacts the detection limits. As a result of the combination, a large number of highly sensitive and reliable biosensors have been developed to date. Depending on the type of bioreceptor-target attachment, amperometric biosensors generate an amperometric signal. A charge transfer between a target molecule and an electron-rich label is capable of producing direct signals [72].

For DENV serotype 2, Bouhemadou and coworkers recently published a new DNA hybridization-based amperometric sensor. Silver-coated interdigitated electrodes based on Cu_2CdSnS_4 alloy nanostructures provide more stable DNA immobilization. 3-aminopropyl-triethoxysilane was used as a covalent linkage between the carboxyl terminated probe DNA and the electrode, and its hybridization with dengue viral DNA altered the electrode characteristics. It is shown that the amperometric signal inversely correlates with target DNA concentration within a range of 100 f–10 nM [73].

Cui et al. developed a glycan-based amperometric biosensor for detecting influenza viruses in vivo with high specificity and sensitivity. Glycans can be selectively released from the influenza virus by the glycoprotein found on its surface. As part of this assay, glucose strips containing dehydrogenase were used to detect galactose released by viruses indirectly via amperometric detection. A notable feature of this sensor is its ability to analyze human nasal swabs and its fast assay time (15 minutes) [74].

Sharma et al. created an amperometric immunosensor to detect HRP2 in human *Plasmodium falciparum* malaria sera. Using multiwall carbon nanotubes and gold nanoparticles, disposable screen-printed electrodes were modified. It was found that Nano-Au/MWCNT/SPE electrodes showed the best immunosensing performance of all electrodes, of detection limit 8 $ng\,ml^{-1}$ [75].

Using bipodal thiolated self-assembled monolayers with reactive N-hydroxysuccinimide ester end groups, Laboria et al. [76] developed a biosensor to detect CEA. It was found significant sensitivity of 3.8 $nA\,ml\,ng^{-1}$ and the limit of detection of 0.2 $ng\,m^{-1}$ demonstrate a linear relationship with CEA concentration over the range of 0–200 $ng\,ml^{-1}$.

A magneto immunoassay-based approach using magnetic micro nanoparticles for detecting *P. falciparum* histidine-rich protein 2 was used for the first time by Castilho et al. [77]. Magneto sensors made of graphite-epoxy composite (m-GEC), which also functioned as electrochemical transducers, were used to capture the modified magnetic particles.

3.4.5 Potentiometric Biosensors

As analytical tools for biomedical applications, potentiometric biocatalytic biosensors are widely used. Potentiometric biosensors are uniquely utilized for measuring urea level. Potentiometric creatinine biosensors are also useful for clinical analysis; however, due to interference concerns, they should be used with caution. The main advantages of these biosensors are their simple operation principles. A variety of biosensors are available, ranging from simple disposable devices to advanced detectors in sophisticated analyzers and monitors. In biomedical, potentiometry is recommended for determining simple ions, most notably pH and physiological electrolytes. In the development of biosensors for biomedical applications, the development of integrated multisensor and biosensor devices that can simultaneously detect all these species using a uniform potentiometric system seems to be an appealing trend. Potentiometric biosensors have two important characteristics, according to IUPAC recommendations: (i) Sensor receptors contain a biological component, recognizing an analyte, and (ii) Analytical signals generated by biosensors are potential sources of information. Although the definition is well-known, it is necessary to recall these two seemingly obvious statements because the term "biosensor" is frequently misused in the analytical literature [78]. Working and reference electrode potential differences vary directly with analyte concentration at zero current flow [45].

A potentiometric biosensor based on chemically modified graphene and aptamers has been reported for the ultrasensitive, rapid detection of *Staphylococcus aureus* (*S. aureus*). The biosensor doesn't require time-consuming pretreatment procedures such as DNA extraction by detecting one colony forming unit (CFU)/ml in a few seconds. This biosensor shows a high potential for detection of difficult microorganisms. The purpose of this study is to develop a new generation of potentiometric aptasensors with a significantly improved performance by taking advantage of the excellent electrochemical and transduction properties of chemically modified graphene [79].

Potentiometric sensors typically use field effect transistors (FETs), gas-sensitive and ion-selective electrodes that are chosen according to the species used for different studies. Researchers have recently focused on the fabrication of biosensors for POC devices, so potentiometric sensors with FET have received great consideration, which could be attributed to the inherent miniaturization potential of FETs [80]. The rapid response, easy construction and high sensitivity of FETs have also increased researchers' interest in this transducer platform. Basically, FET has three terminals: source, drain, and gate. The gate terminal (dielectric material) of biologically sensitive FET contains bioreceptors that are specific to target. By binding charged biomolecules such as proteins, DNA, RNA to the gate terminal, changes in surface charge can change the gate voltage, and therefore the charge transport properties of the FET channel also get affected. The FET-based immunosensor is the most important tool for current scenario. Seo et al. developed an important immunosensor for the extremely pandemic COVID-19 causative virus [81].

3.4.6 Electrochemical Immunosensor

Immunosensors are the biosensor devices that use immunochemical reactions in the serum and other media to detect specific antigens or antibodies. Immunosensors are based on antigen–antibody complexes that are naturally formed by antibodies, it is thus an excellent tool for pathogen detection which is illustrated in Figure 3.4. According to the transducer, immunosensor can be categorized into four types, e.g. thermometric, electrochemical, magnetic, and optical. When the antigen–antibody combination forms in an electrochemical immunosensor, the biological signal is translated into an electrical signal. Electrochemical immunosensors are also gaining popularity in clinical diagnosis techniques. These type of immunosensors had already been effective in detecting certain viruses. Electrochemical immunosensors have recently found viruses such as Plum pox virus, DENV, Fig mosaic virus, and others [3].

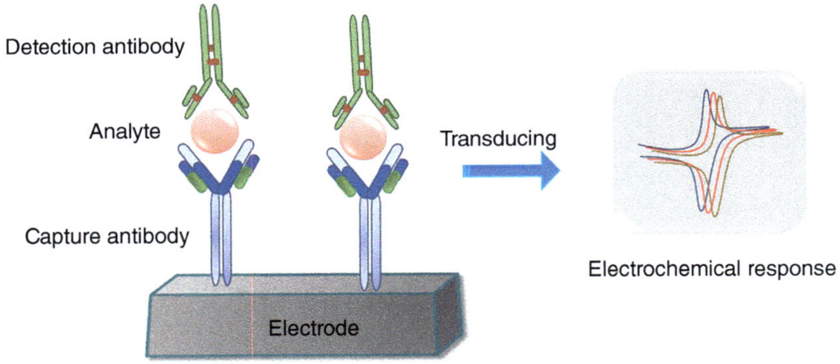

Figure 3.4 Pictorial representation of electrochemical immunosensor.

For the detection of the avian leukovirus subgroup J, Ning and coworkers devised highly sensitive electrochemical immunosensor (ALV-J) [82]. Viruses that cause avian leukosis are endogenous retrovirus that can cause cancer, tumors, and other diseases in doves, poultry, mammals, and birds. The virus can be transmitted to humans through commercially available eggs or poultry but not directly infected by it. Certain disorders can also be detected in the blood or serum by determining tumor markers. These disorders can be detected using electrochemical immunosensors, in which two antibodies, the tracer antibody and the capture antibody are combined on an electrode surface, leading to a sandwich immunoreactivity [4].

Using magnetic Fe_3O_4 nanoparticles as antibody carriers and methylene blue are used to construct an electrochemical immunosensor. With a gold sheet and hemin/G-quadruplex DNAzyme as the signal amplifiers, it proved sensitive in detecting Hepatitis B surface antigen. The proposed immunosensor offered a promising disease diagnosis platform [4]. The disposable immunosensor made by Wu et al. [83] detects CEA (SPCE) by depositing a CEA/colloid Au/chitosan membrane on a carbon screen. According to Pan et al., an electrochemical immunosensor has been developed for detecting the urine tract infection (UTI) biomarker lactoferrin. A SAM of 11-mercaptoundecanoic acid (MUDA) or a mixture of MUDA and 6-mercapto-1-hexanol was applied to the electrode surfaces. Lactoferrin was detected in urine using a sandwich amperometric immunoassay with a detection limit of $145\,pg\,ml^{-1}$ [72].

3.5 Conclusion and Future Direction

Over the last decade, the development of sensors has been one of the most interesting areas of analytical chemistry research. Electrochemical biosensors are gaining traction due to their sensitivity, speed of analysis, low-cost, and their ability to detect a wide range of diseases through their choice of biomarkers and materials. As a result, electrochemical biosensor detection has a lot of potential in terms of complex diseases and in the meantime, biosensors for disease detection have sparked a lot of interest [1]. However, in this chapter, we focus on the impact of electrochemical biosensors, a sub-type of biosensor for such purposes.

Antibiotics have been frequently used to treat COVID-19 in the present pandemic, which can be expected to lead to an increase in bacterial resistance [34]. Their distinct characteristics were showcased. Biomolecules were immobilized onto electrode surfaces using a variety of immobilization methods. In the last few decades, the field of electrochemical biosensors has progressed considerably. As indicator labels, signal amplifiers, and electrodes have all been improved, the linear response range has been extended and the detection limits lowered. Indeed, coating electrodes with nanomaterials and using innovative indicators may improve sensor performance.

Molecular diagnostics remains a strong focus for future research to obtain improved sensitivity and stability. In the meantime, more clinical samples from patients must be processed for analytical validation. The type of the protein, antigen enzyme, or other biomolecules, along with the directed immobilization, must all be taken into account. Future studies will need to look into the best recognition component as well as the best immobilization techniques. But there are still many obstacles to overcome, including the number of proteins (or enzymes) that provide information about the process of transferring electrons on the transducer, as well as determining the kinetics of electron transfer from the recognition element to the electrode surface [4].

References

1 Kim, J.H., Suh, Y.J., Park, D. et al. (2021). Technological advances in electrochemical biosensors for the detection of disease biomarkers. *Biomedical Engineering Letters* 11 (4): 309–334.

2 Anik, Ü. (2017). Electrochemical medical biosensors for POC applications. In: *Medical Biosensors for Point of Care (POC) Applications* (ed. R. Narayan), 275–292. Woodhead Publishing.

3 Khan, M.Z., Hasan, M.R., Hossain, S.I. et al. (2020). Ultrasensitive detection of pathogenic viruses with electrochemical biosensor: state of the art. *Biosensors and Bioelectronics* 166: 112431.

4 Huang, Y., Xu, J., Liu, J. et al. (2017). Disease-related detection with electrochemical biosensors: a review. *Sensors* 17 (10): 2375.

5 Thévenot, D.R., Toth, K., Durst, R.A., and Wilson, G.S. (2001). Electrochemical biosensors: recommended definitions and classification. *Biosensors and Bioelectronics* 16 (1, 2): 121–131.

6 Wen, W. (2016). Introductory chapter: what is chemical sensor? In: *Progresses in Chemical Sensor* (ed. W. Wang), 3–8. IntechOpen.

7 Yang, A. and Yan, F. (2020). Flexible electrochemical biosensors for health monitoring. *ACS Applied Electronic Materials* 3 (1): 53–67.

8 Srivastava, K.R., Awasthi, S., Mishra, P.K., and Srivastava, P.K. (2020). Biosensors/molecular tools for detection of waterborne pathogens. In: *Waterborne Pathogens* (ed. M.N.V. Prasad and A. Grobelak), 237–277.

9 Naresh, V. and Lee, N. (2021). A review on biosensors and recent development of nanostructured materials-enabled biosensors. *Sensors* 21 (4): 1109.

10 Miku, E. (2021). Recent advancements in electrochemical biosensors for Alzheimer's disease biomarkers detection. *Current Medicinal Chemistry* 28 (20): 4049–4073.

11 Valkova, P. and Pohanka, M. (2021). Novel trends in electrochemical biosensors for early diagnosis of Alzheimer's disease. *International Journal of Analytical Chemistry* 2021: 1–13.

12 Ronkainen, N.J., Halsall, H.B., and Heineman, W.R. (2010). Electrochemical biosensors. *Chemical Society Reviews* 39 (5): 1747–1763.

13 Pohanka, M. and Skládal, P. (2008). Electrochemical biosensors – principles and applications. *Journal of Applied Biomedicine* 6 (2): 57–64.

14 Chaubey, A. and Malhotra, B. (2002). Mediated biosensors. *Biosensors and Bioelectronics* 17 (6, 7): 441–456.

15 Pohanka, M., Sklá, P., and Kroè, M. (2007). Biosensors for biological warfare agent detection. *Defence Science Journal* 57 (3): 185.

16 Karbelkar, A.A. and Furst, A.L. (2020). Electrochemical diagnostics for bacterial infectious diseases. *ACS Infectious Diseases* 6 (7): 1567–1571.

17 Grieshaber, D., MacKenzie, R., Vörös, J., and Reimhult, E. (2008). Electrochemical biosensors-sensor principles and architectures. *Sensors* 8 (3): 1400–1458.

18 Elbadawi, M., Ong, J.J., Pollard, T.D. et al. (2021). Additive manufacturable materials for electrochemical biosensor electrodes. *Advanced Functional Materials* 31 (10): 2006407.

19 Săndulescu, R., Tertiş, M., Cristea, C., and Bodoki, E. (2015). New materials for the construction of electrochemical biosensors. *Biosensors-Micro and Nanoscale Applications* 24: 1–36.

20 Cho, I.H., Kim, D.H., and Park, S. (2020). Electrochemical biosensors: perspective on functional nanomaterials for on-site analysis. *Biomaterials Research* 24 (1): 1–2.

21 Rosu, R., Iwasaki, Y., Shimizu, N. et al. (1998). Intensification of lipase performance in a transesterification reaction by immobilization on $CaCO_3$ powder. *Journal of Biotechnology* 66 (1): 51–59.

22 Shahdost-Fard, F. and Roushani, M. (2017). Designing an ultra-sensitive aptasensor based on an AgNPs/thiol-GQD nanocomposite for TNT detection at femtomolar levels using the electrochemical oxidation of Rutin as a redox probe. *Biosensors and Bioelectronics* 87: 724–731.

23 Cao, X., Liu, S., Feng, Q., and Wang, N. (2013). Silver nanowire-based electrochemical immunoassay for sensing immunoglobulin G with signal amplification using strawberry-like ZnO nanostructures as labels. *Biosensors and Bioelectronics* 49: 256–262.

24 Krishnan, S.K., Singh, E., Singh, P. et al. (2019). A review on graphene-based nanocomposites for electrochemical and fluorescent biosensors. *RSC Advances* 9 (16): 8778–8881.

25 Wang, Y., Zhang, Y., Wu, D. et al. (2017). Ultrasensitive label-free electrochemical immunosensor based on multifunctionalized graphene nanocomposites for the detection of alpha fetoprotein. *Scientific Reports* 7 (1): 1.

26 Trindade, E.K., Silva, B.V., and Dutra, R.F. (2019). A probeless and label-free electrochemical immunosensor for cystatin C detection based on ferrocene functionalized-graphene platform. *Biosensors and Bioelectronics* 138: 111311.

27 Patolsky, F., Weizmann, Y., and Willner, I. (2004). Long-range electrical contacting of redox enzymes by SWCNT connectors. *Angewandte Chemie* 116 (16): 2165–2169.

28 Guo, X. (2013). Single-molecule electrical biosensors based on single-walled carbon nanotubes. *Advanced Materials* 25 (25): 3397–3408.

29 Shirsat, M.D., Too, C.O., and Wallace, G.G. (2008). Amperometric glucose biosensor on layer by layer assembled carbon nanotube and polypyrrole multilayer film. *Electroanalysis: An International Journal Devoted to Fundamental and Practical Aspects of Electroanalysis* 20 (2): 150–156.

30 Viswanathan, S., Rani, C., Anand, A.V., and Ho, J.A. (2009). Disposable electrochemical immunosensor for carcinoembryonic antigen using ferrocene liposomes and MWCNT screen-printed electrode. *Biosensors and Bioelectronics* 24 (7): 1984–1989.

31 Li, M., Wang, P., Li, F. et al. (2017). An ultrasensitive sandwich-type electrochemical immunosensor based on the signal amplification strategy of mesoporous core–shell Pd@ Pt nanoparticles/amino group functionalized graphene nanocomposite. *Biosensors and Bioelectronics* 87: 752–759.

32 Benvidi, A., Firouzabadi, A.D., Tezerjani, M.D. et al. (2015). A highly sensitive and selective electrochemical DNA biosensor to diagnose breast cancer. *Journal of Electroanalytical Chemistry* 750: 57–64.

33 Lin, P.Y., Cheng, K.L., McGuffin-Cawley, J.D. et al. (2012). Detection of alpha-methylacyl-CoA racemase (AMACR), a biomarker of prostate cancer, in patient blood samples using a nanoparticle electrochemical biosensor. *Biosensors* 2 (4): 377–387.

34 de la Escosura-Muñiz, A., Maltez-da Costa, M., Sánchez-Espinel, C. et al. (2010). Gold nanoparticle-based electrochemical magnetoimmunosensor for rapid detection of anti-hepatitis B virus antibodies in human serum. *Biosensors and Bioelectronics* 26 (4): 1710–1714.

35 Campuzano, S., Yáñez-Sedeño, P., and Pingarrón, J.M. (2017). Electrochemical biosensing for the diagnosis of viral infections and tropical diseases. *ChemElectroChem* 4 (4): 753–777.

36 Zhang, W. and Li, G. (2004). Third-generation biosensors based on the direct electron transfer of proteins. *Analytical Sciences* 20 (4): 603–609.

37 Shen, M., Rusling, J.F., and Dixit, C.K. (2017). Site-selective orientated immobilization of antibodies and conjugates for immunodiagnostics development. *Methods* 116: 95–111.

38 Zhu, C., Yang, G., Li, H. et al. (2015). Electrochemical sensors and biosensors based on nanomaterials and nanostructures. *Analytical Chemistry* 87 (1): 230–249.

39 Teles, F.R. and Fonseca, L.P. (2008). Applications of polymers for biomolecule immobilization in electrochemical biosensors. *Materials Science and Engineering: C* 28 (8): 1530–1543.

40 Jia, X., Hu, G., Nitze, F. et al. (2013). Synthesis of palladium/helical carbon nanofiber hybrid nanostructures and their application for hydrogen peroxide and glucose detection. *ACS Applied Materials and Interfaces* 5 (22): 12017–12022.

41 Zhao, Y., Zhang, W., Lin, Y., and Du, D. (2013). The vital function of Fe_3 O_4@Au nanocomposites for hydrolase biosensor design and its application in detection of methyl parathion. *Nanoscale* 5 (3): 1121–1126.

42 Zhang, S., Wright, G., and Yang, Y. (2000). Materials and techniques for electrochemical biosensor design and construction. *Biosensors and Bioelectronics* 15 (5, 6): 273–282.

43 Nikolelis, D.P., Hianik, T., and Krull, U.J. (1999). Biosensors based on thin lipid films and liposomes. *Electroanalysis* 11 (1): 7–15.

44 Hu, C., Yang, D.P., Zhu, F. et al. (2014). Enzyme-labeled Pt@ BSA nanocomposite as a facile electrochemical biosensing interface for sensitive glucose determination. *ACS Applied Materials and Interfaces* 6 (6): 4170–4178.

45 Menon, S., Mathew, M.R., Sam, S. et al. (2020). Recent advances and challenges in electrochemical biosensors for emerging and re-emerging infectious diseases. *Journal of Electroanalytical Chemistry* 878: 114596.

46 Sohrabi, H., Kholafazad Kordasht, H., Pashazadeh-Panahi, P. et al. (2020). Recent advances of electrochemical and optical biosensors for detection of C-reactive protein as a major inflammatory biomarker. *Microchemical Journal* 158: 105287.

47 Gao, C. and Huang, X.J. (2013). Voltammetric determination of mercury (II). *TRAC Trends in Analytical Chemistry* 51: 1–2.

48 Yüce, M., Filiztekin, E., and Özkaya, K.G. (2021). COVID-19 diagnosis—A review of current methods. *Biosensors and Bioelectronics* 172: 112752.

49 Kokkinos, C., Prodromidis, M., Economou, A. et al. (2015). Disposable integrated bismuth citrate-modified screen-printed immunosensor for ultrasensitive quantum dot-based electrochemical assay of C-reactive protein in human serum. *Analytica Chimica Acta* 886: 29–36.

50 Kerman, K., Morita, Y., Takamura, Y., and Tamiya, E. (2003). Label-free electrochemical detection of DNA hybridization on gold electrode. *Electrochemistry Communications* 5: 887–891.

51 Azek, F., Grossiord, C., Joannes, M. et al. (2000). Hybridization assay at a disposable electrochemical biosensor for the attomole detection of amplified human cytomegalovirus DNA. *Analytical Biochemistry* 284 (1): 107–113.

52 Kara, P., Meric, B., Zeytinoglu, A., and Ozsoz, M. (2004). Electrochemical DNA biosensor for the detection and discrimination of herpes simplex type I and type II viruses from PCR amplified real samples. *Analytica Chimica Acta* 518 (1, 2): 69–76.

53 Yan, Z., Gan, N., Zhang, H. et al. (2015). A sandwich-hybridization assay for simultaneous determination of HIV and tuberculosis DNA targets based on signal amplification by quantum dots-PowerVision™ polymer coding nanotracers. *Biosensors and Bioelectronics* 71: 207–213.

54 Dong, S., Zhao, R., Zhu, J. et al. (2015). Electrochemical DNA biosensor based on a tetrahedral nanostructure probe for the detection of avian influenza A (H7N9) virus. *ACS Applied Materials and Interfaces* 7 (16): 8834–8842.

55 Ma, X., Jiang, Y., Jia, F. et al. (2014). An aptamer-based electrochemical biosensor for the detection of Salmonella. *Journal of Microbiological Methods* 1 (98): 94–98.

56 Ojeda, I., Barrejón, M., Arellano, L.M. et al. (2015). Grafted-double walled carbon nanotubes as electrochemical platforms for immobilization of antibodies using a metallic-complex chelating polymer: application to the determination of adiponectin cytokine in serum. *Biosensors and Bioelectronics* 74: 24–29.

57 Yan, Y., Qiao, Z., Hai, X. et al. (2021). Versatile electrochemical biosensor based on bi-enzyme cascade biocatalysis spatially regulated by DNA architecture. *Biosensors and Bioelectronics* 174: 112827.

58 BnHamblin, M.R. (2016). Shining light on the head: photobiomodulation for brain disorders. *BBA Clinical* 6: 113–124.

59 Yu, Y., Zhang, L., Li, C. et al. (2014). A method for evaluating the level of soluble β-amyloid (1–40/1–42) in Alzheimer's disease based on the binding of gelsolin to β-amyloid peptides. *AngewandteChemie* 126 (47): 13046–13049.

60 Caifeng, D., Fei, Z., and Rui, R. (2009). An electrochemical biosensor for a-fetoprotein based on carbon paste electrode constructed of room temperature ionic liquid and gold nanoparticlcs. *Talanta* 78 (1): 148.

61 Hou, L., Huang, Y., Hou, W. et al. (2020). Modification-free amperometric biosensor for the detection of wild-type p53 protein based on the in situ formation of silver nanoparticle networks for signal amplification. *International Journal of Biological Macromolecules* 158: 580–586.

62 Dilley, R.J. and Morrison, W.A. (2014). Vascularisation to improve translational potential of tissue engineering systems for cardiac repair. *The International Journal of Biochemistry and Cell Biology* 56: 38–46.

63 Aydın, E.B., Aydın, M., and Sezgintürk, M.K. (2020). Selective and ultrasensitive electrochemical immunosensing of NSE cancer biomarker in human serum using epoxy-substituted poly (pyrrole) polymer modified disposable ITO electrode. *Sensors and Actuators B: Chemical* 306: 127613.

64 Faria, H.A. and Zucolotto, V. (2019). Label-free electrochemical DNA biosensor for Zika virus identification. *Biosensors and Bioelectronics* 131: 149–155.

65 Lee, J., Morita, M., Takemura, K., and Park, E.Y. (2018). A multi-functional gold/iron-oxide nanoparticle-CNT hybrid nanomaterial as virus DNA sensing platform. *Biosensors and Bioelectronics* 102: 425–431.

66 George, A., Amrutha, M.S., Srivastava, P. et al. (2019). Label-free detection of Chikungunya non-structural protein 3 using electrochemical impedance spectroscopy. *Journal of the Electrochemical Society* 166 (14): B1356.

67 Duan, Y., Chen, J., Jin, Y. et al. (2021). Antibody-free determinations of low-mass, soluble oligomers of Aβ42 and Aβ40 by planar bilayer lipid membrane-based electrochemical biosensor. *Analytical Chemistry* 93 (7): 3611–3617.

68 Takeda, K., Kusuoka, R., Inukai, M. et al. (2021). An amperometric biosensor of L-fucose in urine for the first screening test of cancer. *Biosensors and Bioelectronics* 174: 112831.

69 Bogomolova, A., Komarova, E., Reber, K. et al. (2009). Challenges of electrochemical impedance spectroscopy in protein biosensing. *Analytical Chemistry* 81 (10): 3944–3949.

70 Mishra, N.N., Retterer, S., Zieziulewicz, T.J. et al. (2005). On-chip micro-biosensor for the detection of human $CD4^+$ cells based on AC impedance and optical analysis. *Biosensors and Bioelectronics* 21 (5): 696–704.

71 Chowdhury, A.D., Takemura, K., Li, T.C. et al. (2019). Electrical pulse-induced electrochemical biosensor for hepatitis E virus detection. *Nature Communications* 10 (1): 1–2.

72 Ferreira, A.A., Uliana, C.V., de Souza, C.M. et al. (2013). *Amperometric Biosensor for Diagnosis of Disease*, 253–289. Rijeka, Croatia: InTech.

73 Odeh, A.A., Al-Douri, Y., Voon, C.H. et al. (2017). A needle-like Cu_2CdSnS_4 alloy nanostructure-based integrated electrochemical biosensor for detecting the DNA of Dengue serotype 2. *Microchimica Acta* 184 (7): 2211–2218.

74 Cui, X., Das, A., Dhawane, A.N. et al. (2017). Highly specific and rapid glycan based amperometric detection of influenza viruses. *Chemical Science* 8 (5): 3628–3634.

75 Sharma, M.K., Rao, V.K., Agarwal, G.S. et al. (2008). Highly sensitive amperometric immunosensor for detection of *Plasmodium falciparum* histidine-rich protein 2 in serum of humans with malaria: comparison with a commercial kit. *Journal of Clinical Microbiology* 46 (11): 3759–3765.

76 Laboria, N., Fragoso, A., Kemmner, W. et al. (2010). Amperometric immunosensor for carcinoembryonic antigen in colon cancer samples based on monolayers of dendritic bipodal scaffolds. *Analytical Chemistry* 82 (5): 1712–1719.

77 de Souza, C.M., Laube, T., Yamanaka, H. et al. (2011). Magneto immunoassays for *Plasmodium falciparum* histidine-rich protein 2 related to malaria based on magnetic nanoparticles. *Analytical Chemistry* 83 (14): 5570–5577.

78 Koncki, R. (2007). Recent developments in potentiometric biosensors for biomedical analysis. *Analytica Chimica Acta* 599 (1): 7–15.

79 Hernández, R., Vallés, C., Benito, A.M. et al. (2014). Graphene-based potentiometric biosensor for the immediate detection of living bacteria. *Biosensors and Bioelectronics* 54: 553–557.

80 Kaisti, M. (2017). Detection principles of biological and chemical FET sensors. *Biosensors and Bioelectronics* 98: 437–448.

81 Seo, G., Lee, G., Kim, M.J. et al. (2020). Rapid detection of COVID-19 causative virus (SARS-CoV-2) in human nasopharyngeal swab specimens using field-effect transistor-based biosensor. *ACS Nano* 14 (4): 5135–5142.

82 Ning, S., Zhou, M., Liu, C. et al. (2019). Ultrasensitive electrochemical immunosensor for avian leukosis virus detection based on a β-cyclodextrin-nanogold-ferrocene host-guest label for signal amplification. *Analytica Chimica Acta* 1062: 87–93.

83 Wu, J., Tang, J., Dai, Z. et al. (2006). A disposable electrochemical immunosensor for flow injection immunoassay of carcinoembryonic antigen. *Biosensors and Bioelectronics* 22 (1): 102–108.

4

Biosensors for Point-of-Care (POC) Applications

The Flag Bearer of the Modern Medicinal Technology to Tackle Infectious Diseases

Sumit Kumar[1,2], *Garima Rathee*[3], *Gaurav Bartwal*[1,4], *and Pratima R. Solanki*[3]

[1]*University of Delhi, Department of Chemistry, Delhi, 110007, India*
[2]*National Tsing Hua University, Hsinchu, Taiwan*
[3]*Jawaharlal Nehru University, Special Centre for Nanoscience, New Delhi, 110067, India*
[4]*Indian Institute of Technology, Department of Chemistry, Kanpur, 208016, India*

4.1 Introduction

The post-COVID-19 period has provided immense public attention for the fast and rapid detection of diseases that could be infectious and problematic. Although the detection of disease has not been fixed to the post-covid period, the panicked situation has propelled numerous companies and institutes to develop the most effective and easy-to-use methods to detect antigens, antibodies, and other forms, which are associated with the core structure of the virus [1–3]. One critical progress for the fast disease data analysis with an accurate diagnosis has been the point-of-care (POC) detection, which efficiently produced noticeable patient clinical output while allowing patients and healthcare workers to access simple and fast results [4–6].

Diagnosing human illness using POC devices has become possible and realistic due to the timely unveiling and execution of deep-rooted phenomena in various fields such as life science, electrochemistry, molecular biology, optics, and nanotechnology with the noticeable and distinguishable contribution to other aligned fields [7, 8]. Additionally, the versatility of the POC can be exercised by diverse approaches of biosensors such as fluorescence-based biosensors, transistors showing field effects, electrochemical biosensors, and surface plasmon resonance (SPR) biosensors with immense applicability in the field of biochemistry, electrochemistry, electronics, superconductor materials, and integrating interdisciplinary fields by engineering [9]. Since POC devices have evolved with contributions from various fields, many materials have been filtered, shorted, and selected for sensing important biological targets, including biomacromolecules such as proteins, DNA, RNA, and their denatured or detectable forms specific to particular illnesses and pathways [10–15]. The applicability and

Point-of-Care Biosensors for Infectious Diseases, First Edition.
Edited by Sushma Dave and Jayashankar Das.

authority of POC devices over conventional assays can be attributed to detection of biomarker color and structural and spectral changes, with higher selectivity, sensitivity, and applicability even to areas that are yet to be fully assessed and addressed [16, 17]. Moreover, the time-consuming, expensive, and complex handling of biomolecular quantitative techniques such as northern and western blots and ELISA have limited the usability of these methods for immediate responsive applicability, which is needed in early diagnosis [6, 18]. Thus, the gold standard criteria for designing POC devices have been fulfilled by developing novel biosensors that satisfy all the critical factors of the term "ASSURED" set forth by the World Health Organization Special Programme for the formation of ideal diagnostic tools [19]. The vastness of biosensors for POC devices has been optimized and implemented further by substituting alternate materials and connecting POC parts. Generally, the POC devices can be regarded as a complex system having multiple parts which are dexterously combined to give final output in real response time. Target, probe, and sensing tools have been considered central and essential components of the POC devices. Additionally, transducer and signal readout devices play a significant part in the final signal output. Henceforth, the proper working and efficiency of POC devices depend upon the timely technological update of these parts to maintain the selectivity and sensitivity for their ultimate function [20]. One of the eras in which biosensor technology has brought immense change in day-to-day testing is the example of glucose home testing systems that depend on potentiometric sensors and proved efficient for controlling and monitoring diabetes [21]. Other examples of home pregnancy tests based on lateral flow detection tremendously contributed to the development of the biotechnological sector [22].

Similarly, biosensing applications have been improved depending on the biosensors' material and fabrication method. Therefore, varieties of biosensors have microfluidic chip-based, potentiometric-based, field effect transistor (FET)-based, nanoelectromechanical systems or microelectromechanical systems-based, and complementary metal-oxide semiconductor (CMOS)-based biosensors have been engineered to promote the POC devices to next level [23–25]. Various composites can form biosensors, and one such also contains a composite made from graphene oxide (GO) to keep track of disease at various stages, detecting, determining, and analyzing illness for the POC platform. Progressively, significant applications of GO biosensors as a multifunctional platform in biological imaging, cancer sensing, and therapeutic delivery have been demonstrated [26].

With an outlook on the frequency of infectious disease outbreaks in the last two decades, the demand for POC devices has increased exponentially for early *in vitro* diagnostics (IVDs) and monitoring the fast recovery of the large population. Primary outbreaks such as SARS in 2002, MERS in 2012, Ebola (in West Africa) in 2013, and severe acute respiratory syndrome coronavirus (SARS, COVID-19) have culminated in the death of millions of people [27–29]. Thus, the rapid transmission of infectious diseases due to pathogens such as waterborne or airborne viruses, parasites, and bacteria raised the concern of detecting, controlling, and terminating them at an early stage with the assistance of biosensors at the POC level. Therefore, the current chapter comprises the development of various biosensors for better diagnostic strategies to tackle the problems associated with infectious diseases. In detail, applications of various portable devices such as smartphones have also been provided to improve the testing at the POC level for diagnostic purposes.

4.2 Classification of POC Biosensors for Detection of Infectious Diseases

Timely detection, screening, and identification of infectious diseases need a well-advanced system of integrated laboratory techniques with gold standard devices mainly based on culturing techniques, immunoassays, polymerase chain reaction (PCR) methods, and microscopies techniques [30, 31]. During emergency and pandemics situations where highly accurate mass population screening is needed, these techniques have revealed inadequacy in meeting the expectations. Thus, for detection of various infectious diseases, a large number of biosensors have been developed and employed, which are based on electrochemical, fluorescence, SPR, and surface-enhanced Raman scattering (SERS), chemiluminescence, colorimetric, and magnetic biosensors.

4.2.1 Electrochemical-Based Biosensor

Electrochemical biosensors have recognizable potential and applications for diagnosing diseases due to their simplicity, portability, shorter time response, and easy fabrication with POC devices [32]. Therefore, different varieties of electrochemical biosensors such as amperometric/voltammetric [33], impedimetric [34], FET ([35]), and potentiometric [36] have been invented to exercise their applications for POC diagnosis. A detailed description of these biosensors is out of the scope of this chapter; however, essential findings have been chosen to briefly represent the subsequent development in POC diagnosis by these biosensors. For example, to detect the presence of *Mycobacterium tuberculosis* (MTB) in *sputum* by POC and rapid pathway, an immunoassay microtip amperometric biosensor was developed [37]. Another type of electrochemical biosensor is impedimetric biosensors, known for meager amplitude perturbation, and depends on electrochemical impedance spectroscopy (EIS) for target analysis [38].

Furthermore, the detection of infectious disease by impedimetric-based biosensors has been successfully achieved by Cecchetto et al. for the identification and detection of dengue with modification of gold electrode [39]. Similarly, Tarasov et al. developed a novel potentiometric biosensor to detect the Bovine Herpes Virus-1 (BHV-1). This biosensor proved to be highly selective in detecting intervention of animal diseases in humans and based on extended FETs [40]. FET-based biosensors also belong to electrochemical biosensors and have been in demand due to easy production and low cost for the POC diagnosis of infectious diseases. Gao et al. designed a FET-based sensitive graphene nano-biosensor with variable antibody fragments of Lyme single chain for rapid and limited amount detection of an antigen associated with *Borrelia burgdorferi* bacterium [41].

4.2.2 Fluorescence-Based Biosensor

Similarly, fluorescence biosensors have been used for the POC diagnosis of infectious diseases. Fluorescence resonance energy transfer (FRET) from the donor fluorophore to the acceptor fluorophore is the primary method for designing fluorescence biosensors ([42]). Additionally, some materials can act as fluorophores due to their inherent physical, chemical, and electronic properties, making marking fluorophores a crucial step in developing fluorescence biosensors [43]. Successively, fluorescence biosensors can be divided further by directly labeling biosensors based on fluorescence quenching and generation. These are popularly known as "signal-on" and "signal-off" fluorescence biosensors, whereas "signal-on" refers to the proportional change of fluorescence signal to the target quantity, which is binding to the specific ligand; moreover, for "signal-off," the change in fluorescence intensity is inversely dependent on the target quantity. Additionally, depending on the various metal nanoparticles and fluorescent materials, the fluorescence biosensors have been classified and discussed further to consider their preferential applications for diagnostics of infectious diseases at the POC level.

4.2.2.1 Direct Fluorescence Biosensors for Infectious Diseases

Direct fluorescence biosensors have been further classified into various parts depending upon the fluorescence labeling and nature and materials of fluorescent dyes to bind the ligand specifically. Materials such as carbon dots [44, 45], quantum dots [46, 47], metal nanoclusters [48, 49], and up-conversion nanoparticles [50] have shown remarkable fluorescence properties and been recognized for the fabrication of optimum sensors for POC diagnostic of infectious diseases. Xu et al. have detected severe fever symptoms by ultrasensitively detecting the thrombocytopenia syndrome virus (SFTSV) at the POC level by developing a fluorescent lateral flow assay entertaining CDs/SiO_2 nanospheres [44]. Wang et al. also developed a ratiometric fluorescent test paper that was based on Eu(III) functionalized carbon dots (CDs-Eu) and proved to be highly sensitive for POC detection of anthrax biomarker dipicolinic acid (DPA) [45]. Similarly, Hu et al. worked on gold nanoparticles (AuNPs) and fluorescent QDs by encapsulating them into antibodies and thus developed a lateral flow POC detection assay for Ebola virus glycoprotein [46]. The research and development in POC detection of infectious diseases has covered most pathogen- and virus-based illnesses. Sequentially, for the detection of Zika virus nonstructural protein 1 (NS1), a QDs-inspired immunofluorescence biosensor has been developed by Takemura et al. by using the localized surface plasmon resonance (LSPR) from AuNPs [47]. Most POC biosensors are limited to qualitatively detecting single bacteria; however, in some cases, many bacteria have been detected discriminately. Ji et al. fabricated a protein-AuNC-based fluorescence sensor which proved efficient, convenient, and promising for diagnosing many drug-resistant bacteria in distant locations where medical faculty are limited [48]. The competitiveness in developing various types of POC biosensors for infectious diseases has increased exponentially. Therefore, DNA template AgNCs fluorescent and aptamer functionalized UCNPs probes have been developed for the timely, precise, and rapid detection of human immunodeficiency virus type 1 (HIV-1) and Salmonella, respectively [49, 50].

4.2.2.2 Signal-on/off Fluorescent Biosensors for Infectious Disease POC Diagnostics

The progress and application of "signal-on" fluorescence biosensors have been greatly visualized to detect infectious diseases. The detection of the influenza A (H1N1) virus has been achieved by the development of QD-aptamer beacons and 3D photonic crystals [51]. In this biosensor, the fluorescence intensity of the biosensor was proportional to the target H1N1 virus concentration, and the adaption of a low-cost biosensor with a smartphone camera has provided a comprehensive approach to its production. Similarly, He et al. have also developed a very specific and powerful "signal-on" fluorescence biosensor having clustered regularly interspaced short palindromic (CRISPR)-Cas POC detection of African Swine Fever Virus (ASFV) ([52]). "Signal-off" novel fluorescence immunoassay method-based biosensor has been developed by Li et al., which proved to be ultrasensitive for POC detection of avian influenza virus (AIV). In this biosensor, QDs were used as fluorescence signal output due to their luminescent nature, whereas gold ions from AuNPs have been utilized as fluorescence quenchers ([53]).

4.2.3 Surface Plasmon Resonance (SPR)-Based Biosensor

SPR-based biosensors have also been used for POC diagnostics of infectious diseases, owing to their recognizable performance, timely productivity, and real-time detection [54]. However, power consumption, resolution, and sensitivity still require adequate exercise to further develop these biosensors and devices in the form of powerful tools for modern medical facilities [55]. The detection of infectious diseases by SPR-based biosensors depends upon the variation in the refractive index, mainly caused due to molecular interactions of biological macromolecules on the metal surface and competently detected via surface plasmon wave. A number of nanomaterials and microfluidic devices has been assembled efficiently to finalize the integration of SPR-based biosensor systems. Similarly, these biosensors depend on entropy-driven strand displacement reaction to detect HIV [56] and coating CdSeTeS QDs on the AuNPs surface to develop into an LSPR biosensor for the detection of dengue [57]. Additionally, *Escherichia coli* is also being detected using the SPR application of these biosensors that have been achieved by functionalizing antibodies on the gold surface equipped with a microfluidic chip [58].

4.2.4 Surface-Enhanced Raman Scattering (SERS)-Based Biosensor

The lack of multiplex detection and high background signals are the main limitations of the fluorescence-based biosensors that limit their scope, hence favorably opening a trial window for different biosensors. This new kind of biosensor is based on SERS, which proved to be efficient against multiplexing capabilities, single molecule detection, and low background [59]. Although SERS-based biosensors have vivid applications in many medical fields, detection of infectious diseases at the POC, essentially caused by viruses and bacteria, has recently been focused on. The eminently SERS-based biosensor has been developed using aptamer-Fe_3O_4@Au magnetic nanoparticles to detect bacterial cells [60]. Furthermore, paper-based silver nanowires have been assembled in a paper-based biosensor for sensitive determination of *Mycoplasma pneumoniae* DNA. Impressively, integrating the SERS biosensors amplified the detection time for the diagnosis of various viruses and bacteria, thus opening new avenues to continue developing the biosensors integrated with POC devices.

4.2.5 Chemiluminescence-Based Biosensor

Chemiluminescence-based biosensors are one of the popular biosensors for the POC diagnosis of infectious diseases, which ultimately achieve success with the ZstatFlu II test for timely detection of influenza virus ([61]). These biosensors proved efficient for systematical detection of infectious disease through signal-to-noise ratio. Successively, hepatitis B virus (HBV), human immunodeficiency virus (HIV), and hepatitis C virus (HCV) have been detected by dexterously incorporating nucleic acid purification with magnetic separation technology by developing a DNA hybridization-based chemiluminescence biosensor ([62]). Owing to the recent improvement in the design of the chemiluminescence biosensor, the necessity for the automated detection of infectious disease is needed, which can show great potential for biosensing applications.

4.2.6 Colorimetric-Based Biosensors

Colorimetric biosensors proved to be the preferred choice for POC diagnostic applications due to their fast preparation and economical approach [63]. The oxidation of peroxidase and peroxidase-like nanomaterials supported colorimetric detection. Various nanoparticles such as AuNPs, AgNPs, platinum nanoparticles, and magnetic nanoparticles (MNPs) have been used to form colorimetric biosensors [64]. Therefore, by utilizing the SPR property of NPs, a colorimetric POC diagnostic platform can be developed to detect infectious diseases. Subsequently, POC detection of *E. coli* has been achieved by the aggregation of AuNPs and smartphone imaging [65]. The merits of colorimetric biosensors, such as sensitivity and specificity, can be enhanced further for POC detection of bacterial diseases [66].

4.2.7 Magnetic-Based Biosensors

Magnetic phenomena achieve surging importance for POC detection due to specificity, sensitivity, low cost, and improved detection. Over the past few decades, magnetic biosensors have been developed tremendously and classified as magnetic tunnel junction biosensors, giant magnetoresistance (GMR) biosensors, nuclear magnetic resonance (NMR) biosensors, and magnetic particle spectroscopy (MPS) ([67]). Magnetic bioassay simplified the detection process using magnetic particle imaging and conducted spectroscopic studies of superparamagnetic iron oxide nanoparticles (SPIONs), producing a sinusoidal magnetic field with a large amplitude [68]. POC diagnostics requires seamless and perfect integration of these biosensors with various advanced devices to exponentially make the process labor free and smooth.

4.3 Modern Devices for the Detection of Infectious Diseases

Tuberculosis associated with HIV was detected by the first POC device produced by GeneXpert and accepted by WHO in 2010. This device proved to be the perfect combination of multiple steps to detect nucleic acid within two hours. Since then, a noticeable number of POC devices have been fabricated to detect infectious diseases. Some of these devices have detected well-known diseases such as Influenza A and B viruses by GenePoc, HIV by LIAT analyzer, and malaria by Truelab Uno® [69]. Furthermore, the classification of POC devices in lab-on-a-chip devices, Lab-on-a-disc (LOAD) devices, microfluidic paper-based analytical devices, miniaturized PCR devices, and isothermal nucleic acid amplification devices have further advanced to detect the infectious diseases. Progress in software and smartphone technology has also enhanced the usability and detection rate while maintaining the cost and complex operation simply and robustly [70].

4.3.1 Lab-on-a-Chip Devices and Lab-on-a-Disc Devices

LOC devices integrate microfluidic devices to rapidly detect the results for automation of multiple assay functions in a single instrument. These devices have attracted many applications for the on-site detection of infectious diseases caused by viruses, bacteria, and parasites ([71]). LOC devices are a sensitive and accurate alternative to detect the infectious diseases caused by HIV, SARS, Ebola, and COVID-19 viruses. The task's success solemnly depends on detecting these viruses in low concentration, which reaches the limit of 3.37×10^{-3} ng ml^{-1} observed for Ebola RNA without any amplification [72]. Similarly, multidrug-resistant tuberculosis (MDR-TB), proving its havoc, has been considerably detected by POC diagnostic applications. Therefore, to successfully detect the MDR-TB in a more precise way, a sample pretreatment process has to be followed, which itself is associated with many noticeable challenges such as MTB mutations, difficulties in acid-fast MTB bacilli lysis, viscosity, and highly heterogeneous nature of sputum. Some of these challenges have been overcome with the advent of the Lab-on-a-film device, which acts as an independent workstation to effectuate the POC diagnostic of MDR-TB. The future promise of POC devices again lies within the wide detection range in which a broad selection of ranges (1×10^{4}–1×10^{8} copies ml^{-1}) can be detected for MRSA DNA while ascertaining the low-cost molecular diagnostic assay for future biotechnological approaches.

Lab-on-a disc (LOAD) device proved to be better than LOC devices due to the non-necessity of the external pumping system, active values, and high degree of parallel connection owing to the automatically working fluid actuation (Figure 4.1) [73]. LOAD is a type of centrifugal microfluidics device having efficient integration of sample preparation, mixing of reagents, and single device detection and separation. In these devices, various forces such as capillary forces, centrifugal forces, Coriolis, and Euler forces pose together to maintain the sophisticated work of microfluidic design [74]. LOAD devices have successively advanced numerous diagnostics and POC detection of infectious diseases.

4.3.2 Microfluidic Paper-Based Analytical and Lateral Flow Devices

Due to paper-based substrate, microfluidic paper-based analytical devices (μPAD) have proved to be economical, equipment independent, and easily transportable and foldable since their first introduction in 2007, highlighting their porosity and hydrophilic nature [75]. The applicability of μPAD is adverse and can be applied to various biological fluids such as whole blood, plasma, saliva, serum, urine, and saliva without an external source (Figure 4.2)[76]. Hydrophilic/hydrophobic microstructures on specially designed paper substrates enhance the automatic field test, multiplex detection, and sorting, which are ideal for POC diagnostic applications [77]. μPADs are applied to many fields such as environmental protection, food safety, and remote areas' biological detection of infectious diseases. Moreover, further classifying μPADs into 2D and 3D μPADs by distinct chemical and physical hydrophobic boundaries leads to liquid channels [78]. The fabrication becomes complex from 2D μPADs to 3D μPADs. However, repeatedly staking 2D microfluidic paper leads to the development of 3D μPADs. The further advancement in 2D and 3D μPADs leads to more depth division which ultimately gives rise to novel μPADs for the early detection and prevention of the mass spread of infectious diseases.

Lateral flow devices are one of the most widely used POC diagnostic tools for detecting infectious diseases because of their simplicity, easy operation, rapidness, durability, stability, cost-effectiveness, and minimum user intervention. Since the first lateral flow device was developed in the 1980s for pregnancy self-testing, various lateral flow devices have been developed for all applications, including biomarker detection, disease diagnosis, food safety, and environmental monitoring. A typical lateral flow device consists of a nitrocellulose membrane, sample pad, conjugate pad, and absorbent pad with all the required chemicals and reagents prestored in the test strip. The sample pad is designed for sample fluid loading and adsorption; the conjugated pad is impregnated with specific molecule-conjugated bio labels for binding to the analyte and signal generation; the nitrocellulose membrane immobilized with capture molecules is the core component for testing, and the absorbent pad provides the driven capillary forces and collects excess sample fluid. Pre-stored on the conjugate pad and form target-molecule-conjugated-biolabels complexes, which are then captured by the capture molecules immobilized on the test zone and generated test and control lines, which the naked eye can directly read. Therefore, the target concentration is proportional to the test line signal in a sandwich lateral flow device. In a competitive lateral flow device, the analyte target

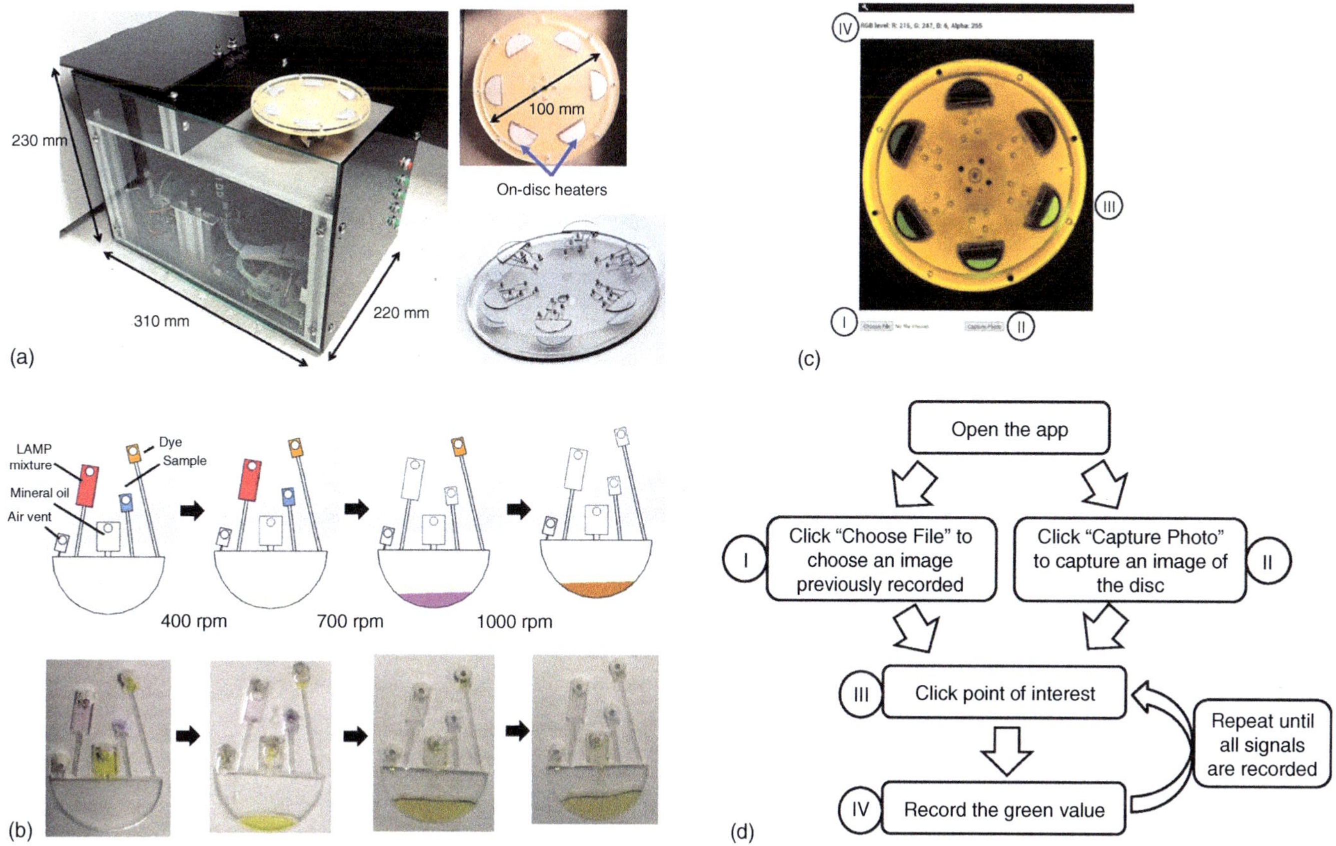

Figure 4.1 The LOAD diagram for detecting GM papaya. The LOAD (100 mm in diameter and 3 mm in thickness) has two layers: the top layer is a disposable microfluidic disc with six identical compartments for GM food screening, and the bottom layer is made up of mechanical and electronic components for onboard heating (a). Spin procedure for the entire assay, with illustrations and photos for each step (b). The LOAD analyzer has an easy-to-use interface (c) and a flow chart of the app's functioning mechanism (d). Source: Loo et al. [73]/with permission of Elsevier.

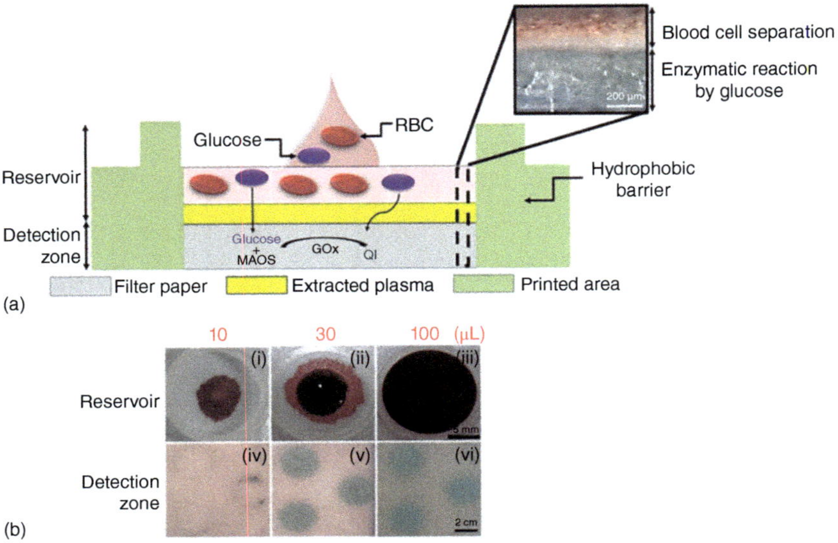

Figure 4.2 Blood cell separation and glucose detection in 3D-µPADs (a) Schematic depicting the separation of blood cells, including RBCs, in the top PSM and the creation of color signals for glucose detection in the separated plasma delivered into the bottom filter paper immobilized with a glucose detection enzyme system, including GOx and MAOS. Quinonimine (QI) generated by GOx in the presence of glucose and MAOS produces cyan hue. Inset: a cross-sectional picture of the reservoir and detecting zone following the introduction of whole blood into the reservoir. (b) Wetting of PSM and color creation of the detecting zone in various volumes (10–100 l) of whole blood containing 5 mM glucose. Source: Reproduced with permission from Park et al. [76], American Chemical Society.

will compete with the capture molecules immobilized on the test zone to bind to the specific molecule-conjugated-biolabels pre-stored on the conjugation pad. Thus, in the presence of the analyte target, the test line will not display the signal, while in the absence of the analyte target, there will be signals at both the test line and control line. Thus, the target concentration is inversely proportional to the test line signal. Both sandwich and competitive lateral flow devices can be used for qualitative and quantitative detection of targets. The difference is that a sandwich lateral flow device is usually utilized to detect targets with multiple antigen epitopes, while a competitive lateral flow device is more suitable for detecting single antigen epitope targets.

4.3.3 Miniaturized PCR and Isothermal Nucleic Acid Amplification Devices

PCR is one of the world-renowned and gold standards for diagnosis to sensitively and specifically detect infectious diseases due to amplifying a unique sequence of RNA or DNA by specific and sensitive [79]. Nowadays, the development of novel microfluidics systems concerning electromechanical systems (MEMS) altered the bulky and complex PCR to a miniaturized form manifesting numerous merits such as high integration, low sample consumption, and full automation [80]. Additionally, miniaturized PCR manifested the quantitative real-time detection and nucleic acid amplification by introducing the new potential of PCR-based NAT despite complex cooling and heating control [81]. Furthermore, essential strategies such as enhancing the heat transfer rate and less sample volume addressed the two challenges: shortening and simplifying conventional PCR processes [81]. Similarly, Zhou et al. engineered a microfluidic chip system for ultimately and sensitively detecting SARS coronavirus and allowing efficient amplification of cDNA by electrophoresis on a glass microchip.

Moreover, isothermal nucleic acid amplification methods have been conducted at a fixed temperature. Thus, POC diagnostics become effortless without a thermal cycle. These isothermal nucleic acid amplification methods are defined as strand displacement amplification (SDA)[82], rolling circle amplification (RCA), helicase-dependent amplification (HDA), LAMP, NASBA, and RPA [83]. Additionally, digital microfluidics (DMF) has addressed the problem of droplet manipulation, which is part of microfluidic devices, and developed INAA devices with tremendous merits for detecting viruses, bacteria, and parasites.

4.4 Scope and Challenges Associated with the Next-Generation POC Devices

Although detecting the results without sample penetration is extremely difficult, integration of POC diagnostics with clinical sample assays may circumvent the sample penetration step for early detection of infectious diseases. The additional challenges of speed, sensitivity, and cost have been progressively coercing to design and incorporate novel materials and devices. The future of POC diagnostics is very opportunistic due to the potential of using applications of fast-growing machine learning, AI, mobile networking system, and the internet of things (IoT) to reach far distant places with personalized diagnostics, timeless and low-cost approaches. The fast growth of the internet network from 4G to 5G, and innovation in AI's deep learning methods, have brought forward the revolution of fast data transfer and easy diagnoses with improving automatic and intelligent medical practices. Additionally, the scope of integrating POC devices with cloud computing would play an efficient and progressive role in speedily and accurately detecting infectious diseases. Despite data safety challenges that cloud computing faces, it might play a substantial role in improving infectious diseases withstanding novel and highly sophisticated encryption methods to improve data safety.

4.5 Conclusion

The recent outbreak of COVID-19 has spread awareness for rapid detection of infections, either in overcrowded cities where medical facilities are better but population pressure is at a peak or in resource-poor regions where medical services are not fully advanced. Herein, in the current chapter, the progress of biosensors integrated with POC devices has been highlighted for the early detection of infectious diseases and, after that, showcased as the flag bearer of the modern healthcare sector due to their fast-growing applications. Biosensors have been explained and summarized depending upon their classifications while explaining the growth of various POC devices in parallel to them. It is appropriate to add that the current chapter represents the rapid growth of biosensors and POC devices to meet the tremendous challenges of accurately and sensitively detecting infectious diseases while achieving the "ASSURED" standards.

References

1 Feng, W., Newbigging, A.M., Le, C. et al. (2020). Molecular diagnosis of COVID-19: challenges and research needs. *Analytical Chemistry* https://doi.org/10.1021/acs.analchem.0c02060.

2 Kevadiya, B.D., Machhi, J., Herskovitz, J. et al. (2021). Diagnostics for Sars-CoV-2 infections. *Nature Materials* https://doi.org/10.1038/s41563-020-00906-z.

3 Yao, L., Zhu, W., Shi, J. et al. (2021). Detection of coronavirus in environmental surveillance and risk monitoring for pandemic control. *Chemical Society Reviews* https://doi.org/10.1039/d0cs00595a.

4 Quesada-González, D. and Merkoçi, A. (2018). Nanomaterial-based devices for point-of-care diagnostic applications. *Chemical Society Reviews* https://doi.org/10.1039/c7cs00837f.

5 Wu, L. and Xiaogang, Q. (2015). Cancer biomarker detection: recent achievements and challenges. *Chemical Society Reviews* 44 (10): https://doi.org/10.1039/c4cs00370e.

6 Zhou, W., Gao, X., Liu, D., and Chen, X. (2015). Gold nanoparticles for in vitro diagnostics. *Chemical Reviews* https://doi.org/10.1021/acs.chemrev.5b00100.

7 Trinh, N., Jolliffe, K.A., and New, E.J. (2020). Dual-functionalisation of fluorophores for the preparation of targeted and selective probes. *Angewandte Chemie International Edition* https://doi.org/10.1002/anie.202007673.

8 Yang, M., Liu, Y., and Jiang, X. (2019). Barcoded point-of-care bioassays. *Chemical Society Reviews* https://doi.org/10.1039/c8cs00303c.

9 Jabbour, C.R., Parker, L.A., Hutter, E.M., and Weckhuysen, B.M. (2021). Chemical targets to deactivate biological and chemical toxins using surfaces and fabrics. *Nature Reviews Chemistry* https://doi.org/10.1038/s41570-021-00275-4.

10 Darwish, N.T., Sekaran, S.D., and Khor, S.M. (2018). Point-of-care tests: a review of advances in the emerging diagnostic tools for dengue virus infection. *Sensors and Actuators B: Chemical* https://doi.org/10.1016/j.snb.2017.09.159.

11 Das, J., Ivanov, I., Safaei, T.S. et al. (2018). Combinatorial probes for high-throughput electrochemical analysis of circulating nucleic acids in clinical samples. *Angewandte Chemie International Edition* 57 (14): https://doi.org/10.1002/anie.201800455.

12 Dhiman, A., Kalra, P., Bansal, V. et al. (2017). Aptamer-based point-of-care diagnostic platforms. *Sensors and Actuators B: Chemical* https://doi.org/10.1016/j.snb.2017.02.060.

13 Furst, A., Landefeld, S., Hill, M.G., and Barton, J.K. (2013). Electrochemical patterning and detection of DNA arrays on a two-electrode platform. *Journal of the American Chemical Society* 135 (51): https://doi.org/10.1021/ja410902j.

14 Nekoueian, K., Amiri, M., Sillanpää, M. et al. (2019). Carbon-based quantum particles: an electroanalytical and biomedical perspective. *Chemical Society Reviews* https://doi.org/10.1039/c8cs00445e.

15 Sharma, A.C., Jana, T., Kesavamoorthy, R. et al. (2004). A general photonic crystal sensing motif: creatinine in bodily fluids. *Journal of the American Chemical Society* 126 (9): https://doi.org/10.1021/ja038187s.

16 Inan, H., Poyraz, M., Inci, F. et al. (2017). Photonic crystals: emerging biosensors and their promise for point-of-care applications. *Chemical Society Reviews* https://doi.org/10.1039/c6cs00206d.

17 Sun, J., Xianyu, Y., and Jiang, X. (2014). Point-of-care biochemical assays using gold nanoparticle-implemented microfluidics. *Chemical Society Reviews* https://doi.org/10.1039/c4cs00125g.

18 Nasseri, B., Soleimani, N., Rabiee, N. et al. (2018). Point-of-care microfluidic devices for pathogen detection. *Biosensors and Bioelectronics* https://doi.org/10.1016/j.bios.2018.05.050.

19 Land, K.J., Boeras, D.I., Chen, X.S. et al. (2019). REASSURED diagnostics to inform disease control strategies, strengthen health systems and improve patient outcomes. *Nature Microbiology* https://doi.org/10.1038/s41564-018-0295-3.

20 Shrivastava, S., Trung, T.Q., and Lee, N.E. (2020). Recent progress, challenges, and prospects of fully integrated mobile and wearable point-of-care testing systems for self-testing. *Chemical Society Reviews* https://doi.org/10.1039/c9cs00319c.

21 Adeel, M., Rahman, M.M., Caligiuri, I. et al. (2020). Recent advances of electrochemical and optical enzyme-free glucose sensors operating at physiological conditions. *Biosensors and Bioelectronics* https://doi.org/10.1016/j.bios.2020.112331.

22 Liu, J., Geng, Z., Fan, Z. et al. (2019). Point-of-care testing based on smartphone: the current state-of-the-art (2017–2018). *Biosensors and Bioelectronics* 132: https://doi.org/10.1016/j.bios.2019.01.068.

23 Dzik, S. (2020). COVID-19 convalescent plasma: now is the time for better science. *Transfusion Medicine Reviews* https://doi.org/10.1016/j.tmrv.2020.04.002.

24 Ho, W.J., Chen, J.S., Ker, M.D. et al. (2007). Fabrication of a miniature CMOS-based optical biosensor. *Biosensors and Bioelectronics* 22 (12): https://doi.org/10.1016/j.bios.2006.12.031.

25 Naresh, V. and Lee, N. (2021). A review on biosensors and recent development of nanostructured materials-enabled biosensors. *Sensors (Switzerland)* https://doi.org/10.3390/s21041109.

26 Campbell, E., Hasan, M.T., Pho, C. et al. (2019). Graphene oxide as a multifunctional platform for intracellular delivery, imaging, and cancer sensing. *Scientific Reports* 9 (1): https://doi.org/10.1038/s41598-018-36617-4.

27 Muyembe-Tamfum, J.J., Mulangu, S., Justin Masumu, J.M. et al. (2012). Ebola virus outbreaks in Africa: past and present. *Onderstepoort Journal of Veterinary Research* 79 (2): https://doi.org/10.4102/ojvr.v79i2.451.

28 WHO Ebola Response Team (2016). After Ebola in West Africa – unpredictable risks, preventable epidemics. *New England Journal of Medicine* 375 (6): 587–596.

29 Yang, Y., Peng, F., Wang, R. et al. (2020). The deadly coronaviruses: the 2003 SARS pandemic and the 2020 novel coronavirus epidemic in China. *Journal of Autoimmunity* https://doi.org/10.1016/j.jaut.2020.102434.

30 Haliga, R.E., Sorodoc, V., Lionte, C. et al. (2021). Acute clinical syndromes and suspicion of Sars-Cov-2 infection: the experience of a single romanian center in the early pandemic period. *Medicina (Lithuania)* 57 (2): https://doi.org/10.3390/medicina57020121.

31 Oyewole, A.O., Barrass, L., Robertson, E.G. et al. (2021). Covid-19 impact on diagnostic innovations: emerging trends and implications. *Diagnostics* 11 (2): https://doi.org/10.3390/diagnostics11020182.

32 Dai, Y. and Liu, C.C. (2019). Recent advances on electrochemical biosensing strategies toward universal point-of-care systems. *Angewandte Chemie International Edition* https://doi.org/10.1002/anie.201901879.

33 Martins, G.V., Marques, A.C., Fortunato, E. et al. (2018). Wax-printed paper-based device for direct electrochemical detection of 3-nitrotyrosine. *Electrochimica Acta* 284: https://doi.org/10.1016/j.electacta.2018.07.150.

34 Reich, P., Preuß, J.A., Bahner, N., and Bahnemann, J. (2020). Impedimetric aptamer-based biosensors: principles and techniques. *Advances in Biochemical Engineering/Biotechnology* 174: https://doi.org/10.1007/10_2019_113.

35 Li, Q., Na, L., Wang, L., and Fan, C. (2018). Advances in nanowire transistor-based biosensors. *Small Methods* https://doi.org/10.1002/smtd.201700263.

36 Mousavi, M.P.S., Ainla, A., Tan, E.K.W. et al. (2018). Ion sensing with thread-based potentiometric electrodes. *Lab on a Chip* 18 (15): https://doi.org/10.1039/c8lc00352a.

37 Hiraiwa, M., Kim, J.H., Lee, H.B. et al. (2015). Amperometric immunosensor for rapid detection of *Mycobacterium tuberculosis*. *Journal of Micromechanics and Microengineering* 25 (5): https://doi.org/10.1088/0960-1317/25/5/055013.

38 Brosel-Oliu, S., Mergel, O., Uria, N. et al. (2019). 3D Impedimetric sensors as a tool for monitoring bacterial response to antibiotics. *Lab on a Chip* 19 (8): https://doi.org/10.1039/c8lc01220b.

39 Cecchetto, J., Carvalho, F.C., Santos, A. et al. (2015). An impedimetric biosensor to test neat serum for dengue diagnosis. *Sensors and Actuators, B: Chemical* 213: https://doi.org/10.1016/j.snb.2015.02.068.

40 Tarasov, A., Gray, D.W., Tsai, M.Y. et al. (2016). A potentiometric biosensor for rapid on-site disease diagnostics. *Biosensors and Bioelectronics* 79: https://doi.org/10.1016/j.bios.2015.12.086.

41 Gao, Z., Ducos, P., Ye, H. et al. (2020). Graphene transistor arrays functionalized with genetically engineered antibody fragments for lyme disease diagnosis. *2D Materials* 7 (2): https://doi.org/10.1088/2053-1583/ab5dce.

42 Yang, D., Liu, M., Jing, X. et al. (2018). Carbon nanosphere-based fluorescence aptasensor for targeted detection of breast cancer cell MCF-7. *Talanta* 185: https://doi.org/10.1016/j.talanta.2018.03.045.

43 Li, B., Qiaoling, Y., and Duan, Y. (2015). Fluorescent labels in biosensors for pathogen detection. *Critical Reviews in Biotechnology* https://doi.org/10.3109/07388551.2013.804487.

44 Xu, L., D., Zhang, Q., Ding, S.N. et al. (2019). Ultrasensitive detection of severe fever with thrombocytopenia syndrome virus based on immunofluorescent carbon dots/SiO_2 nanosphere-based lateral flow assay. *ACS Omega* 4 (25): https://doi.org/10.1021/acsomega.9b03130.

45 Wang, J., Daquan Li, Y., Qiu, X.L. et al. (2020). An europium functionalized carbon dot-based fluorescence test paper for visual and quantitative point-of-care testing of anthrax biomarker. *Talanta* 220: https://doi.org/10.1016/j.talanta.2020.121377.

46 Hu, J., Jiang, Y.Z., Ling Ling, W. et al. (2017). Dual-signal readout nanospheres for rapid point-of-care detection of ebola virus glycoprotein. *Analytical Chemistry* 89 (24): https://doi.org/10.1021/acs.analchem.7b02222.

47 Takemura, K., Adegoke, O., Suzuki, T., and Park, E.Y. (2019). A localized surface plasmon resonance-amplified immunofluorescence biosensor for ultrasensitive and rapid detection of nonstructural protein 1 of Zika virus. *PLoS One* 14 (1): https://doi.org/10.1371/journal.pone.0211517.

48 Ji, H., Li, W., Fang, P. et al. (2018). Point-of-care identification of bacteria using protein-encapsulated gold nanoclusters. *Advanced Healthcare Materials* 7 (13): https://doi.org/10.1002/adhm.201701370.

49 Zhao, H., Wang, L., Zhu, J. et al. (2015). Label-free nucleic acids detection based on DNA templated silver nanoclusters fluorescent probe. *Talanta* 138: https://doi.org/10.1016/j.talanta.2015.02.021.

50 Jin, B., Yang, Y., He, R. et al. (2018). Lateral flow aptamer assay integrated smartphone-based portable device for simultaneous detection of multiple targets using upconversion nanoparticles. *Sensors and Actuators, B: Chemical* 276: https://doi.org/10.1016/j.snb.2018.08.074.

51 Lee, N., Wang, C., and Park, J. (2018). User-friendly point-of-care detection of influenza A (H1N1) virus using light guide in three-dimensional photonic crystal. *RSC Advances* 8 (41): https://doi.org/10.1039/C8RA02596G.

52 He, Q., Dongmei, Y., Bao, M. et al. (2020). High-throughput and all-solution phase African swine fever virus (ASFV) detection using CRISPR-Cas12a and fluorescence based point-of-care system. *Biosensors and Bioelectronics* 154: https://doi.org/10.1016/j.bios.2020.112068.

53 Li, X., Donglian, L., Sheng, Z. et al. (2012). A fast and sensitive immunoassay of avian influenza virus based on label-free quantum dot probe and lateral flow test strip. *Talanta* 100: https://doi.org/10.1016/j.talanta.2012.08.041.

54 Chen, C. and Wang, J. (2020). Optical biosensors: an exhaustive and comprehensive review. *Analyst* https://doi.org/10.1039/c9an01998g.

55 Na, W., Kang, M., Jang, D., and Shin, S. (2018). Miniaturized surface plasmon resonance biosensor with vacuum-driven hydrodynamic focusing. *Sensors and Actuators, B: Chemical* 254: https://doi.org/10.1016/j.snb.2017.07.041.

56 Diao, W., Tang, M., Ding, S. et al. (2018). Highly sensitive surface plasmon resonance biosensor for the detection of HIV-related DNA based on dynamic and structural DNA nanodevices. *Biosensors and Bioelectronics* 100: https://doi.org/10.1016/j.bios.2017.08.042.

57 Chowdhury, A.D., Takemura, K., Khorish, I.M. et al. (2020). The detection and identification of dengue virus serotypes with quantum dot and aunp regulated localized surface plasmon resonance. *Nanoscale Advances* 2 (2): https://doi.org/10.1039/c9na00763f.

58 Tokel, O., Yildiz, U.H., Inci, F. et al. (2015). Portable microfluidic integrated plasmonic platform for pathogen detection. *Scientific Reports* 5: https://doi.org/10.1038/srep09152.

59 Zrimsek, A.B., Chiang, N., Mattei, M. et al. (2017). Single-molecule chemistry with surface- and tip-enhanced Raman spectroscopy. *Chemical Reviews* https://doi.org/10.1021/acs.chemrev.6b00552.

60 Pang, Y., Wan, N., Shi, L. et al. (2019). Dual-recognition surface-enhanced Raman scattering(SERS)biosensor for pathogenic bacteria detection by using vancomycin-SERS tags and Aptamer-Fe_3O_4@Au. *Analytica Chimica Acta* 1077: https://doi.org/10.1016/j.aca.2019.05.059.

61 Wang, C., Xiao, R., Dong, P. et al. (2014). Ultra-sensitive, high-throughput detection of infectious diarrheal diseases by portable chemiluminescence imaging. *Biosensors and Bioelectronics* 57: https://doi.org/10.1016/j.bios.2014.01.016.

62 He, L., Yang, H., Xiao, P. et al. (2017). Highly selective, sensitive and rapid detection of *Escherichia coli* O157:H7 using duplex PCR and magnetic nanoparticle-based chemiluminescence assay. *Journal of Biomedical Nanotechnology* 13 (10): https://doi.org/10.1166/jbn.2017.2422.

63 Sriram, G., Bhat, M.P., Patil, P. et al. (2017). Paper-based microfluidic analytical devices for colorimetric detection of toxic ions: a review. *TrAC Trends in Analytical Chemistry* https://doi.org/10.1016/j.trac.2017.06.005.

64 Alhogail, S., Suaifan, G.A.R.Y., Bikker, F.J. et al. (2019). Rapid colorimetric detection of *Pseudomonas aeruginosa* in clinical isolates using a magnetic nanoparticle biosensor. *ACS Omega* 4 (26): https://doi.org/10.1021/acsomega.9b02080.

65 Zheng, L., Cai, G., Wang, S. et al. (2019). A microfluidic colorimetric biosensor for rapid detection of *Escherichia coli* O157:H7 using gold nanoparticle aggregation and smart phone imaging. *Biosensors and Bioelectronics* 124–125: https://doi.org/10.1016/j.bios.2018.10.006.

66 Yang, H., Xiao, M., Lai, W. et al. (2020). Stochastic DNA dual-walkers for ultrafast colorimetric bacteria detection. *Analytical Chemistry* 92 (7): https://doi.org/10.1021/acs.analchem.9b05149.

67 Wu, K., Diqing, S., Saha, R. et al. (2019). Magnetic particle spectroscopy-based bioassays: methods, applications, advances, and future opportunities. *Journal of Physics D: Applied Physics* https://doi.org/10.1088/1361-6463/ab03c0.

68 Poller, W.C., Löwa, N., Wiekhorst, F. et al. (2016). Magnetic particle spectroscopy reveals dynamic changes in the magnetic behavior of very small superparamagnetic iron oxide nanoparticles during cellular uptake and enables determination of cell-labeling efficacy. *Journal of Biomedical Nanotechnology* 12 (2): https://doi.org/10.1166/jbn.2016.2204.

69 Stevens, W.S., Scott, L., Noble, L. et al. (2017). Impact of the geneXpert MTB/RIF technology on tuberculosis control. *Microbiology Spectrum* 5 (1): https://doi.org/10.1128/microbiolspec.tbtb2-0040-2016.

70 Wood, C.S., Thomas, M.R., Budd, J. et al. (2019). Taking connected mobile-health diagnostics of infectious diseases to the field. *Nature* https://doi.org/10.1038/s41586-019-0956-2.

71 Wang, J., Karnaushenko, D., Medina-Sánchez, M. et al. (2019). Three-dimensional microtubular devices for lab-on-a-chip sensing applications. *ACS Sensors* https://doi.org/10.1021/acssensors.9b00681.

72 Du, K., Park, M., Griffiths, A. et al. (2017). Microfluidic system for detection of viral RNA in blood using a barcode fluorescence reporter and a photocleavable capture probe. *Analytical Chemistry* 89 (22): https://doi.org/10.1021/acs.analchem.7b03527.

73 Loo, J.F., Chuen, G.W., But, C. et al. (2019). A rapid sample-to-answer analytical detection of genetically modified papaya using loop-mediated isothermal amplification assay on lab-on-a-disc for field use. *Food Chemistry* 274: https://doi.org/10.1016/j.foodchem.2018.09.049.

74 Van Nguyen, H., Nguyen, V.D., Nguyen, H.Q. et al. (2019). Nucleic acid diagnostics on the total integrated lab-on-a-disc for point-of-care testing. *Biosensors and Bioelectronics* https://doi.org/10.1016/j.bios.2019.111466.

75 Chinnadayyala, S.R., Park, J., Le, H.T.N. et al. (2019). Recent advances in microfluidic paper-based electrochemiluminescence analytical devices for point-of-care testing applications. *Biosensors and Bioelectronics* https://doi.org/10.1016/j.bios.2018.10.038.

76 Park, C., Kim, H.R., Kim, S.K. et al. (2019). Three-dimensional paper-based microfluidic analytical devices integrated with a plasma separation membrane for the detection of biomarkers in whole blood. *ACS Applied Materials and Interfaces* 11 (40): https://doi.org/10.1021/acsami.9b13644.

77 Akyazi, T., Basabe-Desmonts, L., and Benito-Lopez, F. (2018). Review on microfluidic paper-based analytical devices towards commercialisation. *Analytica Chimica Acta* https://doi.org/10.1016/j.aca.2017.11.010.

78 Li, L., Miao, B., Li, Z. et al. (2019). Sample-to-answer hepatitis B virus DNA detection from whole blood on a centrifugal microfluidic platform with double rotation axes. *ACS Sensors* 4 (10): https://doi.org/10.1021/acssensors.9b01270.

79 Furst, A.L. and Francis, M.B. (2019). Impedance-based detection of bacteria. *Chemical Reviews* https://doi.org/10.1021/acs.chemrev.8b00381.

80 Park, S., Zhang, Y., Lin, S. et al. (2011). Advances in microfluidic PCR for point-of-care infectious disease diagnostics. *Biotechnology Advances* https://doi.org/10.1016/j.biotechadv.2011.06.017.

81 Neuzil, P., Zhang, C., Pipper, J. et al. (2006). Ultra fast miniaturized real-time PCR: 40 cycles in less than six minutes. *Nucleic Acids Research* 34 (11): https://doi.org/10.1093/nar/gkl416.

82 Zhou, W., Li, H., Ying, L. et al. (2018). A CRISPR–Cas9-triggered strand displacement amplification method for ultrasensitive DNA detection. *Nature Communications* 9 (1): https://doi.org/10.1038/s41467-018-07324-5.

83 Obande, G.A. and Singh, K.K.B. (2020). Current and future perspectives on isothermal nucleic acid amplification technologies for diagnosing infections. *Infection and Drug Resistance* https://doi.org/10.2147/IDR.S217571.

5

Organic- and Inorganic-Based Nanomaterials for Healthcare Diagnostics

Komal Kashyap, Maheswata Moharana, Fahmida Khan, and Subrat K. Pattanayak

National Institute of Technology, Department of Chemistry, Raipur, Chhattisgarh 492010, India

5.1 Introduction

Nanomaterials have become increasingly used in a range of fields recently, including healthcare, biomedical, and environment [1]. Nanotechnology has drawn a lot of attention in novel applications using nanomaterials because of its capacity to store and convert energy [2]. These are used in a wide range of technological products, including quantum dots, nanotubes, films, fullerenes, and plates [3]. These forms are available in a variety of shapes and types, such as tubular, irregular, single, fused, aggregated, and single shapes. One of its new features is the physical behavior of nanomaterials, which transitioned particle size. The usage of nanomaterials as catalysts is expanding in a number of industries, including the mechanical, electrical, and environmental remediation fields [4]. Common examples of nanomaterial usage that is acceptable to the environment include on-site wastewater treatment, solar cells made of nanomaterials for increased energy efficiency, and air and water purification [5]. These are typically made by bottom-up procedures including liquid-phase synthesis, self-assembly, carbonization, physical and chemical vapor deposition, and activation. The bulk of these activities uses a lot of energy and resources, which finally leads to the release of pollutants into the air, water, and soil in the form of effluents and emissions. Most nanomaterials research up to this point has been centered on how they work uniquely in a variety of fields and applications, ignoring potential environmental effects throughout their life cycle [6–8]. These materials differ significantly in their fundamental physicochemical characteristics from the bulk of the same materials because the majority of their component atoms or molecules are positioned on the surface as a result of their small size [9]. For instance, identical material nanorods and nanospheres may have highly different properties. Creating artificial nanostructures like quantum dots requires using the quantum phenomena seen in nanoparticles, such as nanorods [10]. Nanoparticles,

Point-of-Care Biosensors for Infectious Diseases, First Edition.
Edited by Sushma Dave and Jayashankar Das.

like these, have magnetic moments because they are made up of many unpaired electron spins from different elements. Because of all of these characteristics, there are several techniques for classifying nanomaterials [11–13]. Nanomaterials are classified into three types depending on their chemical composition, carbon-based allotrope nanomaterials, composed of carbon atoms, and inorganic nanoparticles [14, 15]. These are largely composed of polymeric nanomaterials and organic nanomaterials. Although both of the aforementioned techniques are necessary for the development of nanomaterial-based biosensors, the bottom-up route has greater useful applications. Nanotechnology developments have fueled the creation of faster, less costly, more accurate, and more sensitive diagnostic tests and diagnostic equipment [16, 17]. Nanomaterial-based biosensors integrate molecular engineering, biotechnology, material science, and chemistry [18, 19]. Certain biosensors may presently identify as low as one parasite per microliter of blood due to their extraordinarily high sensitivity. Nanotechnology enables the identification of disease biomarkers at exceedingly low abundances, allowing enabling disease diagnosis at an early stage. This might speed up medical operations for follow-up treatment and routine prognosis to track patient diagnoses. Combining biosensing with nanotechnology is particularly important for point-of-care diagnostics in nations lacking modern medical facilities.

5.2 Nanomaterials Based on Carbon Allotropes in Healthcare

Nanomaterials which are based on carbon allotrope have piqued the curiosity about medical biosensing community in recent years. Because they include diverse forms of carbon allotropes, some of them are fullerenes, graphite, and diamonds, as well as more sophisticated forms such as nanohorns, nanotubes, and graphene. These allotropes differ in that they each have distinct properties. As a result, they are widely used in a variety of biological applications, such as tissue engineering, medicine delivery, cancer therapy, medical diagnostics, biosensing, and bioimaging. These biosensor materials may be used to identify a wide range of substances useful in point-of-care disease analysis and medical diagnostics. Carbon allotrope-based nanomaterials, including fullerenes, graphene, nanotubes, and quantum dots, have played an important part in recent biosensor findings [20].

The use of carbon-based nanomaterials for various biological compounds has increased in recent years. Despite possessing remarkable material characteristics, carbon allotrope-based nanomaterials are hampered by a lack of surface heterogenic reactivity, which is essential for clinically meaningful biomarkers [21]. Precision nanomaterial interface engineering is necessary to improve biomolecule adhesion to the functionalized surface and the consequent protein–protein recognition. The bulk of these nanomaterials requires undergoing covalent or non-covalent changes to achieve this. Many biomolecules can be functionalized to immobilize them, including enzymes, aptamers, and antigens antibodies [22]. Carbon allotrope-based nanomaterials act as transducers in all of these situations by providing the necessary interfaces for the transfer of biorecognition inputs into extremely sensitive and quantified outputs. Quantum dots, fullerenes, graphene sheets, and nanotubes, considering the fact that almost all crystalline and amorphous carbon allotropes have been employed in healthcare biosensing [23]. Water-soluble hydroxyl fullerene preserves the biological function of proteins. Due to their high aspect ratio, substantial surface area, outstanding mechanical strength, outstanding optical and electrical properties, and strong thermal and chemical durability, carbon nanotubes are regarded as advantageous building blocks for biosensors [24]. These are superior to biosensors because of their better signal-to-noise ratio, reduced background, real-time monitoring of wide absorption spectrum, and label-free detection. They are good for optical biosensing since they also have great luminescence qualities and a high luminous intensity. The hexagonally organized carbon atoms make up the relatively new carbon allotrope known as graphene. At ambient temperature, the electrons in graphene have unique features such as high thermal conductivity with higher surface area [25]. The characteristics of graphene may be altered by modifying the number of layers and the stacking sequence. It is also an excellent material for biosensors due to its ability to interact with diverse biomolecules through physisorption. The fluorescence of graphene oxide is one of the fascinating properties of graphene derivatives. Carbon dots (CD) have attracted a lot of scientific interest as the optimal medium for detecting a variety of cancer biomarkers for early-stage diagnosis and tumor growth [26]. The ionization mass spectrometry or surface-enhanced laser desorption is excellent for the gold–CD nanocomposite [27]. Doping techniques have the ability to alter the basic properties of CDs. These have been used to diagnose other dangerous and curable diseases in addition to cancer. Through simple functionalization with amines, thiol group halides, or hydroxyl groups, they may easily create covalent or non-covalent connections with biomolecules [28]. Due to its huge potential for treating serious bone abnormalities and associated disorders, bone tissue engineering has drawn a lot of attention. Due to their superior qualities, such as excellent mechanical strength, large surface area, tunable surface functionalities high biocompatibility as well as their abundant and low-cost nature carbon-based nanomaterials have recently attracted significant interest for their applications as scaffolds [29]. The most prevalent and essential component of life is carbon. Although nonmetallic, it possesses a specific capacity to self-bind and associate with other elements to create structurally complex compounds with a range of different physical and chemical properties. The use of carbon-based nanomaterials in medicine for diagnosis, drug administration, and therapeutic reasons has wide-ranging technical implications. Graphene nanomaterials which belong to the family of carbon-based nanomaterials have special physical and chemical properties that have drawn researchers to use them for biomedical applications [30]. Worldwide, cancer continues to be a serious health concern, and conventional medicinal methods have been ineffective in finding a cure. The complex architecture and various components of tumor microenvironments, which also act as the foundation for tumor growth, and development, pose a special challenge in the field of tumor therapy [31]. Carbon nanomaterials such as graphene–fullerene, carbon nanotubes, and carbon quantum dots have distinct physiochemical features that present chances to overcome obstacles by specifically targeting cancer cells [31]. The variety and use of medications have increased due to the quick advancements in the medical and pharmaceutical professions nowadays. However, there are issues like the overuse of over-the-counter medications and the insensitive use of multiple or combined medications in the treatment of diseases [32]. The performance of the electrochemical sensors with catalytic effects is improved by using carbon nanomaterials such as graphene oxide, carbon nanotubes, carbon nanofibers, and nanodiamonds [32]. Nano-graphene oxide is a derivative of graphene that contains sp^2-bonded carbon atoms arranged in a hexagonal conformation in a two-dimensional (2D) atomic layer along with sp^3 domains that contain carbon atoms connected to oxygen functional groups. The fundamental reason why nano-graphene oxide is superior is due to its unique inherent chemical and physical structure, which confers an amazing degree of chemical diversity, a large aspect ratio, and unique physical properties. In addition to its inherent optical, mechanical, and electrical capabilities, the synergistic effects brought about by the assembly of well-defined structures at the nano-graphene oxide surface enable the development of new multifunctional hybrid materials for cancer therapy [33]. Over the past decades, nanotechnology has advanced significantly and found numerous applications in a variety of fields. Carbon allotropes are among the most efficient nanomaterials because of their ease of functionalization, conductivity, surface area, and electrical activity. They are referred to as "wonder materials." Applications of carbon nanoparticles range from biosensors to the removal of contaminants along with the detection of potentially hazardous substances in food, pharmaceuticals, genes, and drug delivery. The pharmaceuticals have also benefited from the use of allotropes such as carbon nanotubes, graphene, graphene oxide, fullerenes, and CD. With advancements in the construction of electrochemical biosensors, the application of carbon allotropes in the current situation opens up several possibilities in the future. In comparison to other nanomaterials, it is a better solution because of its selectivity, sensitivity, and cost effectiveness [34]. Graphitic carbon nitride is a member of 2D semiconducting materials that are inexpensive, metal-free, and essential for sensing applications. Various literatures reported the applications of graphitic carbon nitride as photocatalyst on energy conversion devices, hydrogen evolution, and photoelectrochemical studies as well as photoluminescence-based optical, fluorescence, and colorimetric sensors [35]. Electrochemical detection techniques have been used in a number of research to confirm the presence of azo dyes [36]. Chemical exfoliation of graphite into graphite oxide or chemical vapor deposition techniques can be used to make graphene. The graphite oxide is subsequently transformed into reduced graphite oxide sheets [36]. There has been an increase in interest in employing carbon nanomaterials for biological applications since the discovery of low-dimensional carbon allotropes. In the biomedical industry, carbon nanomaterials have been used for bioimaging, chemical sensing, targeting, delivery, therapies, catalysis, and energy harvesting. To utilize a desired attribute of the nanoparticles, each application needs a customized surface functionalization [37]. Applications demand for the use of the protein–nanocarbon system in a complicated setting that could impair its functionality or activity [37].

5.3 Inorganic Nanomaterials in Health Diagnosis

Recent developments in nanotechnology have significantly improved biological applications, such as human healthcare diagnosis, monitoring, and therapy. Furthermore, due to their excellent biomedical compatibility, nanomaterials combined with hybrid organic/inorganic polymeric materials have demonstrated exceptional and prospective biomedical applications, particularly in the areas of biosensing, drug delivery, wound healing, cancer therapy, and bioimaging [38]. The discipline of biomedicine has a great deal of potential for inorganic nanomaterials with remarkable physicochemical features (such as catalytic, optical, thermal, electrical, or magnetic performance) that can offer desired functionality. However, the long-term, nonspecific accumulation of these inorganic nanoparticles in healthy tissues can result in toxicity, inhibiting their widespread clinical application [39]. The development of biodegradable and clearable inorganic nanoparticles over the past few decades has provided the opportunity to stop such long-term toxicity. Additionally, the expansion of theranostic applications for various diseases and the advancement of clinical trials depend on a thorough understanding of the design of such nanomaterials and their metabolic routes within the body [39]. Due to their innate characteristics and wide range of therapeutic uses, hybrid organic and inorganic nanomaterials are emerging in the field of health research. In the study of hybrid materials, natural materials can be copied to create brand-new synthetic materials. Designing novel hybrid materials will be made easier with an understanding of the material's mechanism and structure [40]. The incidence of pancreatic cancer is rising alarmingly, and survival rates have not increased significantly over the past three decades. Although significant efforts have been made to identify this disease early and treat it thoroughly, little or no improvement in survival has been seen, necessitating the creation of novel approaches. Inorganic nanomaterials that are currently being developed, such as carbon nanotubes, quantum dots, and mesoporous silica/gold/super magnetic nanoparticles, have been extensively exploited in biomedical research with tremendous promise for the prevention and treatment of cancer [41]. Through the prevention, early detection, diagnosis, treatment, and follow-up of various diseases, the development and application of nanotechnology-related knowledge and tools in modern medicine have been demonstrating great potential in raising human living standards and improving healthcare conditions. The research of nanoparticulate formulations for theranostic uses in a living body has significantly advanced nanomedicine. Inorganic nanoparticles, which are common, have unique physiochemical characteristics and biological consequences that their normal organic counterparts often lack [42]. A significant cause of mortality, particularly in people with diabetes and other diseases, has been identified as chronic wounds. Due to the increasing prevalence of chronic wounds and their associated diseases, as well as the shortcomings of current therapies, there is an urgent need for fresh and creative methods to speed up wound healing [43]. Due to their distinct photophysical, photochemical, and biological characteristics, ruthenium polypyridyl complexes are molecules of great interest in the treatment of cancer. They effectively enter tumor cells for the treatment. They might deliberately increase the cellular uptake of ruthenium-based complexes in tumor cell [44]. At the nanoscale, transitional and noble metals exhibit amazing characteristics. Incompletely filled penultimate or pre-penultimate orbitals in conjunction with additional surface atoms provide unusual quantum phenomena and optical characteristics. They can be combined with other organic and carbon-based materials to create nanocomposites with a combination of preexisting or wholly unique properties. They can also be utilized to create excellent alloys. Bimetallic alloys, core–shell structures, metal–organic frameworks, nanotubes, and nanowire arrays are only a few of the designs that have been characterized. Each of these nanomaterials can enhance the biocompatibility and transmission capabilities of biosensors by providing attractive interface and surface features. Biosensor development has lately received a lot of interest in novel inorganic designs such as nanoshells, nanocages, and nanowires [45]. Nanoshells are a new type of nanomaterials with controlled plasmon resonance. The materials may be designed to match the wavelength for specific applications, such as near-infrared regions that need excellent light penetration into tissue [46]. For cytosensing and other related medical diagnostic purposes, nanowire arrays can be employed because they have the potential to puncture cellular lipid bilayers like nanoneedles. Gold nanoparticles have a huge surface area with unique optical properties. They exhibit metallic lattice-like oscillations. The amount of heat and light dispersion caused by surface plasmonic decay may vary depending on the shape of AuNPs. This parameter is useful in the development of optical immunoassays and biosensors [47]. Multifunctional AuNPs are now commonly used to identify a wide range of diseases, neurological illnesses, diabetes mellitus, nucleic acids, amino acids, hemoglobin, and cancer biomarkers. Similar to AuNPs, silver nanoparticles (AgNPs) are frequently utilized in diagnostic procedures. Their optical characteristics are governed by their aggregation level, shape, and size in the same way. Silver nanoparticles are analogous to AuNPs. However, AgNPs differ from AuNPs in that they are antimicrobial. They are also cheaper than AuNPs and have advantageous electrical characteristics [48]. Due to being less expensive than AuNPs and having superior electrical properties, these are also used to improve the functionality of a biosensing device. Many pharmacological and narcotic medicines have been identified using AgNPs [49]. The porous walls and hollow interior of nanocages, a unique type of nanomaterials, are composed of noble metals. The optical characteristics of these cube-shaped nanoparticles differ from those of spherical nanoparticles [50].

5.4 Organic Nanomaterials in Healthcare Diagnosis

The great majority of organic nanomaterials are polymeric in nature, with a small number of the most powerful nanostructure materials representing an exception [51]. In biomedical activities, such as drug delivery and medical diagnostics, polymeric nanoparticles are being employed significantly more consistently. Because of their biocompatibility, intrinsic inertness, and design flexibility, they are preferred. These nanoparticles are thermally stable and affordable to produce.

Nanostructure films may be recovered even after usage and just require simple handling and processing processes. Their superior electrical, magnetic, and optical properties make them ideal components for a wide range of biosensing systems in light of hyperbranched polymeric–polymeric nanocomposites are great transducers due to their structural variety, exquisite responsibility, and easy fabrication processes. They are affordable, have high signal-to-noise ratios, and may be used in DNA, aptamer, or antibody sensors. Molecular machines are small devices. All biosensor components must be biocompatible to build sensors for sensing diverse biological variables [52]. Materials must be nontoxic as well as impenetrable to bodily tissues and fluids. Furthermore, it should not cause a chronic or acute inflammatory response in the tissues. The hybrid films were employed as transducers in the impedimetric biosensor, and the urease enzyme was covalently immobilized utilizing the hydroxyl groups from polyvinyl alcohol [53]. Enzymes, in particular, and proteins in general, are critical structural constituents in biosensing [54]. Nanogels, generally described as nanostructured hydrogels, were hydrophilic three-dimensional (3D) cross-linked polymeric structures that are nanoscale in dimension and are capable of changing their chemical properties [54]. As a result, they are known as stimuli-responsive smart materials. They transfer swelling effectively, are strong, and have a large surface area. Three categories of nanogel applications may be made based on how they function in a biosensing system [55]. Three types of materials are used for sensory membranes, stimuli-responsive multifunctional materials, and encapsulation carriers. Nanostructured hydrogels are especially useful as optical biosensors because of their huge surface area, which makes them suitable for encasing fluorescent compounds [56]. When cross-linking polymers are polymerized with functional monomers in the presence of a template molecule, synthetic polymers termed molecularly imprinted polymers are formed. A cavity is created after the template is taken out that is identical in shape, size, and function to the template molecule. Numerous bacteria become resistant to antibiotics as a result of bioaccumulation from dairy products as well as food. Knowing how much antibiotics are present in typical biological fluids is crucial since antibiotic resistance makes it more difficult to treat infectious diseases [57]. Molecular nanoparticles called dendrimers are created by the covalent combination of atoms, which are divided into three architectural parts. One of their most outstanding qualities is their capacity to self-assemble into superstructures at material peripheries. Dendrimers are one of the most current forms of macromolecular sensing devices for medical diagnostics due to their unique structural properties such as their high solubility, spheroidal surface, and nanoscopic size [58]. Since dendrimers and proteins are readily attached to surfaces of inorganic biosensors, denaturation can make them unproductive. Immobilizing dendrimers on the biosensor surface is an intriguing way to prevent this from happening [59]. There are structural similarities between dendrimers and hyperbranched polymers (HBP). The irregular arrangement of linear and dendritic units throughout the macromolecular architecture also contributes to the anomalies of the structure. Contrary to dendrimers, HBP is frequently synthesized in a single step. Because of their linearity, availability of functional groups and minimal chain entanglements HBPs are suitable for a variety of biosensing systems [60]. Additionally, the label-free aptasensor was used in *in vitro* studies to develop tests for platelet adhesion, whole blood, and hemolysis, as well as to recognize abnormalities in overall red blood cell shape [61]. Crystalline assemblies of porous polymers are held together by covalent bonds to form covalent organic frameworks (COF) [62]. Strong covalent connections enable them to bind biological building pieces in an atomically precise and organized way. The surface areas of these frameworks are large, they are thermally stable, and they have persistent porosity, and the size of the holes in it may be modified. These are helpful for locating a number of illness biomarkers, such as those that point to malignancy. The enormous-conjugation procedure of the imine-linked COFs was used by the sensor [63]. Polymer nanocomposites (PNCs) give many types of topologies, creative uses, and simple manufacturing procedures. They have several remarkable qualities, which contribute to their environmental stability and biocompatibility [64]. It can interact with a large number of biomolecules – some of them aptamers, enzymes, and proteins to help in biorecognition [65]. They are the best solutions available for signal transduction in biosensing systems because of their substantial surface area, combination of the aforementioned features, and quick electron transfer rate. Molecular machines are also known as nanomotors; they are microscopic devices that can move independently or with the support of an external energy source when placed in a liquid condition [66]. Modern and effective biosensing devices are being made faster due to molecular machinery functionalized with biomolecules [45, 67]. Classifications of nanomaterials and their different healthcare applications are shown in Figure 5.1.

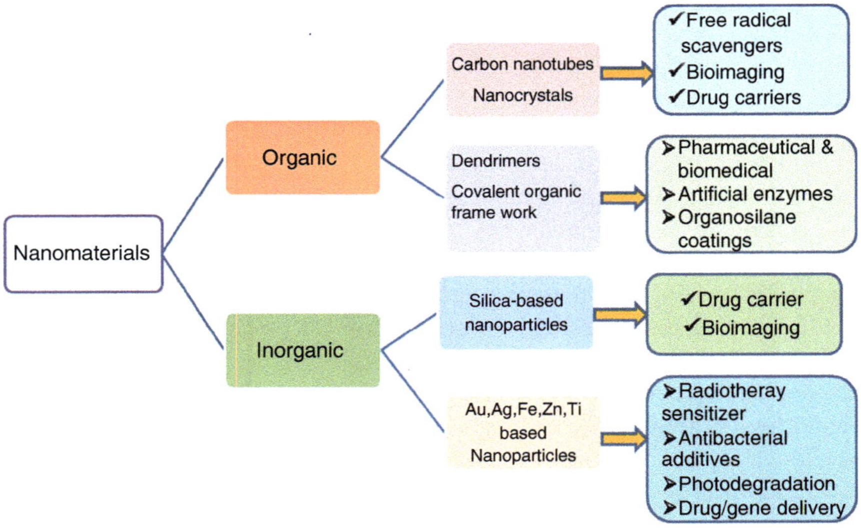

Figure 5.1 Classifications of nanomaterials and their different healthcare applications.

5.5 Future Prospects

Due to their small size and comparatively large surface area, nanomaterials exhibit unique piezoelectric, electromagnetic, and optical capabilities. These characteristics have great potential for bioimaging, healthcare diagnostics, and medical diagnosis. Proteins, nucleic acids, enzymes, antibodies, and other clinically significant substances can all be immobilized using nanomaterials due to their high affinity for biomolecules. This has made it possible to create applications in the field of sensors such as immunosensors, aptasensors, enzyme sensors, sandwich assays, and many more. The latest research in the field of biosensing made use of several innovative nanomaterials, including monomolecular nanomotors and noticeably bigger nanocages. In the near future, it is projected that nanomaterial-based biosensors would replace normally expensive sensing devices due to their quick, affordable, and simple operational processes. Designing novel hybrid materials will be made easier by an understanding of the material's mechanism toward the healthcare diagnostic.

References

1 Sagadevan, S. and Periasamy, M. (2014). Recent trends in nanobiosensors and their applications-a review. *Reviews on Advanced Materials Science* 36 (2014): 62–69.

2 Haleem, A., Javaid, M., Singh, R.P. et al. (2021). Biosensors applications in medical field: a brief review. *Sensors International* 2: 100100.

3 Bacakova, L., Pajorova, J., Tomkova, M. et al. (2020). Applications of nanocellulose/nanocarbon composites: focus on biotechnology and medicine. *Nanomaterials* 10 (2): 196.

4 Kulkarni, M.B. and Goel, S. (2020). Microfluidic devices for synthesizing nanomaterials—a review. *Nano Express* 1 (3): 032004.

5 Ciambelli, P., La Guardia, G., and Vitale, L. (2020). Nanotechnology for green materials and processes. In: *Studies in Surface Science and Catalysis*, vol. 179 (ed. A. Basile, G. Centi, M. De Falco, and G. Iaquaniello), 97–116. Elsevier https://doi.org/10.1016/B978-0-444-64337-7.00007-0.

6 Bauer, C., Buchgeister, J., Hischier, R. et al. (2008). Towards a framework for life cycle thinking in the assessment of nanotechnology. *Journal of Cleaner Production* 16 (8–9): 910–926.

7 Gavankar, S., Suh, S., and Keller, A.F. (2012). Life cycle assessment at nanoscale: review and recommendations. *The International Journal of Life Cycle Assessment* 17 (3): 295–303.

8 Linkov, I., Satterstrom, F.K., and Corey, L.M. (2008). Nanotoxicology and nanomedicine: making hard decisions. *Nanomedicine: Nanotechnology, Biology and Medicine* 4 (2): 167–171.

9 Oberdörster, G., Maynard, A., Donaldson, K. et al. (2005). Principles for characterizing the potential human health effects from exposure to nanomaterials: elements of a screening strategy. *Particle and Fibre Toxicology* 2 (1): 1–35.

10 Whitesides, G.M. (2005). Nanoscience, nanotechnology, and chemistry. *Small* 1 (2): 172–179.

11 Issa, B., Obaidat, I.M., Albiss, B.A., and Haik, Y. (2013). Magnetic nanoparticles: surface effects and properties related to biomedicine applications. *International Journal of Molecular Sciences* 14 (11): 21266–21305.

12 Farzin, A., Etesami, S.A., Quint, J. et al. (2020). Magnetic nanoparticles in cancer therapy and diagnosis. *Advanced Healthcare Materials* 9 (9): 1901058.

13 Johnson, C.J., Zhukovsky, N., Cass, A.E., and Nagy, J.M. (2008). Proteomics, nanotechnology and molecular diagnostics. *Proteomics* 8 (4): 715–730.

14 Zaytseva, O. and Neumann, G. (2016). Carbon nanomaterials: production, impact on plant development, agricultural and environmental applications. *Chemical and Biological Technologies in Agriculture* 3 (1): 1–26.

15 Madima, N., Mishra, S.B., Inamuddin, I., and Mishra, A.K. (2020). Carbon-based nanomaterials for remediation of organic and inorganic pollutants from wastewater. A review. *Environmental Chemistry Letters* 18 (4): 1169–1191.

16 Wang, C., Liu, M., Wang, Z. et al. (2021). Point-of-care diagnostics for infectious diseases: from methods to devices. *Nano Today* 37: 101092.

17 Syedmoradi, L., Daneshpour, M., Alvandipour, M. et al. (2017). Point of care testing: the impact of nanotechnology. *Biosensors and Bioelectronics* 87: 373–387.

18 Malhotra, B.D., Srivastava, S., Ali, M., and Singh, C. (2014). Nanomaterial-based biosensors for food toxin detection. *Applied Biochemistry and Biotechnology* 174 (3): 880–896.

19 Tiwari, J.N., Vij, V., Kemp, K.C., and Kim, K.S. (2016). Engineered carbon-nanomaterial-based electrochemical sensors for biomolecules. *ACS Nano* 10 (1): 46–80.

20 Niazvand, F., Wagh, P.R., Khazraei, E. et al. (2021). Application of carbon allotropes composites for targeted cancer therapy drugs: a review. *Journal of Composites and Compounds* 3 (7): 140–151.

21 Pirzada, M. and Altintas, Z. (2019). Nanomaterials for healthcare biosensing applications. *Sensors* 19 (23): 5311.

22 Wang, C.F., Sun, X.Y., Su, M. et al. (2020). Electrochemical biosensors based on antibody, nucleic acid and enzyme functionalized graphene for the detection of disease-related biomolecules. *Analyst* 145 (5): 1550–1562.

23 Kour, R., Arya, S., Young, S.J. et al. (2020). Recent advances in carbon nanomaterials as electrochemical biosensors. *Journal of the Electrochemical Society* 167 (3): 037555.

24 Mallakpour, S. and Khadem, E. (2016). Carbon nanotube–metal oxide nanocomposites: fabrication, properties and applications. *Chemical Engineering Journal* 302: 344–367.

25 Tang, H., Hessel, C.M., Wang, J. et al. (2014). Two-dimensional carbon leading to new photoconversion processes. *Chemical Society Reviews* 43 (13): 4281–4299.

26 Iannazzo, D., Espro, C., Celesti, C. et al. (2021). Smart biosensors for cancer diagnosis based on graphene quantum dots. *Cancers* 13 (13): 3194.

27 Abdelhamid, H.N., Talib, A., and Wu, H.F. (2017). One pot synthesis of gold–carbon dots nanocomposite and its application for cytosensing of metals for cancer cells. *Talanta* 166: 357–363.

28 Khan, R. and Nishina, Y. (2021). Covalent functionalization of carbon materials with redox-active organic molecules for energy storage. *Nanoscale* 13 (1): 36–50.

29 Peng, Z., Zhao, T., Zhou, Y. et al. (2020). Bone tissue engineering via carbon-based nanomaterials. *Advanced Healthcare Materials* 9 (5): 1901495.

30 Singhal, S., Gupta, M., Alam, M.S. et al. (2022). Carbon allotropes-based nanodevices: graphene in biomedical applications. In: *Nanotechnology*, 241–269. CRC Press.

31 Saleem, J., Wang, L., and Chen, C. (2018). Carbon-based nanomaterials for cancer therapy via targeting tumor microenvironment. *Advanced Healthcare Materials* 7 (20): 1800525.

32 Cetinkaya, A., Kaya, S.I., Ozcelikay, G. et al. (2022). Carbon nanomaterials-based novel hybrid platforms for electrochemical sensor applications in drug analysis. *Critical Reviews in Analytical Chemistry* 1–16.

33 Gonçalves, G., Vila, M., Portolés, M.T. et al. (2013). Nano-graphene oxide: a potential multifunctional platform for cancer therapy. *Advanced Healthcare Materials* 2 (8): 1072–1090.

34 Sridharan, R., Monisha, B., Kumar, P.S., and Gayatahri, K.V. (2022). Carbon nanomaterials and its applications in pharmaceuticals: a brief review. *Chemosphere* 294: 133731.

35 Vinoth, S., Devi, K.S., and Pandikumar, A. (2021). A comprehensive review on graphitic carbon nitride based electrochemical and biosensors for environmental and healthcare applications. *TrAC Trends in Analytical Chemistry* 140: 116274.

36 Gupta, A. K., Dutta, K., Dutta, A., and Rana, S. (2021). Graphene based functional Nano-materials for bio-chemical azo dyes. *2021 5th International Conference on Electronics, Materials Engineering & Nano-Technology (IEMENTech)*. IEEE, pp. 1–8.

37 Oliveira, S.F., Bisker, G., Bakh, N.A. et al. (2015). Protein functionalized carbon nanomaterials for biomedical applications. *Carbon* 95: 767–779.

38 Bairagi, P.K., Rajbanshi, P., and Khare, P. (2022). Hybrid organic or inorganic nanomaterials for healthcare diagnostics. In: *Advanced Nanomaterials for Point of Care Diagnosis and Therapy* (ed. S. Dave, J. Das, and S. Ghosh), 275–312. Elsevier https://doi.org/10.1016/B978-0-323-85725-3.00014-3.

39 Wang, X., Zhong, X., Li, J. et al. (2021). Inorganic nanomaterials with rapid clearance for biomedical applications. *Chemical Society Reviews* 50 (15): 8669–8742.

40 Sharma, I. and Bhattacharjee, S. (2022). Hybrid organic or inorganic nanomaterials for healthcare diagnostics. In: *Advanced Nanomaterials for Point of Care Diagnosis and Therapy* (ed. S. Dave, J. Das, and S. Ghosh), 203–210. Elsevier https://doi.org/10.1016/B978-0-323-85725-3.00024-6.

41 Yang, F., Jin, C., Subedi, S. et al. (2012). Emerging inorganic nanomaterials for pancreatic cancer diagnosis and treatment. *Cancer Treatment Reviews* 38 (6): 566–579.

42 Huang, H., Feng, W., Chen, Y., and Shi, J. (2020). Inorganic nanoparticles in clinical trials and translations. *Nano Today* 35: 100972.

43 Nethi, S.K., Das, S., Patra, C.R., and Mukherjee, S. (2019). Recent advances in inorganic nanomaterials for wound-healing applications. *Biomaterials Science* 7 (7): 2652–2674.

44 Soliman, N., Gasser, G., and Thomas, C.M. (2020). Incorporation of Ru (II) polypyridyl complexes into nanomaterials for cancer therapy and diagnosis. *Advanced Materials* 32 (47): 2003294.

45 Yao, J., Yang, M., and Duan, Y. (2014). Chemistry, biology, and medicine of fluorescent nanomaterials and related systems: new insights into biosensing, bioimaging, genomics, diagnostics, and therapy. *Chemical Reviews* 114 (12): 6130–6178.

46 Altınoğlu, E.İ. and Adair, J.H. (2010). Near infrared imaging with nanoparticles. *Wiley Interdisciplinary Reviews: Nanomedicine and Nanobiotechnology* 2 (5): 461–477.

47 Amendola, V., Pilot, R., Frasconi, M. et al. (2017). Surface plasmon resonance in gold nanoparticles: a review. *Journal of Physics: Condensed Matter 29* (20): 203002.

48 Lee, S.H. and Jun, B.H. (2019). Silver nanoparticles: synthesis and application for nanomedicine. *International Journal of Molecular Sciences* 20 (4): 865.

49 Kim, D.M., Go, M.J., Lee, J. et al. (2021). Recent advances in micro/nanomaterial-based aptamer selection strategies. *Molecules* 26 (17): 5187.

50 Kale, S.N., Kitture, R., Ghosh, S. et al. (2017). Nanomaterials as enhanced antimicrobial agent/activity-enhancer for transdermal applications: a review. *Antimicrobial Nanoarchitectonics* 279–321.

51 Pryjmaková, J., Kaimlová, M., Hubáček, T. et al. (2020). Nanostructured materials for artificial tissue replacements. *International Journal of Molecular Sciences* 21 (7): 2521.

52 Rocchitta, G., Spanu, A., Babudieri, S. et al. (2016). Enzyme biosensors for biomedical applications: strategies for safeguarding analytical performances in biological fluids. *Sensors* 16 (6): 780.

53 Nemiwal, M., Zhang, T.C., and Kumar, D. (2022). Enzyme immobilized anomaterials as electrochemical biosensors for detection of biomolecules. *Enzyme and Microbial Technology* 110006.

54 Sasaki, Y. and Akiyoshi, K. (2010). Nanogel engineering for new nanobiomaterials: from chaperoning engineering to biomedical applications. *The Chemical Record* 10 (6): 366–376.

55 Kabanov, A.V. and Vinogradov, S.V. (2009). Nanogels as pharmaceutical carriers: finite networks of infinite capabilities. *Angewandte Chemie International Edition* 48 (30): 5418–5429.

56 Eslahi, N., Abdorahim, M., and Simchi, A. (2016). Smart polymeric hydrogels for cartilage tissue engineering: a review on the chemistry and biological functions. *Biomacromolecules* 17 (11): 3441–3463.

57 Manyi-Loh, C., Mamphweli, S., Meyer, E., and Okoh, A. (2018). Antibiotic use in agriculture and its consequential resistance in environmental sources: potential public health implications. *Molecules* 23 (4): 795.

58 Bharti, J.P., Prajapati, S.K., Jaiswal, M.K., and Yadav, R.D. (2011). Dendrimer multifunctional nano-device: a review. *International Journal of Pharmaceutical Sciences and Research* 2 (8): 1947.

59 Noriega-Luna, B., Godínez, L.A., Rodríguez, F.J. et al. (2014). Applications of dendrimers in drug delivery agents, diagnosis, therapy, and detection. *Journal of Nanomaterials* 2014.

60 Tande, B.M. (2003). The rheology, microstructure, and phase behavior of dendritic and hyperbranched polymers. University of Delaware.

61 Gheorghiu, M. (2020). Carbonic anhydrases: hematologic relevance and a iosensing perspective. *Blood and Genomics 4* (1): 19–30.

62 Waller, P.J., Gándara, F., and Yaghi, O.M. (2015). Chemistry of covalent organic frameworks. *Accounts of Chemical Research* 48 (12): 3053–3063.

63 Niu, F., Shao, Z.W., Zhu, J.L. et al. (2021). Structural evolution of imine-linked covalent organic frameworks and their NH_3 sensing performance. *Journal of Materials Chemistry C* 9 (27): 8562–8569.

64 Wu, H., Fahy, W.P., Kim, S. et al. (2020). Recent developments in polymers/polymer nanocomposites for additive manufacturing. *Progress in Materials Science* 111: 100638.

65 Ahmed, S.F., Mofijur, M., Ahmed, B. et al. (2022). Nanomaterials as a sustainable choice for treating wastewater. *Environmental Research* 214: 113807.

66 Sokolov, I.L., Cherkasov, V.R., Tregubov, A.A. et al. (2017). Smart materials on the way to theranostic nanorobots: molecular machines and nanomotors, advanced biosensors, and intelligent vehicles for drug delivery. *Biochimica et Biophysica Acta (BBA) - General Subjects* 1861 (6): 1530–1544.

67 Arshak, K., Velusamy, V., Korostynska, O. et al. (2009). Conducting polymers and their applications to biosensors: emphasizing on foodborne pathogen detection. *IEEE Sensors Journal* 9 (12): 1942–1951.

6

CRISPR/Cas System

Applications in Diagnosis of Infectious Diseases

Deepak Kumar Sahel[1] *and Mohd Azhar*[2]

[1]*Bhiwani, 127041, Haryana, India*
[2]*Research and Development, Tata Medical and Diagnostics Limited, Mumbai, 400001, Maharastra, India*

6.1 Introduction

The diseases which are mainly caused by microorganisms, such as bacteria, viruses, fungi, or worms/helminths, are called infectious diseases [1]. Depending on the infection-causing agent, the repercussions of an infectious disease can vary from mild inflammatory phase to life-threatening conditions. Infectious proteins such as Prions are transmitted either by inheritance or acquired via eating/receiving contaminated meat or other biological products and is one of the common causes of infectious diseases [1]. Although infectious diseases have common symptoms including fever, diarrhea, fatigue, muscle aches, sneezing, and coughing, sometimes the severity of the symptoms depends majorly on the nature of the antigen or pathogen. Therefore, an infectious disease is different from normal infection. From conception to death, people are attacked by a plethora of different living species, all of which compete for a position in the same environment. According to World Health Organization (WHO) 2019 report, lower respiratory infections, and diarrheal diseases are listed as two of the top 10 causes of death worldwide [2]. A multitude of pathogenic pathogens can be the cause of each of these disorders. Other examples of life-threatening diseases include AIDS, pneumonia, and meningitis, which are some of the most common causes of death in the past two decades. Additionally, some infectious diseases are linked to some debilitating diseases at the chronic stage [3].

Point-of-Care Biosensors for Infectious Diseases, First Edition.
Edited by Sushma Dave and Jayashankar Das.

For instance, the human papillomavirus (HPV) has been found to linked with cervical cancer, helicobacter pylori with stomach cancer as well as peptic ulcers, and both hepatitis B and C to liver cancer, etc. Over the past five years, the WHO has recorded over 1000 infectious disease outbreaks, including avian flu, swine flu, polio, cholera etc. The most prominent example is COVID-19, which is caused by the SARS-CoV-2 virus and predicted as the leading cause of death in 2020 [4]. Epidemiologically, COVID-19 was reported as the third leading cause of death in the United States, as per the data released by the Center for Disease Control and Prevention (CDC) for 2020 [5]. Over 533 million cases and over 6.3 million deaths have been confirmed by WHO worldwide due to COVID-19 [6]. Respiratory failure, which was associated with burst release of inflammatory cytokines (IL10. IL7, IL2, TNFα, GCSF, IP10, MIP1A, etc.) and death has been linked to the chronic stage of the COVID infection [7]. A virus may also be detected in the stool, in severe cases, and in the blood. Therefore, an early-stage diagnosis of the infections is very important to avoid a pandemic situation. It must be assumed that the multiplex PCR panels, which are currently available, do not include COVID-19 [8]. At the early stage, chest X-ray (CXR) results show normal architecture but bilateral infiltrates have been observed in later stages. Although CT imaging is more sensitive and specific, it typically ground glass opacities to display infiltrates, and sub-segmental consolidation [9]. However, in some cases abnormal CT scans have been utilized to detect COVID-19; the majority of these individuals got positive molecular tests on repeat testing [10]. Therefore. It is very important to develop a universal, versatile, and precise diagnostic tool for the fast detection of any infectious disease. However, ample diagnostics have been already developed employing molecular markers including direct or indirect entities. In recent years, nucleic acids have gained much attention in terms of their role in developing next-generation biosensors for early-stage detection of infections.

6.2 Nucleic Acids: Role in the Diagnosis

Early recognition of the infectious disease is usually crucial for improving patient prognosis and restricting transmission of disease. As a disease worsens, treatment strategies frequently become more restricted. Interestingly, nucleic acids play a crucial role in infection diagnosis, monitoring disease severity, and therapy. Technological advancements in nucleic acid analysis have enabled the development and clinical use of nucleic acid as disease biomarker. Nucleic acids are frequently utilized as biomarkers for a variety of disorders, including cancer, neurological diseases, and infectious diseases. This makes them a valuable tool for diagnosis and monitoring of disease progression. In particular, they have proven useful in detecting and monitoring the presence of infectious agents in patients. The versatility and sensitivity of nucleic acid biomarkers make them a critical component in the fight against infectious diseases. Various nucleic acid biomarkers are discussed below in detail [11].

6.2.1 Deoxyribonucleic Acids

A lot of different kinds of medical treatments use DNA analysis for the detection of genotypic changes such as single nucleotide polymorphisms (SNPs), translocations, deletions, short tandem repeats, and insertions. Mitochondrial DNA and pathogenic DNA are two further types of DNA biomarkers, which can be detected in cells or as cell-free DNA (cfDNA). When necrosis or apoptosis happens, cfDNA is often released into the bloodstream that can be detected in the blood or other biological fluids. But it is hard to detect DNA, especially cfDNA, in several circumstances [12]. Because there are only two copies of DNA in each cell, tools used to study DNA may need to be very sensitive. Also, the amount of target DNA is reduced even more when only one copy of a gene is affected. Also, there is not much cfDNA in biological fluids like blood because the cfDNA is very diluted by the time it gets to the biological fluid. It has been reported that the total amount of cfDNA in plasma is less than $10\,\text{ng}\,\text{ml}^{-1}$ [13]. Another problem is that the amount of cfDNA from the target cell may not be as high as the amount of cfDNA from other cells. Because of this, the methods used to analyze cfDNA may need to be very specific. Even with these problems, cfDNA analysis has a lot of potential for liquid biopsies, which are a less invasive way to diagnose [14].

6.2.2 Ribonucleic Acids

The most common RNA biomarkers are messenger RNA (mRNA), noncoding RNAs like microRNAs, and long noncoding RNAs (lncRNAs) [15]. Intracellular mRNA analysis can be used to profile gene expression and to figure out what's wrong with a cell. Recent research suggests that differences in gene expression levels may be a sign of illness [16]. So, the levels of gene expression can be used as diagnostic and prognostic biomarkers. Single-cell gene expression profiling has also been used to find out more about diseases like cancer. For analyzing gene expression levels, it is helpful to have methods that can reliably and sensitively find mRNA in many different samples. Because there are only a few hundred copies of mRNA per cell, ultrasensitive methods are needed [17]. MicroRNAs and other noncoding RNA biomarkers have a lot of potential in the field of diagnostics. MicroRNAs are RNAs that are 17–25 nucleotides long and modulate gene expression. It has been shown that pathogenic processes are linked to higher or lower levels of certain microRNAs [18]. For example, microRNA expression profiles are a good way to screen for cancer. Conventional technologies do have a hard time detecting microRNAs as their sequences are so short. Also, microRNAs are often found in very small amounts in many different body fluids. As per literature, the amount of microRNA in plasma is in the femtomolar range. (lncRNAs) are another type of RNA biomarker that regulates transcription, intercellular transport, chromosomal remodeling, and various other cellular activities [19].

6.3 Nucleic Acid Biomarkers in Infectious Diseases

Because virus and bacteria possess genes that encode nucleic acids, using them as biomarkers in the context of infectious diseases poses an interesting area of research. A disease marker is a component of genetic information that is particularly specific to the condition being studied and can be utilized in its detection. This genotypic specificity is usually not always mirrored at the phenotypic level, which can lead to cross-reactivity and inaccurate diagnosis when detecting biomarkers such as proteins. One prominent example of this is the spike protein that is produced by SARS-CoV, which is identical in both SARS-CoV-1 and SARS-CoV-2. The capacity to distinguish genetically between closely related diseases is especially crucial for bacterial or viral infections, which may rapidly evolve and produce new strains. Another advantage of nucleic acid biomarkers in infectious disease is that being exogenous they may be spotted in the body soon after the infection. Therefore, nucleic acid could play a vital role in early detection and can help slow or stop disease transmission. The ability to create tests quickly following the introduction of a novel infection is perhaps the most significant benefit of nucleic acid testing in the context of the diagnosis of infectious diseases [20]. This agility is due to the relative ease of identifying nucleic acid biomarkers and developing disease-specific targeting ligands. Infectious disorders can also be diagnosed using genetic markers. For instance, the peripheral expression of lncRNAs namely nuclear-enriched abundant transcript 1 (NEAT1) was examined in patients with sepsis and further explored for its clinical value in the field of diagnosis of sepsis. In this study, NEAT1 was identified as an ideal candidate for the early-stage diagnosis of sepsis. Interestingly, the level of circulating miRNAs is also linked with the onset of infectious disease and therefore providing ample opportunities to develop diagnostics employing miRNAs as a biomarker [21]. Additionally, the potential of miRNA, as a biomarker has been proven in various infectious diseases such as Hendra virus, HIV, TB, malaria, and Ebola (Table 6.1). Changes in miRNA profiles were identified in several investigations early before the pathogen could be explicitly detected and before the commencement of seroconversion. MicroRNAs level has also been linked to influenza and rhinovirus infections. The active release of extracellular miRNAs lends credence to the idea that they may function as "hormones" for cell-to-cell communication. Regardless of their particular roles, the primary value of miRNAs in prognostics is predicated on the assumption that distinct miRNA expression profiles are associated with different disease states. With this in mind, it is worth noting that a variety of viral disorders have recently been the subject of research evaluating circulating miRNAs as biomarkers [64].

Table 6.1 Different nucleic acids as biomarkers in infectious diseases.

S. No.	Infectious disease	Type of nucleic acid	Nucleic acid biomarker	Source of sample	References
1	Hendra virus	RNA	miRNA	Blood	[22]
2	HIV	RNA	miRNA	Blood	[23]
3	Tuberculosis	RNA	miRNA	Blood	[24]
4	P. falciparum malaria	RNA	miRNA	Blood	[25]
5	P. vivax malaria	RNA	miRNA	Blood	[26]
6	Ebola	RNA	miRNA	Blood	[27]
7	Influenza infections	RNA	miRNA	Bronchoalveolar lavage fluid	[28]
8	Rhinoviruses	RNA	miRNA	Nasal airway samples	[29]
9	Rabies	RNA	miRNA	Blood	[30]
10	Varicella (chickenpox)	RNA	miRNA	Blood	[31]
11	Pertussis (whooping cough)	RNA	miRNA	Blood	[32]
12	Measles	RNA	miRNA	Blood	[33]
13	Enteroviral infection	RNA	miRNA	Blood	[34]
14	Hepatitis B	RNA	miRNA	Blood	[35]
15	Hepatitis C	RNA	miRNA	Blood	[36]
16	Schistosomiass	RNA	miRNA	Blood	[37]
17	Japanese encephalitis	RNA	miRNA	Blood	[38]
18	SARS-CoV-2	RNA	RNA	Bronchoalveolar-lavage fluid samples	[39]
19	Tuberculosis	RNA	Ribosomal RNA	Sputum	[40]
20	Lupus erythematosus	RNA	Circular RNA	Blood	[41]
21	Rheumatoid arthritis	RNA	Circular RNA	Blood	[42]
22	Tuberculosis	RNA	Circular RNA	Blood	[43]
23	Multiple sclerosis	RNA	Long non-coding RNAs	Blood	[44]
24	Tuberculosis infection	RNA	Long non-coding RNAs	Blood	[45]
25	Malaria	RNA	Non-coding RNAs (ncRNAs)	Blood	[46]
26	Sepsis	RNA	Long non-coding RNAs	Blood	[47]
27	Paracoccidioidomycosis	RNA	Circulating miRNA	Blood	[48]

(Continued)

Table 6.1 (Continued)

S. No.	Infectious disease	Type of nucleic acid	Nucleic acid biomarker	Source of sample	References
28	Cutaneous leishmaniasis	RNA	mRNA	Blood	[49]
29	Viral respiratory infections	RNA	mRNA	Nasal swabs	[50]
30	Cancers	RNA	miRNA and mRNA	Blood	[51]
31	SARS-CoV-2	RNA	Cell-free RNAs	Blood	[52]
32	Sepsis	DNA	cfDNA	Blood	[53]
33	Anti-NMDAR encephalitis	DNA	Cell-free mitochondrial DNA	Cerebrospinal fluid	[54]
34	Chronic HBV infection	DNA	Cell-free circulating mitochondrial DNA	Blood	[55]
35	Leptomeningeal disease	DNA	cfDNA	Cerebrospinal fluid	[56]
36	Staphylococcus aureus bacteremia	DNA	cfDNA	Blood	[57]
37	Autoimmune rheumatic diseases	DNA	cfDNA	Blood	[58]
38	Sepsis	DNA	Plasma microbial cell-free DNA (mcfDNA)	Blood	[59]
39	SARS-CoV-2	DNA	cfDNA	Body fluids	[60]
40	Viral infections (HIV, HBV, and HCV)	DNA	Cell-free mitochondrial DNA	Blood	[61]
41	Invasive fungal disease	DNA	cfDNA	Blood	[62]
42	Tuberculosis	DNA	cfDNA	Urine and blood samples	[63]

6.4 Nucleic Acid Detection and Limitations

Initially, nucleic acids were recognized mostly by gene cloning approaches and hybridization methods, which are tedious and quite time-consuming and are limited to scientific outcomes [65] Since nucleic acid has emerged as a prominent biomacromolecule due to its potential as a biomarker (as stated in Section 6.3), it is possible to collect more precise information when doing viral monitoring [66]. As a result, a straightforward and universal method for nucleic acid detection was needed, which led to the development of the PCR technique. In a typical PCR procedure, DNA duplex templates are melted at high temperatures, and then oligonucleotides (primers) that are complementary to the flanking gene target sequence are annealed at a temperature that is precisely dependent on the primer sequence and length [67]. Using a set of oligonucleotides to identify distinct organisms or variations in a single reaction on a biological material is one variation of the approach [68]. Additionally, a nested and hemi/semi-nested PCR can be performed utilizing an initial PCR result as a template to boost sensitivity and specificity [69]. Quantitative reverse transcription-PCR (qRT-PCR) technique have been utilized to a great extent for coronaviruses, most notably in COVID-19 detection. In this scenario, the target viral RNA is first transformed to complementary DNA (cDNA), which is necessary for amplification of target gene using primers as occurs in PCR reaction [70]. To identify the molecular targets produced within the virus's single-stranded RNA genome, specialized primers must be designed first and then tested. These molecular targets include open reading frame 1a/b or 8 (ORF1a/b or ORF8) sections, N-, S-, and E-genes, as well as RNA-dependent RNA polymerase (RdRP) and aid in identifying viral components using these primers [71]. Nucleic acid amplification of the product was able to be achieved due to the advancement of PCR in the presence of fluorescent dye which leads to the detection of products via amplification cycle and real-time PCR [72]. For real-time PCR, thermocyclers are accompanied by fluorescence detection systems and software to facilitate data interpretation [73]. Conventional PCR is capable of amplification of DNA fragments up to 20 Kb, but real-time PCR only produces DNA fragments no longer than 150 bp, which further could not be used for POS–PCR [20]. Using Carl Wittwer's melting curve analysis or dual fluorescently labeled probes determines the specificity of real-time PCR products [74]. Real-time PCR is allowed for the detection of genotypes as well as quantification of amplification product to determine gene copy numbers of pathogenic microorganisms, and detecting genes related to reactivation, virulence, and genetic modification [20]. Although the PCR-based diagnosis of nucleic acid is a hallmark and standard method in the field of diagnostics, there are several limitations exist with this technology, such as time consuming process, low sensitivity, high-cost, false positive/negative results etc., which sometimes forces us to think about the development of novel, more precise, sensitive, and universal type nucleic acid diagnosis techniques [75]. Various limitations such as time consumption, low sensitivity, high cost, false positive/negative results, etc. As a result, there is scope for further improvement in the diagnosis rate of infections using RT-qPCR [76]. Furthermore, RT-qPCR has limitations such as biological hazards, time-consuming nucleic detection processes, and extended waiting times for outcomes. Although amendments have been done to PCR to improve its sensitivity and accuracy, more advancement is still required [77].

For the detection of microbial specifies in biological samples, fluorescence in situ hybridization (FISH) is used, which is considered a non-amplification cytogenic method. In this method, a specific complementary target sequence is hybridized with synthetic fluorochrome-labeled oligonucleotide probes, and the presence of the target sequence is detected using fluorescence microscopy. The FISH method is primarily utilized in diagnostic applications specifically for the identification of chromosomal alterations for the diagnosis of rRNA associated with pathogenic microorganisms. However, this technique has many confines such as the requirement of dedicated instruments, time-consuming procedures, the requirement of costly oligonucleotide probes, and high limit of detection (LOD) [78]. Clinical molecular laboratories generally utilize amplification-based molecular techniques to overcome the LOD regarding limitations, since they are regarded as the "gold standard" in this field. The molecular techniques based on PCR allow us to easily overcome the low sensitivity usually based on fluorescence, however, this reaction requires a thermostable DNA polymerase, designed primers, thermocyclers, nucleotides, and related buffers [79]. Further, the PCR bridges with different other molecular biology techniques viz., electrophoresis, DNA microarray, and RT-PCR, which further require sophisticated instruments for their operation.

Unfortunately, the amplification step is difficult to accomplish in portable devices despite being necessary for quick, accurate, and sensitive identification. Several attempts have been made in recent years to develop an innovative and affordable isothermal amplification tool, including exponential amplification reaction (EXPAR), loop-mediated isothermal amplification (LAMP), nucleic acid sequence-based amplification (NASBA), and strand displacement amplification (SDS). These approaches showed lower sensitivity and specificity than conventional PCR, and they were unable to bridge the gap between isothermal-based laboratory methods or POC devices and PCR-based laboratory diagnosis methods [80].

6.5 CRISPR/Cas System

Clustered Randomly Interspaced Short Palindromic Repeats (CRISPR) system was explored by Doudna and Charpentier and their colleagues in 2012 as an adaptive immune system in the bacteria and archaea against the invading phages. The system comprises a CRISPR repeat array and a set of CRISPR-associated genes. In brief, the CRISPR repeat array consist of transcripts, crRNA (CRISPR RNA) and tracrRNA (trans-activating RNA), and CRISPR-associated gene encode a Cas effector (endonucleases) having DNA binding and cleavage properties. Interestingly, the DNA binding and cleavage property of the Cas9 effector is directed by the guide RNA (i.e. crRNA & tracrRNA), and therefore the CRISPR system is also called as RNA guided DNA cleavage system. When prokaryotes are invaded by foreign genetic material, Cas proteins can cut the invading DNA into short fragments, then incorporate them into the CRISPR array as new spacers. When the same invader returns, a signal of cascades leads to generate the crRNA which anneals to tracrRNA, and due to its complementarity to the foreign materials, guides the Cas protein to specifically bind and break specific regions of foreign DNA, safeguarding the host [81] As per the latest classification by Makarova, et al. in 2020, the CRISPR/Cas system has 2 classes, 6 types, and 33 subtypes [82] (Figure 6.1).

The Cas9 effector is the heart of CRISPR technology. It has an endonuclease activity that is controlled by a single guide RNA (sgRNA). The Cas9 effector has two nuclease domains called HNH and RuvC. Each of these domains cuts one strand of

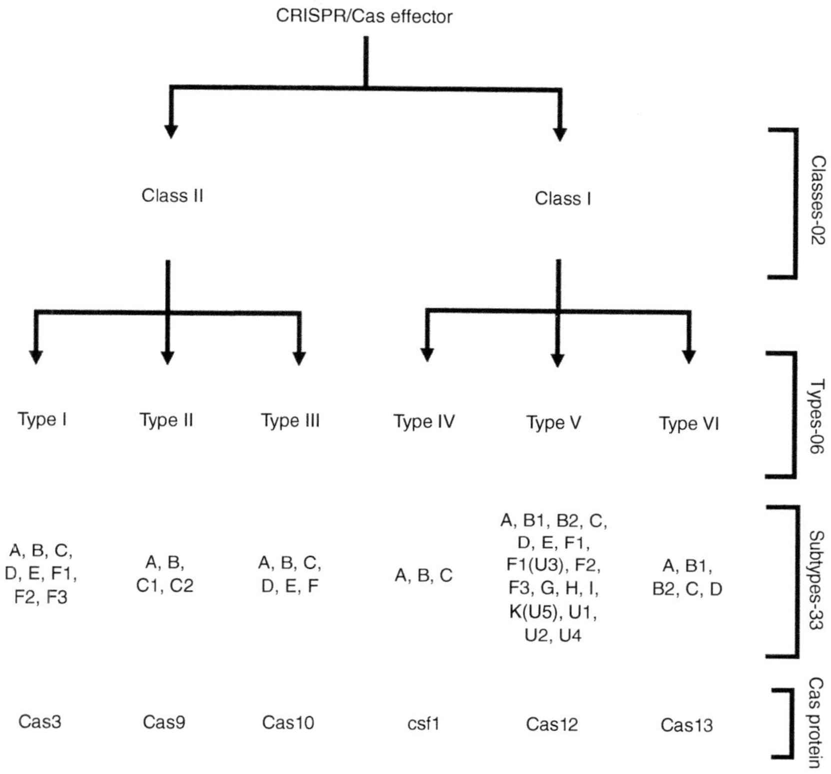

Figure 6.1 Latest classification of CRISPR/Cas system.

the dsDNA target. A single-guide RNA, or sgRNA, is a shorter version of both tracrRNA and crRNA. A Cas9 ribonucleoprotein (RNP), which is made up of Cas9 nuclease and sgRNA, can bind to and cut a specific DNA target that is complementary to the sgRNA [83] Also, a protospacer adjacent motif (PAM) sequence is an important part that is needed for Cas9 protein to find and bind to the target dsDNA. In short, the sgRNA tells the Cas9 effector to bind to a specific gene sequence that matches the sgRNA sequence to make a double-strand break (DSB). As shown in Figure 6.2, the DSB could be fixed by either nonhomologous end joining (NHEJ) or HDR (homology-directed repair). According to the research, HDR only works during the S/G2 phase of cell cycle and requires donor template, while NHEJ works throughout the cell cycle and does not depend on the donor template. Because of this, most DSBs are fixed by NHEJ pathways. Random insertions or deletions (indels) at cleavage sites can happen because of NHEJ. This can cause frameshift mutations or premature stop codons in target genes, which turn off the target gene [84]. Alternatively, utilizing a homologous DNA repair template, HDR can make precise genomic modifications at the target site. Furthermore, by utilizing several sgRNAs targeting one or more sequences within a gene, extensive fragment deletions and simultaneous

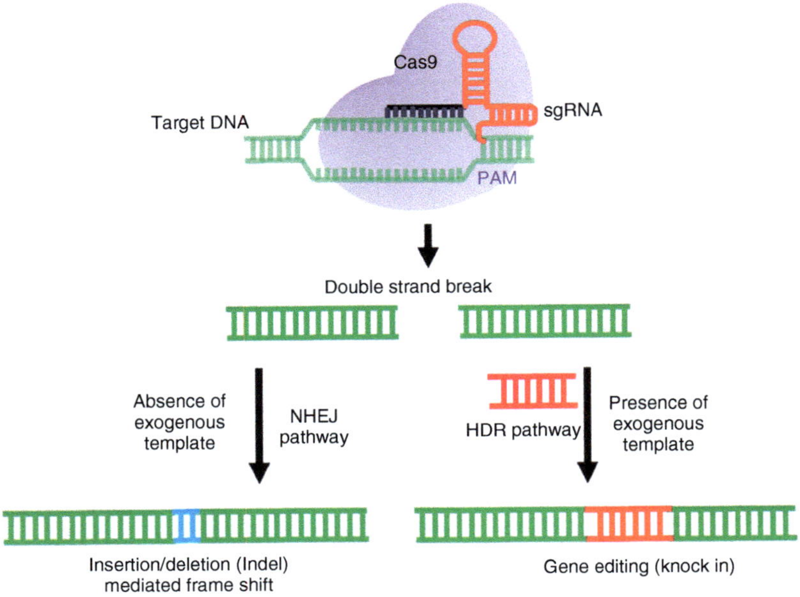

Figure 6.2 Schematic representation of the mechanism of CRISPR/Cas9 mediated gene editing.

knockout of numerous sequences might be accomplished [85]. A detailed mechanism of gene editing by the CRISPR system is shown in Figure 6.2. In 2013, Zhang and Church pioneered CRISPR/Cas system for gene editing in mammalian cells [86]. Till then, CRISPR/Cas system has evolved as a next-generation gene editing tool along with various other applications such as epigenome editing, biosensing, diagnostics, etc.

6.5.1 Characteristics Features of Different Cas Effectors

Cas9 effector possesses HNH and RuvC-like nuclease domains that are responsible for the DSB. Both complementary and non-cDNA strands undergo site-specific cleavage at the same location, composed of three base pairs upstream of the PAM region [87]. In comparison to Cas9, dead Cas9, the catalytically inactive Cas9, lacks cleavage activity due to two mutations in the RuvC1 and HNH nuclease domains. The dCas9/sgRNA complex, on the other hand, retains a specific binding affinity for target DNA and has been put to use in a variety of applications, including live-cell imaging, gene expression regulation, and others [88]. In Cas12a, a single RuvC domain is sufficient to cleave both strands of target DNA with the assistance of a single crRNA, which results in a PAM-distal dsDNA break with staggered cut at 5′ and 3′ ends. Interestingly, Cas12a identifies target dsDNA through a short PAM sequence that is rich in T nucleotides. Furthermore, after binding to a target, Cas12a can degrade nonspecific single-stranded DNA (ssDNA), a process known as "trans cleavage" activity [89]. Similarly, after binding to the target ssRNA, Cas13a also has a *trans cleavage* activity for nonspecific nucleic acids, just like Cas12a. Therefore, the major difference between Cas12a and Cas13a is the target DNA, which is DNA in the case of Cas12a and ssRNA in the case of Cas13a [90]. Cas13a, on the other hand, employs two conserved higher eukaryotic and prokaryotic nucleotide-binding (HEPN) domains to cut target ssRNA at uracil sites and entirely cuts non-target ssRNA following the formation of a guide-target ssRNA duplex. In addition, instead of a PAM sequence, Cas13a detects the target ssRNA via a 3′ protospacer flanking site (PFS) [91] Cas effectors may precisely bind target nucleic acids when guided by a 20–30 bp guide RNA (sgRNA or crRNA). And PAM sequence recognition is required for Cas effectors to identify target dsDNA, which causes the target dsDNA helix to unwind, allowing the guide RNA to hybridize it [92]. However, Cas12a can also cut guide RNA-complementary ssDNA, which increases Cas12a's trans cleavage activity in the absence of a PAM sequence. A 3′ PFS (non-G) is necessary to activate Cas13a's cleavage activity for ssRNA detection [89]. Cas9, dCas9, and Cas12a are particularly sensitive to around ten base pairs that are located close to the PAM sequence inside the duplex of guide RNA and target nucleic acids. On the other hand, Cas13a is highly sensitive to the base pairs that are located in the center of the duplex [93]. The cleavage efficiency of Cas effectors is dramatically diminished when base pair mismatches occur in the sensitive area known as the "seed region," [94].

6.5.2 CRISPR in Diagnostics

In 2018, CRISPR diagnostic technology was nominated in the top ten science and technology advancements. To date, the CRISPR/Cas9 system has evolved its empire specifically in the field of diagnostics due to the discovery of new Cas effectors, mainly the ones with trans cleavage activity [95]. Currently, Cas effectors, including Cas9, dCas9, Cas12a, and Cas13a, Cas14, CasX are commonly used and play an important role in nucleic acid detection. The Cas9 showed the diagnostic potential via a binding-based mechanism, on the other hand, Cas effectors Cas12/Cas13/Cas14 showed *collateral* or *trans* cleavage activity-based mechanism, which can induce cleavage of nearby nontarget RNAs after cleavage of the target sequence. The discovery of *trans cleavage* activity of some class 2 Cas effectors makes them more powerful. For instance, Feng Zhang et al. created Specific High Sensitivity Enzymatic Reporter UnLOCKing (SHERLOCK), an *in vitro* nucleic acid detection technology based on Cas13a's "collateral cleavage" activity [96]. The SHERLOCK system comprises a Cas13a, sgRNA targeting specific RNA sequences, and a fluorescent RNA reporter. Once the Cas13a RNPs recognize the target RNA, they will also cut the reporter RNA, resulting in a fluorescent signal as an indicator of target recognition. Further, this method was explored to detect viruses and bacteria in different biological samples. Collateral cleavage activity was discovered in Cas12 enzymes as well as Cas13. Firstly, Cas12a (often called Cpf1) was used by Doudna et al. to construct a nucleic acid detection system called DETECTR (DNA endonuclease-targeted CRISPR trans reporter). For the validation of DETECTR, the Human cell lines infected with HPV or clinical samples from patients with cervical cancer have been used to detect HPV16 and

HPV18 [91, 97]. Further, employing the newly found Cas14 and CasX effectors in molecular diagnosis, Doudna et al. also try to develop CRISPR-based next-generation biosensors. The high sensitivity and single-base specificity of CRISPR molecular diagnostic technology make it ideal for the early detection of pathogenic genes. CRISPR diagnostics offer better specificity without a specialized device with a low cost, short turnaround time, and reaction temperature [98]. Many firms are currently working to provide CRISPR diagnostic kits for the detection of HIV, rabies, Toxoplasma Gondi, and other diseases. CRISPR-based diagnostics developed for infectious diseases have been thoroughly discussed in the next section (Table 6.2).

6.5.2.1 Cas9-Based Detection

Cas9 effector has been well explored for its specificity toward the dsDNA for (as explained in Section 6.5) gene-editing applications. Despite the absence of trans cleavage activity, Cas9 has been explored in the area of diagnosis through impressive advancement. For instance, a next-generation diagnostic namely finding low abundance sequences by hybridization (FLASH) was developed by Quan et al. by employing targeted endonuclease activity of the Cas9 effector [99]. In detail, firstly the phosphatase treatment was given to the DNA followed by exposure to the sgRNA-guided Cas9 effector leading to the cleavage in such a way that the resulting fragment ligates to a universal type sequencing aptamer. The target sequences will be amplified over the background by employing the sequencing aptamer for selective amplification and will be prepared for further sequencing. Thus, offering high multiplexing, precision, and identification of the target nucleic acid sequence. Therefore, the FLASH technique has a unique ability to identify the sequence of the target DNA and this property makes this technique advantageous over the trans cleavage-based CRISPR diagnostic methods. This technique made it possible to find antimicrobial resistance genes in a variety of clinical samples, including dried blood and respiratory fluid samples. Another relevant example of Cas9 effector-based diagnosis is FnCas9 editor linked uniform detection assay (FELUDA), which utilized the Cas9 effector originating from *Francisella novicida,* named FnCas9 for the detection of viral genome in biological samples. The method is independent of the *trans cleavage* activity and when employed in the lateral flow readout, exhibits100% and 97% of sensitivity and specificity respectively for all the clinical samples containing viral load within the time range of one hour [100].

6.5.2.2 Cas12-Based Detection

The Cas12 effector possesses both a crRNA-guided dsDNA binding affinity and a trans cleavage activity toward a non-specific nucleic acid reporter, as can be seen in Figure 6.3. Recently, large number of new approaches have been created or established by employing the trans cleavage activity of Cas12. A summarized view of these examples is discussed in this section.

Liu et al. have developed a global CRISPR/Cas12a-based electrochemical biosensor to detect dsDNA from parvovirus B19 and human papillomavirus 16. The system showed a LOD of approx. 50 pM. Electrochemically, a dye, i.e. methylene blue (MB)

Table 6.2 CRISPR/Cas-based diagnostics methods for nucleic acid detection.

Cas effector	Detection assay	Disease/ pathogen	Sample types	Sensitivity	References
Cas9	FLASH	Malaria	Respiratory fluid and dried blood spots	—	[99]
	FELUDA	COVID – 19	Respiratory fluids	97%	[100]
Cas12	E-CRISPR	Human papillomavirus 16 and parvovirus B19	Blood sample	LOD-50 pM	[101]
	CRISPR/Cas system coupling with electrochemical detection			LOD-10–100 fM	[102]
	A portable laser-induced fluorescence system	African swine fever virus	Plasma sample	LOD-1 pM	[103]
	CRISPR-Cas12a-Assisted Nanopores (SCAN)	HIV-1		LOD-10 nM	[104]
Cas13	Cas13a-powered catalytic electrochemical biosensor	Lung cancer	Serum sample	LOD-10 CFU	[105]
	SHERLOCK	Zika and dengue strains	Serum sample	LOD-2 aM	[96]
	HUDSON	Zika virus	Body fluid	LOD as low as 1 copy per microliter	[106]
	SHERLOCKv2	Dengue or zika virus	Patient liquid biopsy samples and saliva samples	LOD-2 aM	[107]
Cas14	Cas14-pMOFs fluorescence sensor	Microcystin-LR (MC-LR) determination.	Tape water and drinking water	LOD-50 $pg\,ml^{-1}$ to 1 $\mu g\,ml^{-1}$	[108]
	Cas14a1-mediated nucleic acid detection platform (CMP)	detect *Streptococcus pyogenes* and *Eberthella typhi*	Milk samples	LOD-10^6 and 10^7 CFU ml^{-1}	[109]

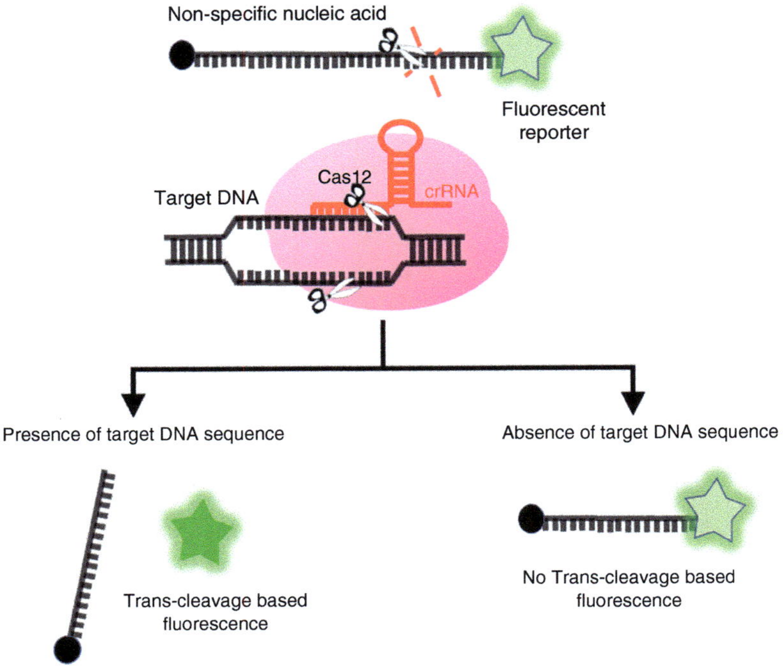

Figure 6.3 Cas12 effector mediated *trans cleavage* activity in the detection of target DNA.

was attached and tagged to a nonspecific ssDNA reporter that was further attached to the electrode surface of the biosensor. Square wave voltammetry showed that the MB signal was high when there was no target dsDNA. Once the presence of target dsDNA activated Cas12a's trans cleavage activity, the MB-ssDNA was cut away from the electrode surface. This made the MB signal much more vulnerable. But it must stop the electrode surface from getting a lot of ssDNA reporter molecules on it. If not, the steric hindrance will make it harder for Cas12a to access the ssDNA reporter [101]. Zhang et al. used a hairpin DNA reporter instead of a linear DNA reporter to solve this problem. The loop structure of the hairpin DNA reporter makes it less likely that a steric hindrance will happen. When compared to the linear DNA reporter, the electron tunneling distance of the hairpin DNA reporter was also shorter. So, when Cas12a cut the reporter, there was a bigger change in the signal. This made it easier to find dsDNA so that as little as 30 pM could be found in 60 minutes time duration. Both of the above methods change the nonspecific ssDNA on the surface of the electrode. Even if there is a base mutation in the target sequence, the target concentration can still be lowered and the same output signal can be made. Because of this, it might be hard to tell the point mutation apart. Multiple researchers have used a hairpin ssDNA reporter that is the same as the target ssDNA to change the electrode. For mutation analysis, it could produce both a signal caused by a change in the structure of the target sequence and a signal caused by cleavage by the CRISPR/Cas system. When the target sequence was there, the hairpin probe hybridized with the targets to move MB away from the electrode surface. This was the first signal change. The *cis cleavage* of Cas9 or Cas12a for the matching probe got rid of all the MB on the electrode, which caused the second change in the signal. So, the single mutation could be found by comparing the changes in the two signals, and the sensitivity was further increased by doubling the signal (10–100 fM). But these methods are all signal-off assays because they use the CRISPR/Cas system in combination with electrochemical detection. When looking for low-concentration targets, it is hard to tell the difference between positive and negative results because the signal change is so narrow. It might be a good approach to replace the electrochemical tag with spatial enzymes like catalase, and glucose oxidase that can be linked to magnetic beads by single-stranded DNA. When the Cas effector cuts the ssDNA, the enzyme will be released into the solution to catalyze substrates. This produces an electrochemical signal, which is what a signal-on essay is all about. The CRISPR/Cas systems can also detect target DNA by combining an ssDNA probe with a fluorophore. After the targets hybridize with the Cas effectors, the probe will be cut, and a fluorescence-sensing unit can read the fluorescence signal. The traditional fluorescence microscope, on the other hand, is big and expensive, even though it is very sensitive. This makes it hard to use for on-site detection [110]. With a parabolic mirror and a small spectrometer and laser, Du et al. made a small, portable, sensitive, and economical fluorescence detection device capable of 40x objective lens light-gathering efficiency. With the CRISPR/Cas12a system and the fluorescence detection platform, the African Swine Fever Virus can be found directly in porcine plasma without any sample preparation. Within 120 minutes, 1 pM of target dsDNA can be detected. It could be even more sensitive if the reaction time is longer or if stronger fluorescent reporters are used. Some nanoparticles that glow, like quantum dots and up-conversion nanoparticles, can also be used as signal reporters to make the signal even stronger [111]. Li et al. used a primer extension reaction triggered by target binding to make a universal CRISPR/Cas12a-targetable system. It can find many types of target ssDNA by using a fixed guide RNA sequence. Two partially complementary (6-nt) primers with a melting temperature of 17 °C were used in the method to precisely recognize the target ssDNA. Target ssDNA made it possible for the two primers to pair up at 37 °C, and the primer extension reaction made a DNA sequence that the guide RNA could recognize. To make the primers work better, a nicking recognition domain was added. After primer extension, a nicking endonuclease cut the ssDNA, and the second round of primer extension give more substrates to activate the Cas12a effector. The LOD could go up to 1 fM. To reduce the background signal made by primer reactions the primer sequences could be improved. For example, the primers can be made with a stem-loop structure (like a molecular beacon) that is more stable [112]. In this way, ample detection methods for pathogenic diseases have been crafted using CRISPR/Cas system. Another example is Solid-State CRISPR-Cas12a-Assisted Nanopores (SCAN), a way to find HIV-1, which can find at least 10 nM of target DNA in one hour without the need for pre-amplification [104]. Using fluorescent readout, a report on detecting HBV-based DETECTR showed high sensitivity within 13 minutes. However, the LOD of the lateral flow test strip method was not revealed but it also takes 20 minutes time for detection [113].

6.5.2.3 Cas13-Based Detection

Cas13 effector belongs to the Class II, Type VI of CRISPR/Cas system and was previously known as C2c2. As of now, there are currently four Cas13 proteins (i.e Cas13a, Cas13b, Cas13c, and Cas13d) have been discovered. Cas13 effector has evolved as a nucleic acid detection tool by the means of its unique RNase activity provided by two HEPN domains. As discussed in section 5.1, the two HEPN domains are required for the maturation of crRNA. Further, the Cas13 identifies the shot hairpin in crRNA and forms an RNPs complex, which is further able to bind the target ssRNA sequence with aid of a spacer via PFS sequence. Similar to Cas12, the Cas13 possesses collateral or *trans cleavage* activity, which makes it able to cut ssRNA nonspecifically (Figure 6.4). Next-generation nucleic acid biosensors have been developed to detect debilitating infectious diseases using *trans cleavage* or collateral activity of the Cas13 effector. An overview of these techniques is included in this section [114].

A Cas13a assay that uses a new aptamer makes it possible to mix and read live pathogenic bacteria without reverse transcription, nucleic acid amplification, or chemical labeling. For Bacillus cereus, the assay showed a LOD of 10 CFU [115]. To detect a particular nucleic acid within a collection of transcripts, the target RNA binding-based collateral cleavage activity of Cas13 was utilized. The attachment of active Cas13 to the target RNA activates the collateral activity of Cas13, causing

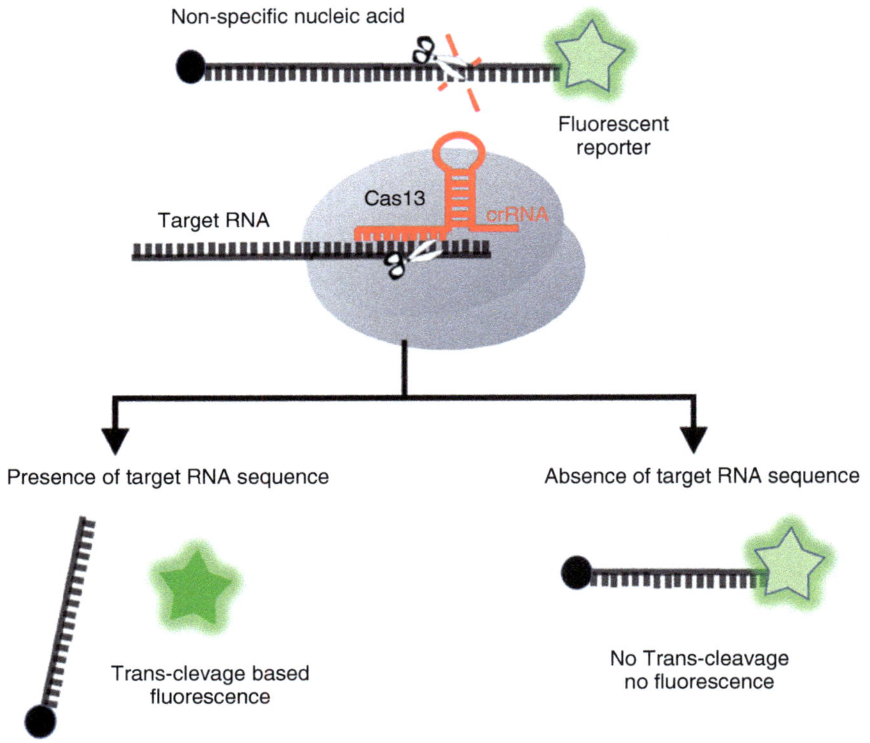

Figure 6.4 Cas13 effector mediated *trans cleavage* activity in the detection of target RNA.

reporter RNAs to be cut and the signal to be produced. This can be accomplished through the utilization of reporter RNAs that emits a fluorescent signal when they are cut. This shows that the target RNA is present. Based on this secondary effect, the SHERLOCK method was made. SHERLOCK harnesses the nonspecific activity of Cas13, i.e. cleavage of quenched fluorescent reporter RNA molecule upon target recognition. The SHERLOCK system features to find specific strains of the Zika and Dengue viruses, inform pathogenic bacteria apart, genotype human DNA with high accuracy, and also find mutations in tumor DNA [116]. The power of SHERLOCK was expanded by creating HUDSON (heating unextracted diagnostic samples to obliterate nucleases), which kills nucleases in diagnostic samples that have not been extracted by heating. This makes SHERLOCK find pathogenic nucleic acid in biological fluids at significantly lower concentrations (as low as 1 copy/μl). More work has been done to make SHERLOCK better to solve the diagnosing problems, which led to the establishment of SHERLOCKv2. Later, Feng Zhang and his colleagues.

made changes to the SHERLOCK system and changed its name to SHERLOCKv2, which can find four viruses at once. By using the different specificities of different Cas proteins, like Cas12 and Cas13, SHERLOCKv2 was able to find at least four different targets in a single nucleic acid detection reaction. Also, the system's sensitivity was increased by using an extra CRISPR enzyme called Csm6 that amplifies the signal when Cas13 does the collateral activity. To make it portable and easy to use, FAM-biotin reporters were used to make a lateral flow system that does not need any equipment. In the lateral flow system, an RNA or DNA reporter with fluorescein and biotin at the end is used, depending on the Cas effector. At one end of the reporter, depending on the Cas effector, with fluorescein and biotin at the separate end is used. At one end of the reporter, streptavidin binds to biotin, and at the other end, gold nanoparticle-labeled anti-fluorescein antibodies bind to the fluorescein [96]. Upon activation of collateral Cas activity, the RNA reporter will be cut, and a gold nanoparticle-labeled antibody will flow to a test line in the lateral flow device. The test line has an anti-species secondary antibody that will further react with the gold nanoparticle-labeled antibody to make a colored product. This shows that the target is present. This system is used to find ssRNA viruses and mutations from Zika and Dengue in liquid biopsy samples from patients. As stated in section 3, the miRNAs have been linked to various infectious diseases and their detection at an early stage could provide enough information about the severity of the infections. Interestingly, Cas13 has been employed to detect the miRNA directly in the biological samples with excellent sensitivity and accuracy. The different techniques employing Cas13 for miRNA detection are shown in Table 6.3.

6.5.2.4 Other Cas Effectors-Based Detection

New Cas variants have been also identified and explored for their diagnostic applications. Specifically, Cas14a or Cas12f1 was employed for its potential to detect pathogenic bacteria. The diagnostic system was independent of PAM with a sensitivity of 1 cfu or 1 attomol [122, 123]. For instance, the new Cas variantsi.e.Cas14a1 by utilizing its unique collateral cleavage activity can detect bacterial pathogens.

Table 6.3 Different techniques employing Cas13 effectors for miRNA detection.

Cas effector	Technique	miRNA	Sensitivity (LOD)	Detection method	References
Cas13a	vCas	miR-10b	1 fM to 10 pMol	Visual method	[117]
Cas13a	CRISPR/ CHDC Assay	miRNA-21	2.6 fM	Electrochemical	[118]
Cas13a	EM-CRISPR	miR-19b and miR-20a	10 pMol	Electrochemical microfluidic	[119]
Cas13a	PECL-CRISPR	miR-17	Approx. 1×10^{-12} M	Electrochem-iluminescence	[120]
Cas13a	COMET	miR-17,miR-19b miR-155, and miR-210	50 fM	Electrochemical	[105]
LbuCas13a		miR-17	4.5 amol	Fluorescence signals	[121]

A bioanalysis technique employing CRISPR/Cas namely CMP (Cas14a1 mediated nucleic acid detection platform) associated with molecular amplification which is developed and validated for nucleic acid detection. As per the reported observations, the CAMP offers a highly specific, accurate, and rapid detection of ample species of pathogens in biological samples such as milk. Interestingly, the CMP was found to have excellent sensitivity with an improved LOD of 10^6–10^7 CFU ml^{-1} of pathogens. In brief, the CMP utilized the *trans cleavage* activity of the Cas14a effector to produce a fluorescent signal as an indication of ssDNA target detection. In addition, the detection by Cas14a1 is PAM independent, and therefore Cas14a1 has portability since the redesigning of the crRNA sequence could aid in the detection of other pathogens. Hence, this method offers ample advantages such as universal nature, reduced analysis time, and economic cost, over existing nucleic acid detection techniques [109]. A unique and versatile fluorescence sensor was developed by integrating the 2D porphyrin metal–organic framework nanosheets (2D-pMOFs) with the CRISPR/Cas14a system for Microcystin-LR (MC-LR) determination, which is important to detect because of its high toxicity. Unbound cDNA, which showed a positive correlation with MC-LR content, served as the basic principle for the functioning of the Cas14-pMOFs diagnosis system. Furthermore, the pre-absorbed FAM-labeled single-stranded DNA (ssDNA-FAM) on Cu-TCPP(Fe) nanosheets were indiscriminately cleaved by the activated Cas14a protein. The pre-quenched fluorescence signal was revived again due to the denaturation of FAM-labeled ssDNA. Along with the high specificity, the Cas14-pMOFs showed a LOD of 0.12 nM for cDNA and 19 pg/ml for MC-LR. Overall, the Cas14-pMOFs diagnosis system was found feasible and suitable for the indirect detection of the non-nucleic acid target [108].

6.6 Conclusion and Prospects

Early-stage diagnosis is very crucial when it comes to infectious diseases to prevent the pandemic situation, COVID-19 is the foremost example of the same. Despite viruses, other infections that can be diagnosed at an early stage are much easier to treat rather than in the chronic stage. Although existing diagnostics techniques are improving in terms of sensitivity and accuracy but still a better option is a need of the hour. Recently, several attempts have been made to evolve the diagnostics area and EXPAR, LAMP, NASBA, SDS, RPA, and RCA are the results of the same. Being precise to the target nucleic acid CRISPR/Cas effector has shown ample advantages in the field of diagnosis including high sensitivity (up to attomolar), excellent selectivity, lesser false-positive/negative results, versatile nature, multiplex detection, and advanced readout methods. The reported literature provides strong consolidated evidence for employing CRISPR/Cas effectors in the mainstream next-generation biosensors. Although, there are several limitations associated with the inherent property of the CRISPR/Cas effector such as stability, complexity, and availability. However, it will be interesting to watch the progress in the area of CRISPR-based biosensors, since many companies showing interest in the deployment of CRISPR in the field of diagnosis of infections in clinical practices.

References

1 Cairncross, S. and Feachem, R. (2018). *Environmental Health Engineering in the Tropics: Water, Sanitation and Disease Control*. Routledge.

2 Abbafati, C. et al. (2020). Global burden of 369 diseases and injuries in 204 countries and territories, 1990–2019: a systematic analysis for the global burden of disease study 2019. *The Lancet* 396 (10258): https://doi.org/10.1016/S0140-6736(20)30925-9.

3 Perin, J. et al. (2022). Global, regional, and national causes of under-5 mortality in 2000–19: an updated systematic analysis with implications for the sustainable development goals. *The Lancet Child and Adolescent Health* 6 (2): https://doi.org/10.1016/S2352-4642(21)00311-4.

4 Piret, J. and Boivin, G. (2021). Pandemics throughout history. *Frontiers in Microbiology* 11: https://doi.org/10.3389/fmicb.2020.631736.

5 Woolf, S.H., Chapman, D.A., and Lee, J.H. (2021). COVID-19 as the leading cause of death in the United States. *Journal of the American Medical Association* 325 (2): https://doi.org/10.1001/jama.2020.24865.

6 WHO. 2020 World Health Organization: Coronavirus disease (COVID-2019) situation reports. https://www.who.int/emergencies/diseases/novel-coronavirus-2019/situation-reports/. Retrieved on 4 April, 2020. World Health Organization, vol. 2019, no. March.

7 Hojyo, S. et al. (2020). How COVID-19 induces cytokine storm with high mortality. *Inflammation and Regeneration* 40 (1): https://doi.org/10.1186/s41232-020-00146-3.

8 Singhal, T. (2020). A review of coronavirus disease-2019 (COVID-19). *Indian Journal of Pediatrics* 87 (4): https://doi.org/10.1007/s12098-020-03263-6.

9 Carotti, M. et al. Chest CT features of coronavirus disease 2019 (COVID-19) pneumonia: key points for radiologists. *Radiologia Medica* 125 (7): 2020. https://doi.org/10.1007/s11547-020-01237-4.

10 Gietema, H.A. et al. (2020). CT in relation to rt-PCR in diagnosing covid-19 in the Netherlands: a prospective study. *PLoS ONE* 15 (7): https://doi.org/10.1371/journal.pone.0235844.

11 Suea-Ngam, A., Bezinge, L., Mateescu, B. et al. (2020). Enzyme-assisted nucleic acid detection for infectious disease diagnostics: moving toward the point-of-care. *ACS Sensors* 5 (9): https://doi.org/10.1021/acssensors.0c01488.

12 Mahdieh, N. and Rabbani, B. (2013). An overview of mutation detection methods in genetic disorders. *Iranian Journal of Pediatrics* 23 (4): 375.

13 Zhong, Y., Fan, Q., Zhou, Z. et al. (2020). Plasma cfDNA as a potential biomarker to evaluate the efficacy of chemotherapy in gastric cancer. *Cancer Management and Research* 12: https://doi.org/10.2147/CMAR.S243320.

14 Bronkhorst, A.J., Ungerer, V., and Holdenrieder, S. (2019). The emerging role of cell-free DNA as a molecular marker for cancer management. *Biomolecular Detection and Quantification* 17: https://doi.org/10.1016/j.bdq.2019.100087.

15 Ratti, M. et al. (2020). MicroRNAs (miRNAs) and long non-coding RNAs (lncRNAs) as new tools for cancer therapy: first steps from bench to bedside. *Targeted Oncology* 15 (3): https://doi.org/10.1007/s11523-020-00717-x.

16 Singh, K.P., Miaskowski, C., Dhruva, A.A. et al. (2018). Mechanisms and measurement of changes in gene expression. *Biological Research for Nursing* 20 (4): https://doi.org/10.1177/1099800418772161.

17 Hijazo-Pechero, S. et al. (2021). Gene expression profiling as a potential tool for precision oncology in non-small cell lung cancer. *Cancers* 13 (19): https://doi.org/10.3390/cancers13194734.

18 O'Brien, J., Hayder, H., Zayed, Y., and Peng, C. (2018). Overview of microRNA biogenesis, mechanisms of actions, and circulation. *Frontiers in Endocrinology* 9: https://doi.org/10.3389/fendo.2018.00402.

19 Cortez, M.A., Bueso-Ramos, C., Ferdin, J. et al. (2011). MicroRNAs in body fluids-the mix of hormones and biomarkers. *Nature Reviews Clinical Oncology* 8 (8): https://doi.org/10.1038/nrclinonc.2011.76.

20 Sperança, M.A., Suzuki, R.B., Cabral, A.D., and Carmo, A.D. (2016). Nucleic acid-based diagnosis and epidemiology of infectious diseases. In: *Nucleic Acids - From Basic Aspects to Laboratory Tools*. https://doi.org/10.5772/61965.

21 Huang, S., Qian, K., Zhu, Y. et al. (2017). Diagnostic value of the lncRNA NEAT1 in peripheral blood mononuclear cells of patients with sepsis. *Disease Markers* 2017: https://doi.org/10.1155/2017/7962836.

22 Stewart, C.R. et al. (2013). Promotion of Hendra virus replication by microRNA 146a. *Journal of Virology* 87 (7): https://doi.org/10.1128/jvi.01342-12.

23 Biswas, S., Haleyurgirisetty, M., Lee, S. et al. (2019). Development and validation of plasma miRNA biomarker signature panel for the detection of early HIV-1 infection. *eBioMedicine* 43: https://doi.org/10.1016/j.ebiom.2019.04.023.

24 Zhang, H. et al. (2014). Identification of serum microRNA biomarkers for tuberculosis using RNA-seq. *PLoS ONE* 9 (2): https://doi.org/10.1371/journal.pone.0088909.

25 Li, J.J. et al. (2018). Identification of potential whole blood MicroRNA biomarkers for the blood stage of adult imported falciparum malaria through integrated mRNA and miRNA expression profiling. *Biochemical and Biophysical Research Communications* 506 (3): https://doi.org/10.1016/j.bbrc.2018.10.072.

26 Kaur, H., Sehgal, R., Kumar, A. et al. (2018). Screening and identification of potential novel biomarker for diagnosis of complicated plasmodium vivax malaria. *Journal of Translational Medicine* 16 (1): https://doi.org/10.1186/s12967-018-1646-9.

27 Duy, J. et al. (2016). Circulating microRNA profiles of Ebola virus infection. *Scientific Reports* 6: https://doi.org/10.1038/srep24496.

28 Scheller, N. et al. (2019). Proviral MicroRNAs detected in extracellular vesicles from Bronchoalveolar lavage fluid of patients with influenza virus-induced acute respiratory distress syndrome. *Journal of Infectious Diseases* 219 (4): https://doi.org/10.1093/infdis/jiy554.

29 Hasegawa, K. et al. (2018). RSV vs. rhinovirus bronchiolitis: difference in nasal airway microRNA profiles and NF B signaling. *Pediatric Research* 83 (3): https://doi.org/10.1038/pr.2017.309.

30 Han, M.G. et al. (2011). Serum MicroRNA expression profiling in mice infected with rabies virus. *Osong Public Health and Research Perspectives* 2 (3): https://doi.org/10.1016/j.phrp.2011.11.043.

31 Qi, Y. et al. (2014). Dysregulated microRNA expression in serum of non-vaccinated children with varicella. *Viruses* 6 (4): https://doi.org/10.3390/v6041823.

32 Cui, L. et al. (2013). Serum microRNA expression profile as a biomarker for the diagnosis of pertussis. *Molecular Biology Reports* 40 (2): https://doi.org/10.1007/s11033-012-2176-9.

33 Haralambieva, I.H. et al. (2018). Differential miRNA expression in B cells is associated with inter-individual differences in humoral immune response to measles vaccination. *PLoS ONE* 13 (1): https://doi.org/10.1371/journal.pone.0191812.

34 Zheng, Z. et al. (2013). Human MicroRNA hsa-miR-296-5p suppresses enterovirus 71 replication by targeting the viral genome. *Journal of Virology* 87 (10): https://doi.org/10.1128/jvi.02655-12.

35 Wang, J. et al. (2019). Hepatitis B virus induces autophagy to promote its replication by the axis of miR-192-3p-XIAP through NF kappa B signaling. *Hepatology* 69 (3): https://doi.org/10.1002/hep.30248.

36 Abdel Motaleb, F.I., Nabih, E.S., Mohamed, S.M., and Abd Elhalim, N.S. (2017). Up-regulation of circulating miRNA146a correlates with viral load via IRAK1 and TRAF6 in hepatitis C virus-infected patients. *Virus Research* 238: https://doi.org/10.1016/j.virusres.2017.05.026.

37 Cai, P., Mu, Y., Olveda, R.M. et al. (2018). Circulating miRNAs as footprints for liver fibrosis grading in schistosomiasis. *eBioMedicine* 37: https://doi.org/10.1016/j.ebiom.2018.10.048.

38 Baluni, M., Ghildiyal, S., Singh, D. et al. (2018). Increased serum microRNA-29b expression and bad recovery in Japanese encephalitis virus infected patients; a new component to improve the disease recovery. *Journal of Neuroimmunology* 323: https://doi.org/10.1016/j.jneuroim.2018.07.014.
39 Patrucco, F. et al. (2021). SARS-CoV-2 detection on Bronchoalveolar lavage: an Italian multicenter experience. *Respiration* 99 (11): https://doi.org/10.1159/000511964.
40 Musisi, E. et al. (2021). Reproducibility of the ribosomal RNA synthesis ratio in sputum and association with markers of mycobacterium tuberculosis burden. *Microbiology Spectrum* 9 (2): https://doi.org/10.1128/spectrum.00481-21.
41 Miao, Q. et al. (2019). RNA-seq of circular RNAs identified circPTPN22 as a potential new activity indicator in systemic lupus erythematosus. *Lupus* 28 (4): https://doi.org/10.1177/0961203319830493.
42 Ouyang, Q. et al. (2017). Microarray expression profile of circular RNAs in peripheral blood mononuclear cells from rheumatoid arthritis patients. *Cellular Physiology and Biochemistry* 42 (2): https://doi.org/10.1159/000477883.
43 Qian, Z. et al. (2018). Potential diagnostic power of blood circular RNA expression in active pulmonary tuberculosis. *eBioMedicine* 27: https://doi.org/10.1016/j.ebiom.2017.12.007.
44 Santoro, M., Nociti, V., Lucchini, M. et al. (2016). Expression profile of long non-coding RNAs in serum of patients with multiple sclerosis. *Journal of Molecular Neuroscience* 59 (1): https://doi.org/10.1007/s12031-016-0741-8.
45 He, J. et al. (2017). Differential expression of long non-coding RNAs in patients with tuberculosis infection. *Tuberculosis* 107: https://doi.org/10.1016/j.tube.2017.08.007.
46 Lodde, V., Floris, M., Muroni, M.R. et al. (2022). Non-coding RNAs in malaria infection. *Wiley Interdisciplinary Reviews: RNA* 13 (3): https://doi.org/10.1002/wrna.1697.
47 Zhang, T.N. et al. (2017). Non-coding RNA: a potential biomarker and therapeutic target for sepsis. *Oncotarget* 8 (53): https://doi.org/10.18632/oncotarget.21766.
48 Singulani, J.D.L. et al. (2017). Preliminary evaluation of circulating microRNAs as potential biomarkers in paracoccidioidomycosis. *Biomedical Reports* 6 (3): https://doi.org/10.3892/br.2017.849.
49 Khouri, R. et al. (2014). SOD1 plasma level as a biomarker for therapeutic failure in cutaneous leishmaniasis. *Journal of Infectious Diseases* 210 (2): https://doi.org/10.1093/infdis/jiu087.
50 Yahya, M., Rulli, M., Toivonen, L. et al. (2017). Detection of host response to viral respiratory infection by measurement of messenger RNA for MxA, TRIM21, and Viperin in nasal swabs. *Journal of Infectious Diseases* 216 (9): https://doi.org/10.1093/infdis/jix458.
51 Cheung, K.W.E. et al. (2019). The potential of circulating cell free RNA as a biomarker in cancer. *Expert Review of Molecular Diagnostics* 19 (7): https://doi.org/10.1080/14737159.2019.1633307.
52 Wang, Y. et al. (2022). Plasma cell-free RNA characteristics in COVID-19 patients. *Genome Research* 32 (2): https://doi.org/10.1101/GR.276175.121.

53 Hamaguchi, S. et al. (2015). Origin of circulating free DNA in sepsis: analysis of the CLP mouse model. *Mediators of Inflammation* 2015: https://doi.org/10.1155/2015/614518.
54 Peng, Y. et al. (2019). Cell-free mitochondrial DNA in the CSF: a potential prognostic biomarker of anti-NMDAR encephalitis. *Frontiers in Immunology* 10: https://doi.org/10.3389/fimmu.2019.00103.
55 Li, L. et al. (2016). Cell-free circulating mitochondrial DNA content and risk of hepatocellular carcinoma in patients with chronic HBV infection. *Scientific Reports* 6: https://doi.org/10.1038/srep23992.
56 White, M.D. et al. (2021). Detection of leptomeningeal disease using cell-free DNA from cerebrospinal fluid. *JAMA Network Open* https://doi.org/10.1001/jamanetworkopen.2021.20040.
57 Guimaraes, A.O. et al. (2019). Prognostic power of pathogen cell-free DNA in staphylococcus aureus bacteremia. *Open Forum Infectious Diseases* 6 (4): https://doi.org/10.1093/ofid/ofz126.
58 Duvvuri, B. and Lood, C. (2019). Cell-free DNA as a biomarker in autoimmune rheumatic diseases. *Frontiers in Immunology* 10: https://doi.org/10.3389/fimmu.2019.00502.
59 Wang, L. et al. (2021a). Plasma microbial cell-free DNA sequencing technology for the diagnosis of sepsis in the ICU. *Frontiers in Molecular Biosciences* 8: https://doi.org/10.3389/fmolb.2021.659390.
60 Stawski, R., Nowak, D., and Perdas, E. (2022). Cell-free DNA: potential application in COVID-19 diagnostics and management. *Viruses* 14 (2): https://doi.org/10.3390/v14020321.
61 Ali, Z., Waseem, S., Anis, R.A., and Anees, M. (2020). Assessment of cell free mitochondrial DNA as a biomarker of disease severity in different viral infections. *Pakistan Journal of Medical Sciences* 36 (5): https://doi.org/10.12669/pjms.36.5.2476.
62 Hong, D.K., Blauwkamp, T.A., Kertesz, M. et al. (2018). Liquid biopsy for infectious diseases: sequencing of cell-free plasma to detect pathogen DNA in patients with invasive fungal disease. *Diagnostic Microbiology and Infectious Disease* 92 (3): https://doi.org/10.1016/j.diagmicrobio.2018.06.009.
63 Fernández-Carballo, B.L., Broger, T., Wyss, R. et al. (2019). Toward the development of a circulating free DNA-based in vitro diagnostic test for infectious diseases: a review of evidence for tuberculosis. *Journal of Clinical Microbiology* 57 (4): https://doi.org/10.1128/JCM.01234-18.
64 Tribolet, L. et al. (2020). MicroRNA biomarkers for infectious diseases: from basic research to biosensing. *Frontiers in Microbiology* 11: https://doi.org/10.3389/fmicb.2020.01197.
65 Wang, W., Zheng, G., and Lu, Y. (2021b). Recent advances in strategies for the cloning of natural product biosynthetic gene clusters. *Frontiers in Bioengineering and Biotechnology* 9: https://doi.org/10.3389/fbioe.2021.692797.
66 Liu, R., Ye, X., and Cui, T. (2020). Recent progress of biomarker detection sensors. *Yosetsu Gakkai Ronbunshu/Quarterly Journal of the Japan Welding Society* 38 (3): https://doi.org/10.2207/QJJWS.38.193.

67 Kuslich, C.D., Chui, B., and Yamashiro, C.T. (2019). Overview of PCR. *Current Protocols in Essential Laboratory Techniques* 18 (1): https://doi.org/10.1002/cpet.27.

68 Hughes, R.A. and Ellington, A.D. (2017). Synthetic DNA synthesis and assembly: putting the synthetic in synthetic biology. *Cold Spring Harbor Perspectives in Biology* 9 (1): https://doi.org/10.1101/cshperspect.a023812.

69 Sun, M.T., Gu, M.M., Zhang, J.Y. et al. (2022). Meta-analysis of variable-temperature PCR technique performance for diagnosising Schistosoma japonicum infections in humans in endemic areas. *PLoS Neglected Tropical Diseases* 16 (1): https://doi.org/10.1371/JOURNAL.PNTD.0010136.

70 Udugama, B. et al. (2020). Diagnosing COVID-19: the disease and tools for detection. *ACS Nano* 14 (4): https://doi.org/10.1021/acsnano.0c02624.

71 Afzal, A. (2020). Molecular diagnostic technologies for COVID-19: limitations and challenges. *Journal of Advanced Research* 26: https://doi.org/10.1016/j.jare.2020.08.002.

72 Kralik, P. and Ricchi, M. (2017). A basic guide to real time PCR in microbial diagnostics: definitions, parameters, and everything. *Frontiers in Microbiology* 8: https://doi.org/10.3389/fmicb.2017.00108.

73 Taylor, S.C., Nadeau, K., Abbasi, M. et al. (2019). The ultimate qPCR experiment: producing publication quality, reproducible data the first time. *Trends in Biotechnology* 37 (7): https://doi.org/10.1016/j.tibtech.2018.12.002.

74 Zhang, H., Gaňova, M., Yan, Z.Q. et al. (2020). PCR multiplexing based on a single fluorescent channel using dynamic melting curve analysis. *ACS Omega* 5 (46): https://doi.org/10.1021/acsomega.0c04766.

75 Martinez-Dominguez, M.V. et al. (2021). Current technologies for RNA-directed liquid diagnostics. *Cancers* 13 (20): https://doi.org/10.3390/cancers13205060.

76 Tombuloglu, H., Sabit, H., Al-Suhaimi, E. et al. (2021). Development of multiplex real-time RT-PCR assay for the detection of SARS-CoV-2. *PLoS ONE* 16 (4): https://doi.org/10.1371/journal.pone.0250942.

77 Hwang, K.A., Ahn, J.H., and Nam, J.H. (2018). Diagnosis of viral infection using real-time polymerase chain reaction. *Journal of Bacteriology and Virology* 48 (1): https://doi.org/10.4167/jbv.2018.48.1.1.

78 Shakoori, A.R. (2017). Fluorescence in situ hybridization (FISH) and its applications. *Chromosome Structure and Aberrations* https://doi.org/10.1007/978-81-322-3673-3_16.

79 Dwivedi, S. et al. (2017). Diseases and molecular diagnostics: a step closer to precision medicine. *Indian Journal of Clinical Biochemistry* 32 (4): https://doi.org/10.1007/s12291-017-0688-8.

80 Bonini, A. et al. (2021). Advances in biosensing: the CRISPR/Cas system as a new powerful tool for the detection of nucleic acids. *Journal of Pharmaceutical and Biomedical Analysis* 192: https://doi.org/10.1016/j.jpba.2020.113645.

81 Sahel, D.K., Mittal, A., and Chitkara, D. (2019). CRISPR/Cas system for genome editing: progress and prospects as a therapeutic tool. *Journal of Pharmacology and Experimental Therapeutics* 370 (3): https://doi.org/10.1124/jpet.119.257287.

82 Makarova, K.S. et al. (2015). An updated evolutionary classification of CRISPR-Cas systems. *Nature Reviews Microbiology* 13 (11): https://doi.org/10.1038/nrmicro3569.

83 Anders, C., Niewoehner, O., Duerst, A., and Jinek, M. (2014). Structural basis of PAM-dependent target DNA recognition by the Cas9 endonuclease. *Nature* 513 (7519): https://doi.org/10.1038/nature13579.

84 Song, B., Yang, S., Hwang, G.H. et al. (2021). Analysis of NHEJ-based DNA repair after CRISPR-mediated DNA cleavage. *International Journal of Molecular Sciences* 22 (12): https://doi.org/10.3390/ijms22126397.

85 Wu, X., Kriz, A.J., and Sharp, P.A. (2014). Target specificity of the CRISPR-Cas9 system. *Quantitative Biology* 2 (2): https://doi.org/10.1007/s40484-014-0030-x.

86 Cong, L. et al. (1979). Multiplex genome engineering using CRISPR/Cas systems. *Science* 339 (6121): 2013. https://doi.org/10.1126/science.1231143.

87 Chen, H., Choi, J., and Bailey, S. (2014). Cut site selection by the two nuclease domains of the Cas9 RNA-guided endonuclease. *Journal of Biological Chemistry* 289 (19): https://doi.org/10.1074/jbc.M113.539726.

88 Dagdas, Y.S., Chen, J.S., Sternberg, S.H. et al. (2017). A conformational checkpoint between DNA binding and cleavage by CRISPR-Cas9. *Science Advances* 3 (8): https://doi.org/10.1126/sciadv.aao0027.

89 Chen, J.S. et al. (2018). CRISPR-Cas12a target binding unleashes indiscriminate single-stranded DNase activity. *Science* 360 (6387): 436–439. https://doi.org/10.1126/SCIENCE.AAR6245.

90 Feng, W. et al. (2021). CRISPR technology incorporating amplification strategies: molecular assays for nucleic acids, proteins, and small molecules. *Chemical Science* 12 (13): https://doi.org/10.1039/d0sc06973f.

91 Gong, S., Zhang, S., Lu, F. et al. (2021b). CRISPR/Cas-based in vitro diagnostic platforms for cancer biomarker detection. *Analytical Chemistry* 93 (35): https://doi.org/10.1021/acs.analchem.1c02533.

92 Gleditzsch, D. et al. (2019). PAM identification by CRISPR-Cas effector complexes: diversified mechanisms and structures. *RNA Biology* 16 (4): https://doi.org/10.1080/15476286.2018.1504546.

93 Wang, J., Zhang, C., and Feng, B. (2020). The rapidly advancing class 2 CRISPR-Cas technologies: a customizable toolbox for molecular manipulations. *Journal of Cellular and Molecular Medicine* 24 (6): https://doi.org/10.1111/jcmm.15039.

94 Liu, X. et al. (2022). Programmable biosensors based on RNA-guided CRISPR/Cas endonuclease. *Biological Procedures Online* 24 (1): https://doi.org/10.1186/s12575-021-00163-7.

95 Xu, Y. and Li, Z. (2020). CRISPR-Cas systems: overview, innovations and applications in human disease research and gene therapy. *Computational and Structural Biotechnology Journal* 18: https://doi.org/10.1016/j.csbj.2020.08.031.

96 Kellner, M.J., Koob, J.G., Gootenberg, J.S. et al. (2019). SHERLOCK: nucleic acid detection with CRISPR nucleases. *Nature Protocols* 14 (10): https://doi.org/10.1038/s41596-019-0210-2.

97 Gong, J. et al. (2021a). A simple and rapid diagnostic method for 13 types of high-risk human papillomavirus (HR-HPV) detection using CRISPR-Cas12a technology. *Scientific Reports* 11 (1): https://doi.org/10.1038/s41598-021-92329-2.

98 Kaminski, M.M., Abudayyeh, O.O., Gootenberg, J.S. et al. (2021). CRISPR-based diagnostics. *Nature Biomedical Engineering* 5 (7): https://doi.org/10.1038/s41551-021-00760-7.

99 Quan, J. et al. (2019). FLASH: a next-generation CRISPR diagnostic for multiplexed detection of antimicrobial resistance sequences. *Nucleic Acids Research* 47 (14): https://doi.org/10.1093/NAR/GKZ418.

100 Azhar, M. et al. (2021). Rapid and accurate nucleobase detection using FnCas9 and its application in COVID-19 diagnosis. *Biosensors and Bioelectronics* 183: 113207. https://doi.org/10.1016/J.BIOS.2021.113207.

101 Dai, Y. et al. (2019). Exploring the trans-cleavage activity of CRISPR-Cas12a (cpf1) for the development of a universal electrochemical biosensor. *AngewandteChemie – International Edition* 58 (48): https://doi.org/10.1002/anie.201910772.

102 Zhang, Y., Tang, Z., Wang, J. et al. (2010). Hairpin DNA switch for ultrasensitive spectrophotometric detection of DNA hybridization based on gold nanoparticles and enzyme signal amplification. *Analytical Chemistry* 82, no. 15: https://doi.org/10.1021/ac1006238.

103 Tao, D. et al. (2020). Application of CRISPR-Cas12a enhanced fluorescence assay coupled with nucleic acid amplification for the sensitive detection of African swine fever virus. *ACS Synthetic Biology* 9 (9): https://doi.org/10.1021/acssynbio.0c00057.

104 Nouri, R., Jiang, Y., Tang, Z. et al. (2021). Detection of SARS-CoV-2 with solid-state CRISPR-Cas12a-assisted Nanopores. *Nano Letters* 21 (19): https://doi.org/10.1021/acs.nanolett.1c02974.

105 Sheng, Y. et al. (2021). A CRISPR/Cas13a-powered catalytic electrochemical biosensor for successive and highly sensitive RNA diagnostics. *Biosensors and Bioelectronics* 178: https://doi.org/10.1016/j.bios.2021.113027.

106 Myhrvold, C. et al. (1979). Field-deployable viral diagnostics using CRISPR-Cas13. *Science* 360 (6387): 2018. https://doi.org/10.1126/science.aas8836.

107 Gootenberg, J.S., Abudayyeh, O.O., Kellner, M.J. et al. (2018). Multiplexed and portable nucleic acid detection platform with Cas13, Cas12a, and Csm6. *Science* 360 (6387): 439–444. https://doi.org/10.1126/SCIENCE.AAQ0179.

108 Wu, P. et al. (2022). A novel CRISPR/Cas14a system integrated with 2D porphyrin metal-organic framework for microcystin-LR determination through a homogeneous competitive reaction. *Journal of Hazardous Materials* 424: https://doi.org/10.1016/j.jhazmat.2021.127690.

109 Ge, X. et al. (2021). Cas14a1-mediated nucleic acid detectifon platform for pathogens. *Biosensors and Bioelectronics* 189: https://doi.org/10.1016/j.bios.2021.113350.

110 Zhang, J. et al. (2021). Recent improvements in CRISPR-based amplification-free pathogen detection. *Frontiers in Microbiology* 12: https://doi.org/10.3389/fmicb.2021.751408.

111 He, Q. et al. (2020). High-throughput and all-solution phase African swine fever virus (ASFV) detection using CRISPR-Cas12a and fluorescence based point-of-care system. *Biosensors and Bioelectronics* 154: https://doi.org/10.1016/j.bios.2020.112068.

112 Li, Y., Mansour, H., Tang, Y., and Li, F. (2019). A universal proximity CRISPR Cas12a assay for ultrasensitive detection of nucleic acids and proteins. *bioRxiv* .

113 Ding, R. et al. (2021). Crispr/cas12-based ultra-sensitive and specific point-of-care detection of hbv. *International Journal of Molecular Sciences* 22 (9): https://doi.org/10.3390/ijms22094842.

114 Granados-Riveron, J.T. and Aquino-Jarquin, G. (2021). Crispr/cas13-based approaches for ultrasensitive and specific detection of micrornas. *Cell* 10 (7): https://doi.org/10.3390/cells10071655.

115 Zhang, T. et al. (2021). Light-up RNA aptamer signaling-CRISPR-Cas13a-based mix-and-read assays for profiling viable pathogenic bacteria. *Biosensors and Bioelectronics* 176: https://doi.org/10.1016/j.bios.2020.112906.

116 Gootenberg, J.S. et al. (2017). Nucleic acid detection with CRISPR-Cas13a/C2c2. *Science* 356 (6336): 438–442. https://doi.org/10.1126/SCIENCE.AAM9321.

117 Zhou, T., Huang, M., Lin, J. et al. (2021). High-Fidelity CRISPR/Cas13a trans-cleavage-triggered rolling circle amplified DNAzyme for visual profiling of MicroRNA. *Analytical Chemistry* 93 (4): https://doi.org/10.1021/acs.analchem.0c03708.

118 Cui, Y. et al. (2021). Ultrasensitive electrochemical assay for microRNA-21 based on CRISPR/Cas13a-assisted catalytic hairpin assembly. *Talanta* 224: https://doi.org/10.1016/j.talanta.2020.121878.

119 Bruch, R. et al. (2019). CRISPR/Cas13a-powered electrochemical microfluidic biosensor for nucleic acid amplification-free miRNA diagnostics. *Advanced Materials* 31 (51): https://doi.org/10.1002/adma.201905311.

120 Zhou, T., Huang, R., Huang, M. et al. (2020). CRISPR/Cas13a powered portable electrochemiluminescence chip for ultrasensitive and specific MiRNA detection. *Advanced Science* 7 (13): https://doi.org/10.1002/advs.201903661.

121 Shan, Y., Zhou, X., Huang, R., and Xing, D. (2019). High-fidelity and rapid quantification of miRNA combining crRNA programmability and CRISPR/Cas13a trans-cleavage activity. *Analytical Chemistry* 91 (8): https://doi.org/10.1021/acs.analchem.9b00073.

122 Kim, D.Y. et al. (2022). Efficient CRISPR editing with a hypercompact Cas12f1 and engineered guide RNAs delivered by adeno-associated virus. *Nature Biotechnology* 40 (1): https://doi.org/10.1038/s41587-021-01009-z.

123 Song, F. et al. (2021). Combining tag-specific primer extension and magneto-DNA system for Cas14a-based universal bacterial diagnostic platform. *Biosensors and Bioelectronics* 185: https://doi.org/10.1016/j.bios.2021.113262.

7

Role of Piezoelectric Biosensors

Jaykishon Swain[1], Subrat Swain[1], Durgesh Singh[2], Anirudha Jena[1], Raghabendra Samantaray[2], and Rojalin Sahu[1]

[1]*Department of Chemistry, School of Applied Sciences, Kalinga Institute of Industrial Technology Deemed to be University, Patia, Bhubaneswar, 751024, India*
[2]*KIIT School of Biotechnology, Kalinga Institute of Industrial Technology Deemed to be University, Patia, Bhubaneswar, 751024, India*

7.1 Introduction

By definition, a sensor is a device that converts physical inputs into functionally related outputs, typically an optical or electrical output which can either be viewed or detected by the user or by electronic devices. Different physical and chemical properties of compounds can be detected and measured by sensors, e.g. pH, odor, temperature, force, pressure, light intensity, position, flow, and presence of special chemicals [1]. In general, a sensor is characterized as being either (a) purely dependent on the physical or chemical quantity that it is measuring, or (b) independent of all other parameters, which are supposed to come across in its use. When operated, the chemical and/or physical properties of the input do not change. When the physical or chemical property changes, so does its sensitivity, and the magnitude of output also varies accordingly. While selecting a sensor, it is important to check certain aspects such as its sensitivity, selectivity, accuracy, resolution, calibration range, and cost-effectiveness [2].

Sensors can be classified into physical and chemical categories based on the properties of the substance being measured:

(i) The physical sensors measure physical parameters such as pressure, temperature, conductivity, refractive index, force, magnetic field, mass change, and absorbance [3].
(ii) Chemical sensors respond selectively to a specific analyte through a chemically selective layer [4].

An analysis of a total composition can include determining the concentration of a particular component and how the prevailing surrounding parameters affect it. After all, the components have been collected, they are then converted into signals that can be analyzed, such as change in conductance, voltage, light, current, or

Point-of-Care Biosensors for Infectious Diseases, First Edition.
Edited by Sushma Dave and Jayashankar Das.

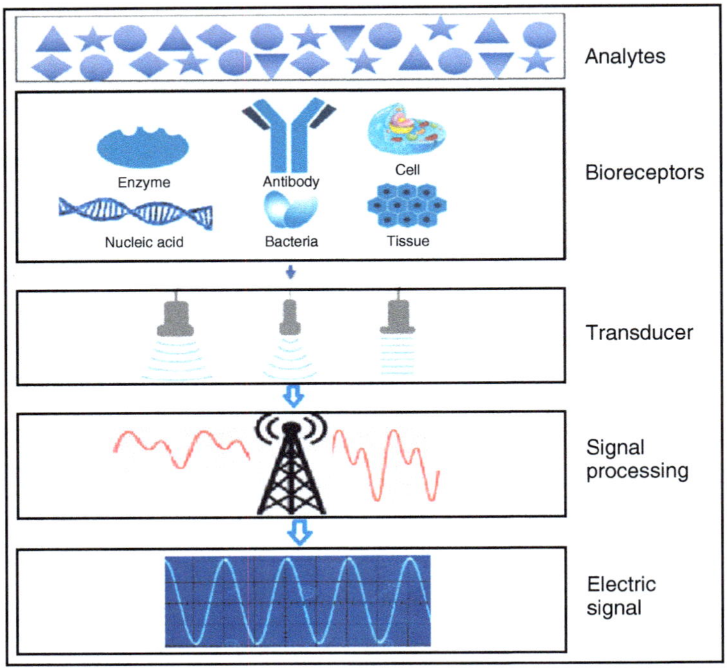

Figure 7.1 Schematic representation of working of biosensors.

sound. A widespread range of industrial, clinical, agricultural, and environmental applications can be achieved with chemical sensors among the currently commercially available ones.

Biosensors usually include a biological component, a receptor, a physicochemical detector, and a transducer. Bioreceptors are biomolecules that are able to detect or identify analytes such as antibodies, aptamers, enzymes, nucleic acids, or cells. Due to their specific structure, these sensors provide a high level of selectivity to the target analyte (biorecognition). The biosensor signal is not interfered by signals from other substances due to this specific interaction. At last, using a transducer, a bioreceptor converts the recognized signal into a measurable one as shown in Figure 7.1.

Immobilizing the bioreceptor using a reversible or irreversible technique at the surface of the transducer is paramount to a stable biosensor. There are several strategies to accomplish this, which are classified according to such criteria as the samples involved, the desired selectivity, the difficulty, and the range in size. Surface adsorption is classified as covalent bonding, cross-linking, metal binding or chelation, entrapment (beads or fibers), etc. When a physical change takes place, a transducer converts one form of energy into another, as part of a process known as "signalization," which is the process of converting detection into an observable signal. In addition to optoelectronics, electrochemistry, quartz–crystal piezoelectric, calorimetric (the amount of heat energy absorbed or emitted by a reaction), and thermal types of transducers are also available. For the first time in 1962, Clark and Lyons used glucose oxidase for sensing glucose [5].

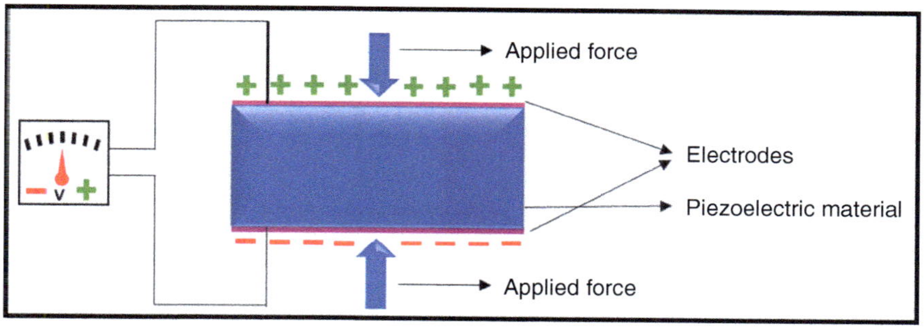

Figure 7.2 Mechanism of electricity generation from a piezoelectric material.

The word "biosensor" was proposed for the first time by Rechnitz et al. in 1977 [6]. Biosensors got much attention in various areas such as drug delivery, treatment of diabetic and cardiac diseases, and forensic investigations in food industries, agricultural and environmental detection systems. A major component of biosensor commercialization is to provide improved properties such as increased selectivity, stability, sensitivity, reproducibility, and portability.

Piezoelectric effect or piezoelectricity is the ability of a material to generate voltage when subjected to a mechanical stress. The mechanism of piezoelectric effect is shown in Figure 7.2.

Conversely, when alternating voltage is applied to surface of a piezoelectric material, it causes mechanical stress or oscillation. Some examples of piezoelectric materials are: aluminum nitride (AlN), aluminum phosphate (berlinite), crystallized topaz, lead titanate, barium titanate, quartz (SiO_2), polyvinylidene fluoride, gallium orthophosphate, polylactic acid, etc. Since the nineteenth century, the piezoelectric effect is not a new concept. It was discovered by Jacques Curie and Pierre Curie, two famous physicists. A piezoelectric material is well suited for the construction of physical sensors and biosensors from an analytical chemistry perspective. To achieve this, the surface of the sensor is subjected to excite by external electric voltage supplied by two electrodes. When alternating voltage is applied to crystal, the crystal oscillates mechanically, so the frequency of the oscillations can be measured when the crystal is kept in oscillation circuit [7]. A variety of oscillation patterns are possible depending on the material, electrical contacts, and crystal shape. A typical oscillation occurs as an adiabatic wave, which is also spread out over the mass. In this chapter, the types of piezoelectric biosensors, their applications in immunosensors, molecularly imprinted polymer, genetic information, etc. have been illustrated.

7.2 Types of Piezoelectric Biosensors

7.2.1 Inorganic Piezoelectric Material

Piezoelectricity in inorganic materials is due to ions being displaced inside crystals. Piezoelectric materials undergo structural changes when stressed, which shift the ion balance in the crystal and result in dipole moments. For net polarization to

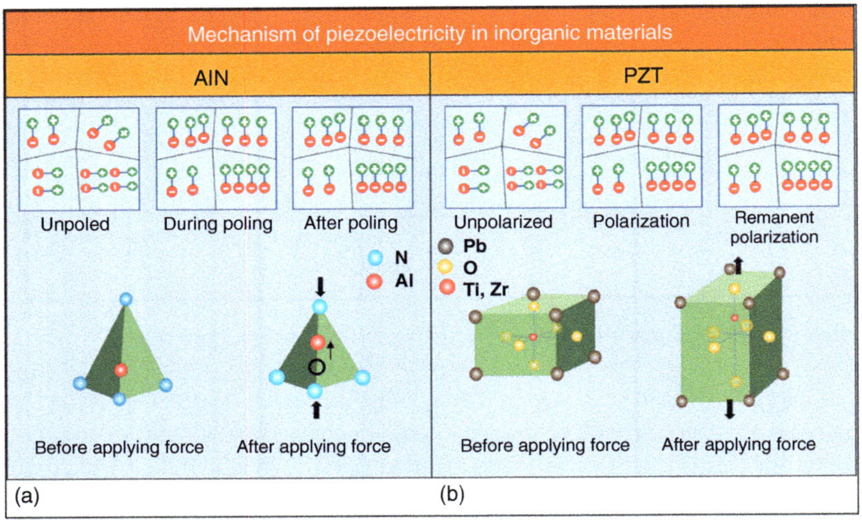

Figure 7.3 Mechanisms of piezoelectricity in inorganic materials. (a) Polarization process of AlN and the effect of piezoelectricity on the tetrahedral coordinate of AlN crystal. (b) Polarization process of Lead Zirconate titanate (PZT) and stress-induced phase transition inPZT.

emerge, dipoles created in the unit cell cannot cancel each other out. For piezoelectricity to exist, the atomic structure must be non-centrosymmetric, i.e. there should be no center of symmetry within the crystal. Some centrosymmetric minerals (such as hydroxyapatite) undergo symmetry breaking in nonequilibrium situations or at nanoscale dimensions, making them piezoelectric [8].

Piezoelectricity (the process of inducing polarization) is illustrated in Figure 7.3 for various inorganic materials. There are four Al atoms surrounding each Al atom in AlN, a tetrahedrally linked semiconductor (Figure 7.3a).

As there is no center of symmetry in each interstice and in response to a stress, dipole moments are produced by the motion of the central atom [9]. High-frequency resonators, filters, sensors, optical devices, acousto-optic devices, surface acoustic wave (SAW) devices, and bulk acoustic wave (BAW) devices are all made from AlN. The benign AlN substance is still a popular choice for biosensors and other biodevices. Because of its chemical stability, large bandgap, and high acoustic velocity, AlN has a wide range of uses. AlN thin films are highly sought after for gigahertz devices and have good electromechanical coupling characteristics, making them ideal for sensor applications [10]. In PZT, before mechanical stress, each unit cell of the crystal has a net nonzero charge. After stress within the unit cell, titanium ions shift positions, changing electrical polarity. Thus, the unit cell is converted into an electric dipole. PZT and AlN are examples of inorganic materials that exhibit piezoelectric behavior when poled under a strong electric field and heated to high temperatures. The piezoelectric coefficient of a crystalline AlN film is largely dependent on the crystal orientation of the film, which cannot be modified after deposition. The internal dipoles of PZT, on the other hand, can be reoriented by applying an external electric field, leaving a residual polarization at zero applied electric field [11]. PZT has a higher operating temperature and greater sensitivity than previously identified metallic oxide-based piezoelectric materials. PZT is a chemically inert, physically strong, and generally inexpensive material to produce [12, 13].

Owing to its enormous piezoelectric charge constant, PZT has outstanding electromechanical characteristics, making it particularly appealing for biomedical applications. Even though PZT, AlN, and other inorganic materials are less flexible than their organic counterparts, they can still exhibit considerable mechanical flexibility when arranged as thin films or nanowires [14, 15]. Biological sensors must be biocompatible for both *in vivo* and *in vitro* applications. The biocompatibility of PZT must be researched extensively, primarily because the material includes lead, which has negative effects on organisms [16]. Several types of research also focus on converting PZT to a biocompatible ceramic. For example, by coating the surface of PZT with titanium, a metal that is commonly used in orthopedic metallic implants, researchers attempted to increase the biocompatibility of the device [17]. Their results show that titanium-coated PZT has a good proliferation rate, but other methods, such as immersing a metal into blood, are still needed. Apart from PZT, biocompatible perovskite piezoelectric materials include $LiNbO_3^-$, $BaTiO_3^-$, lead zirconate niobate–lead titanate (PZN–PT), barium zirconate titanate–barium calcium titanate (BZT–BCT), lead magnesium niobate–lead titanate (PMN–PT), bismuth–sodium titanate (BNT), and potassium–sodium niobate (KNN).

Among other piezoelectric materials, ZnO possesses high electron mobility, making it a unique inorganic piezoelectric material [18], exceptional transparency [19], and biocompatibility [20]. As an active material in strain sensing and transient electronics, as well as for flexible mechanical energy harvesting systems, ZnO can also be employed [21]. There are several advantages of zinc oxide, but it must be processed at extremely high temperatures (400–500 °C) to produce ZnO-based electronics, which limits its application to certain fields [22].

The piezoelectric material gallium nitride (GaN) is biocompatible. GaN has a huge bandgap energy of 3.4 eV [23], strong electron mobility [24], great chemical stability [25], and small electrical drift in ionic solutions, among other benefits. Gallium nitride has no biofunctional impact on cellular surroundings, which is a crucial consideration for future biosensing applications [26].

7.2.2 Organic Piezoelectric Biosensors

Piezoelectricity refers to the reorientation of molecular dipoles within bulk polymers. This can be achieved by stretching or by applying a strong electrical field (drawing). Graphene, collagen, self-assembled diphenylalanine peptide nanotubes (PNTs), and glycine have also been studied in recent years as organic piezoelectric materials. As an example, polyvinylidene fluoride (PVDF) is shown in Figure 7.4a. The five crystal phases ($\alpha, \beta, \gamma, \delta, \varepsilon$) of PVDF has been recorded from which the α and β phases being the most typically used [27]. It consists of molecular dipoles arranged in phase in a unit cell, leading to the formation of nonpolar crystals and antiparallel molecular dipoles. This phase has the most favorable piezoelectric characteristics

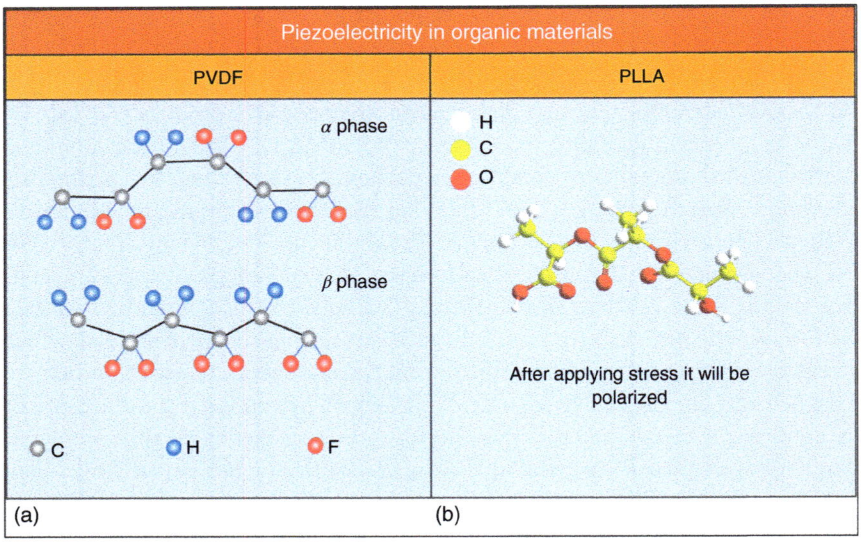

Figure 7.4 Mechanisms of piezoelectricity in inorganic materials. (a) Various PVDF structures including piezoelectric (α-phase) and non-piezoelectric (β-phase) (b) Structure of PLLA chain with orientation of CO dipoles in all directions.

because all the dipoles are parallel and contribute to the maximum dipole moment per unit cell [28]. PVDF has a negative d_{33} value, unlike other piezoelectric materials like PZT. Under the influence of an applied electric field, shifting of charged atomic nuclei, self-consistent quantum redistribution of electron molecular orbitals, and dipole rearrangement results in a negative piezoelectric effect [29]. In the same electric field that causes PZT to expand physically, PVDF will compress [30]. Applications like biological sensors and energy harvesting could benefit from the negative piezoelectric effect. Due to its inertness, PVDF makes an excellent material for surgical meshes and sutures, while its piezoelectricity makes it suitable for wound healing [31]. PVDF has been combined with other materials in an attempt to create composites that would overcome these disadvantages [32, 33].

Figure 7.4d shows poly(L-lactic acid) (PLLA) in its α-crystalline state (conformation is thermodynamically stable), with the C=O dipoles, arranged randomly along the main chain. During piezoelectricity, chains are thermally stretched from their α-crystalline to β-crystalline states. As a result, molecular chains aligned along the stretched direction have changed from randomly oriented chains [34]. C=O bond can be aligned by the electrospinning technique, in a single direction to make piezoelectric PLLAs. Plant-derived polymeric materials like PLLA are flexible and translucent, making them ideal for use in mobile devices as an environmentally benign, transparent, flexible piezoelectric thin film [35]. Its practical application, however, is limited due to its lower piezoelectric constant than inorganic piezoelectric materials like PZT. It is important to note that PLLA films do not spontaneously polarize, unlike poled polymers, such as PVDF, but they still have high piezoelectric constants under high shear loads [36]. Thermal annealing can control the degree of crystallinity in PLLA, despite its complicated higher-order structure with crystalline and amorphous elements intermixed. A PLLA sheet's piezoelectric constant can be tailored and increased by enhancing crystallinity and molecule orientation. In the future, this biodegradable and biocompatible film could be used in biosensors and actuators.

Depending on the crystallization conditions, glycine can crystallize into three different forms (α, β, and γ structures) [37]. A non-piezoelectric compound is α-glycine because it is centrosymmetric [38]. The acentric structures of α- and β-glycine provide shear piezoelectricity. β-glycine's voltage constants and piezoelectric strain were found to be around $g_{16} = 8\,\text{V m N}^{-1}$ and $d_{16} = 190\,\text{pm V}^{-1}$, respectively [38]. Due to glycine's high piezoelectricity and biocompatibility, it could be a beneficial piezoelectric organic material for biotechnology and biomedicine. Collagen's piezoelectric effect is caused by the molecule's polar and charged groups [39]. Collagen molecules reorient their dipole moments toward the long axis when mechanical force is applied, and the amplitude of the dipole moments changes. The overall piezoelectric effect in collagen is the consequence of several effects working together. There is a shear piezoelectric constant of $d_{14} = 0.1\,\text{pm V}^{-1}$ in collagen [40], however recent studies have shown that adding chitosan [41] or shifting the pH from acidic to neutral can boost this value [42].

Silk is recognized for having both amorphous and crystalline phases [43]. A high degree of silk II, crystalline orientation, and β-sheet crystallinity give silk its piezoelectricity [44]. Shear piezoelectric coefficients of $d_{14} = 1.5\,\text{pC N}^{-1}$ are found in silk films with a draw ratio of 2.7 [44]. Tissue engineering [45], regenerative medicine [46], and bone fracture repair [47] have all used silk fibroins. Piezoelectric silk could pave the way for new piezoelectric silk-based BioMEMS devices and wearable sensors.

Shear piezoelectricity is also visible in self-assembled PNTs. PNTs, which are produced from amino acids, are attractive prospects for future "green" piezoelectric materials and piezo devices. PNT has a shear piezoelectric constant of $d_{15} = 60\,\text{pm V}^{-1}$ [48]. The use of 2D materials like graphene in biomedicine is fast expanding [49, 50]. Due to the unique features of these materials, such as great stretchability and flexibility, the recent discovery of piezoelectric effects in graphene has opened a new path for piezoelectric actuators and sensors [51]. Inorganic elements (high piezoelectricity) and organic elements (low piezoelectricity) are blended into new classes of composite piezoelectric materials (flexibility). There is a wide range of applications for piezoelectric nanocomposite-based biomedical devices [52, 53]. Organic–inorganic hybrid nanogenerators hold great promise for a wide range of applications, including flexible wearable electronics and human–machine interfaces [53].

7.3 Application of Piezoelectric Biosensor Devices

7.3.1 Immunosensors Based on Piezoelectric Material

An immunosensor based on piezoelectric material is a type of analytical device which that detects various kinds of macromolecules and microorganisms. It is important to note that, piezoelectric immunosensors are biosensors that contain antibodies for biorecognition. The specificity of the entire immunosensor is dependent on the specificity of antibody. It is possible for immunosensors to show the opposing reaction to antigens despite they often contain immobilized antibodies. In other words, immunosensors can have immobilized antigens, which can be used for antibody recognition. Thus, the device containing immobilized antigens is used as an expedient for the diagnosis of infectious diseases.

Piezoelectric immunosensors have been explored for a long time for the development of various applications. Due to the fact that quartz crystal microbalances (QCMs) are extensively used in electronic equipment and are commercially available, the use of these materials in immunosensor construction is quite common. Muratsugu et al. [54] established label-free assay of albumin which is present in urea (i.e. albuminuria). They used immobilized antibody with QCM against human albumin serum and they determined albumin in the range of 0.1–100 $\mu g\,ml^{-1}$ thanks to assay. The complement component 4 (C4) was determined by QCM-type immunosensor and in that study, nafion membrane electrode was used and an antibody was used to detect C4 [55]. With a standard deviation of 5%, C4 was experimentally determined in 0.08–1.6 $\mu g\,ml^{-1}$ range. As an example, Funari et al. [56] demonstrated the versatility of QCM immunosensors by immobilizing antibodies which were spatially oriented for gluten on gold electrodes and achieved a 4 ppm detection limit for gluten where the sensitivity ranged between 7.5–15. Apparently, piezoelectric immunosensors cause a higher reduction in oscillation frequency when used for analytes having high molecular weight, as shown in the cited papers. This method also has some limitations, since antibodies immobilized on a piezoelectric device cannot directly recognize low-molecular-weight analytes. In recent years, new types of piezoelectric immunosensors have been developed to analyze microorganisms directly [57–59]. Although, piezoelectric immunosensors are not ideal for general health protection purposes, they can detect terrorist activity and other military misuse of biological warfare agents and provide early warning of such activities. Through the immobilization of a biorecognition element, i.e. bacteriophage, Olsen and his group [60] developed a biosensor based on acoustic wave for *Salmonella typhimurium* which could detect 100 cells ml^{-1}. Further, immunosensor based on antibodies was prepared on QCMs of 10 MHz to check the presence of bacteria [61]. This immunosensor contained antibody which detected *Francisella tularensis* type of bacteria and this immunosensor had a detection limit of 10^5 colony-forming units per ml in five minutes.

By coating nanoparticles with antibodies against analytes, the surface of the sensor can be increased in mass and the bacteria can be crosslinked into a firm layer. This makes the bacterial assay better than before. Some researchers have successfully tested an assay that involves a process in which nanoparticle-bound antibody is sandwiched between a crystal surface-bound antibody and an antibody immobilized on a crystal surface. For improving signals, Salam et al. [62] encapsulated gold nanoparticles with antibody and detected *S. typhimurium* type bacteria by forming a QCM immunosensor. Their group achieved the detection limit of 10–20 colony-forming units/ml and they could test the presence of bacteria in food samples. As an amplifier of signal, Guo et al. [63] used antibody-functionalized gold nanoparticles to detect *Escherichia coli* O157: H7 by QCM immunosensors. Their limit of detection was up to 10 colony-forming units/ml. Due to the presence of iron oxides in magnetic nanoparticles, which possess a high density, modified nanoparticles can be used to amplify signal. An ideal piezoelectric assay requires just a high specific weight. Particle magnetism is not required for the assay and is rarely used, although it can be beneficial when organizing the assay microfluidically. In a study of tumor necrosis factor α detection using magnetic nanoparticles covered with antibodies, the research article described the concept of using magnetic nanoparticles with a limit of detection of 1.62 $pg\,ml^{-1}$. The paper also described the full correlation with the standard immunological method [64]. Since the immunosensors are suitable for a label-free assay, piezoelectric immunosensors can provide similar analytical parameters in contrast to enzyme-linked immunosorbent assay (ELISA) without the need for specific reagents. Hence, it can be said that piezoelectric immunosensors have the potential to tackle with ELISA which is currently available in the market. Apart from that, lateral flow immunoassays also take part in the race with piezoelectric immunosensors. An example of lateral flow immunoassay is a pregnancy test in woman where human chorionic gonadotropin (HCG) hormone is detected. The limitation of the lateral flow immunoassays is the poor estimation of concentration of analytes quantitatively. It can be concluded from the comparison that piezoelectric immunosensors can match ELISA in terms of quality while being more straightforward and simpler as compared to tests like lateral flow immunoassays.

7.3.2 Piezoelectric Device with Molecularly Imprinted Polymers

As part of a biosensor, molecularly imprinted polymers can replace antibodies or antigens. In the presence of the target molecule or template, the molecularly imprinted polymers have been synthesized. When the template has the same exact structural characteristics as the analyte, it can be made of the same compound. However, templates with a lower molecular weight can also be used if they contain the same specific structural characteristics. To achieve a functional membrane, the template must be removed as part of the molecularly imprinted polymer construction. The smaller templates can be removed more easily from the polymer and the larger templates can be removed using physical or chemical processes that destroy the membrane. In the past decades, both inorganic and organic membrane materials have been explored for the formation of molecularly imprinted polymers. Among the materials with excellent template retaining capability and resistance to template removal are acrylamides and acrylates, chitosan, dextrin, sol–gel, and organometallic composites [65–67]. A piezoelectric biosensor based on molecularly imprinted polymers is analogous to a piezoelectric immunosensor on the basis of their usage. They interact with the analyte directly due to an affinity reaction which reduces the measured frequency of oscillation. An amplification reagent is not usually required in the assay, however, the frequency of oscillation of low molecular weight compounds is also limited by the assay because the assay is not dependent on the reagent.

Following examples illustrate the usefulness of molecularly imprinted polymer for analytical purposes. The electropolymerization of 3-thiophene acetic acid covers a

QCM sensor with an imprint of melphalan, a cytostatic drug [68]. QCM sensors were electropolymerized with gold electrodes and repeated cycles of cyclic voltammetry were used to control the process. The assay had shown the limit of detection of $5.4\,mg\,ml^{-1}$ which can be compared with general immunoassays. Piezoelectric immunosensors can also measure larger analytes. As an analyte, Gupta et al. [69] used *Neisseria meningitidis*, bacterium responsible for invasive meningococcal disease, and developed the molecularly imprinted polymer on a QCM sensor using methacrylic acid, ethylene glycol dimethacrylate, and azoisobutyronitrile. This sensor was fully capable of detecting proteins in blood serum originating from *N. meningitidis* MC58. An electrochemically fabricated QCM sensor is imprinted with taurine and I-methionine [70]. For a taurine solution, this sensor provided a detection limit of $0.12\,\mu mol\,l^{-1}$. A piezoelectric sensor with molecularly imprinted polymers successfully analyzed low molecular weight compounds as well. As examples of 10 MHz QCM sensors, chloramphenicol can be determined using trimethylolpropane trimethacrylate polymer [71], and copolymerization of ethylene glycoldimethacrylate and methacrylic acid with azobis (isobutyronitrile) as an initiator can determine caffeine [72]. Furthermore, molecularly imprinted polymers can be used as a tool to recognize certain genetic sequences. By using imprinted polymers, Bartold and coworkers are able to recognize single nucleotide polymorphisms [73].

7.3.3 Piezoelectric Biosensors for Genetic Information

Various biosensors are used for the employment of genetic information. DNA and RNA strains with single strands are suitable for constructing biosensors as examples of genetic information. However, there are also specific interactions that require the entire chromosome [74]. In addition to detecting genetic information from pathogens, these biosensors can also detect it from human tissue or blood. The biorecognition part and an analyte of the biosensor may be interacting with carcinogens that interact with immobilized chromosomes or double-stranded DNA chains [75, 76]. DNA and RNA when used as biorecognition elements, they provide two major practical advantages. The first one is that the chains can be simply synthesized using polymerase chain reactions, and thus it is possible for small companies to fabricate biosensors without involving feasible organisms or harmful materials. The second one is that an improved biosensor can be achieved by extending the chain used. In case of false positives, interference must be suppressed and the second step can occur.

As a result of the excellent properties of piezoelectric biosensors, interaction equilibrium and capture of single-stranded chains from solution are quickly established. According to the study of Kirimli et al. [77], immobilization of DNA probes on an $8\,\mu m$ thick lead magnesium niobite–lead titanate piezoelectric plates can detect target DNA in urine samples in 30 minutes with a limit of detection range of $10^{-19}\,mol\,l^{-1}$. Microorganisms can be identified with the help of piezoelectric immunosensors that embed genetic information. According to the study of Lian et al. [78], an aptamer coupled to an interdigitated gold electrode on graphene was used in a piezoelectric biosensor to detect *Staphylococcus aureus* using a crosslinking reaction. For that biosensor, the detection limit was $41\,CFU\,ml^{-1}$. It is very vital to

Table 7.1 Some other piezoelectric biosensors.

Sl. No.	Biosensor	Purpose	Limit of detection	References
1	QCM-based DNA sensor	Beta-thalassemia gene mutations in codon DC17	$2.6\,\text{nmol}\,\text{l}^{-1}$	[81]
2	Lectin affixed with QCM	Hematopoietic cells of leukemia	—	[82]
3	Lead zirconate–lead titanate glass piezoelectric micro cantilever	Epidermal growth factor receptor 2 (Her2)	$0.06–0.6\,\text{nmol}\,\text{l}^{-1}$	[83]
4	QCM modified with ZnO nanorods	Breast cancer biomarker CM15.3	$0.5\,\text{U}\,\text{ml}^{-1}$	[84]
5	Piezoelectric aptasensor	Leukemic cell	$1200\,\text{cells}\,\text{ml}^{-1}$	[85]
6	QCM-based sensor coupled to cyclone air sampler	The aerosol form of *Escherichia coli* (safe strains)	$1.5 \times 10^4\,\text{CFU}\,\text{l}^{-1}$	[86]
7	Labeling of the surface immunocomplex with magnetic particles on quartz crystal	Tumor necrosis factor-alpha (TNF)	$25\,\text{ng}\,\text{ml}^{-1}$	[87]
8	QCM-based immunosensor with scanning approach	Bacteria	$10^2–10^7\,\text{CFU}\,\text{mL}^{-1}$	[88]

control food which can be contaminated by Shiga toxin-producing bacteria such as *E. coli* O157: H7. According to the study of Rijal et al. [79], a biosensor was developed which detects *E. coli* O157: H7 using a piezoelectric-excited cantilever sensing platform containing a genetic probe that recognizes the Stx2 gene. In their experiment, they found that the detection limit for the biosensor was $2500\,\text{cells}\,\text{ml}^{-1}$. Highly infective microorganisms and viruses like dengue can be detected earlier by QCM-based DNA biosensors. A modified gold nanoparticle containing oligonucleotides was utilized to enhance signals produced by interaction between dengue virus and QCM's surface, with the viral genome actually serving as a bridge for the layer-by-layer accumulation of modified gold nanoparticles [80].

Some other papers based on piezoelectric materials for biosensor applications are cited in Table 7.1.

7.4 Conclusion

Piezoelectric biosensors are well suited for rapid, simple, reagentless tests of viruses, bacteria, proteins, and nucleic acids, as well as pharmaceuticals. In the contemporary era, piezo sensors are valuable tools for studying the morphology, adhesion, and apoptosis of eukaryotic cells, as well as testing the physiological effects of new drugs and toxicity and biocompatibility tests. Directly monitoring affinity interactions with piezoelectric sensors in real-time without labels presents a much more economical solution than the other often overpriced options. It is easy and quick to obtain the valuable characteristics of binding reactions. It remains an open platform for researchers with moderate skill levels to construct their own experimental devices from the shell components. In turn, this leads to a wide variety of applications. In addition, innovative scientific advancements are possible through fruitful combinations with other sensing technologies.

References

1 Huang, H., Su, S., Wu, N. et al. (2019). Graphene-based sensors for human health monitoring. *Frontiers in Chemistry* 399.

2 Vetelino, J.F. (2010). A lateral field excited acoustic wave sensor platform. *Proceedings of the IEEE Ultrasonics Symposium* Oct 11 (pp. 2269–2272).

3 Gründler, P. (2007). *Chemical Sensors: An Introduction for Scientists and Engineers*. Springer Science & Business Media.

4 Janata, J. (2010). *Principles of Chemical Sensors*. Springer Science & Business Media.

5 Clark, L.C. Jr. and Lyons, C. (1962). Electrode systems for continuous monitoring in cardiovascular surgery. *Annals of the New York Academy of Sciences* 102 (1): 29–45.

6 Rechnitz, G.A., Kobos, R.K., Riechel, S.J., and Gebauer, C.R. (1977). A bio-selective membrane electrode prepared with living bacterial cells. *Analytica Chimica Acta* 94 (2): 357–365.

7 García-Martinez, G., Bustabad, E.A., Perrot, H. et al. (2011). Development of a mass sensitive quartz crystal microbalance (QCM)-based DNA biosensor using a 50 MHz electronic oscillator circuit. *Sensors* 11 (8): 7656–7664.

8 Lang, S.B., Tofail, S.A., Kholkin, A.L. et al. (2013). Ferroelectric polarization in nanocrystalline hydroxyapatite thin films on silicon. *Scientific Reports* 3 (1): 1–6.

9 Que, M., Zhou, R., Wang, X. et al. (2016). Progress in piezo-phototronic effect modulated photovoltaics. *Journal of Physics. Condensed Matter* 28 (43): 433001.

10 Shih, W.C., Chen, Y.C., Chang, W.T. et al. (2014). Design and fabrication of nanoscale IDTs using electron beam technology for high-frequency saw devices. *Journal of Nanomaterials* 1: 2014.

11 Ramadan, K.S., Sameoto, D., and Evoy, S. (2014). A review of piezoelectric polymers as functional materials for electromechanical transducers. *Smart Materials and Structures* 23 (3): 033001.

12 Swallow, L.M., Luo, J.K., Siores, E. et al. (2008). A piezoelectric fibre composite based energy harvesting device for potential wearable applications. *Smart Materials and Structures* 17 (2): 025017.

13 Paul, D. and Roy, A. (2015). Piezoelectric effect: smart roads in green energy harvesting. *International Journal of Engineering and Technical Research (IJETR)* 3 (2): 112.

14 Zang, Y., Zhang, F., Di, C.A., and Zhu, D. (2015). Advances of flexible pressure sensors toward artificial intelligence and health care applications. *Materials Horizons* 2 (2): 140–156.

15 Pan, C., Dong, L., Zhu, G. et al. (2013). High-resolution electroluminescent imaging of pressure distribution using a piezoelectric nanowire LED array. *Nature Photonics* 7 (9): 752–758.

16 Jaishankar, M., Tseten, T., Anbalagan, N. et al. (2014). Toxicity, mechanism and health effects of some heavy metals. *Interdisciplinary Toxicology* 7 (2): 60.

17 Sakai, T., Hoshiai, S., and Nakamachi, E. (2006). Biochemical compatibility of PZT piezoelectric ceramics covered with titanium thin film. *Journal of Optoelectronics and Advanced Materials* 8 (4): 1435.

18 Janotti, A. and Van de Walle, C.G. (2009). Fundamentals of zinc oxide as a semiconductor. *Reports on Progress in Physics* 72 (12): 126501.

19 Wei, Q. (2009). Emerging approaches to the surface modification of textiles. *Surface Modification of Textiles* 1: 318–323.

20 Zhang, Y., Nayak T, R., Hong, H., and Cai, W. (2013). Biomedical applications of zinc oxide nanomaterials. *Current Molecular Medicine* 13 (10): 1633–1645.

21 Lu, J., Zhang, L., Takagi, H. et al. (2014). Hybrid piezoelectric MEMS resonators for application in bio-chemical sensing. *Journal of Applied Science and Engineering* 17 (1): 17–24.

22 Kumar, R., Al-Dossary, O., Kumar, G., and Umar, A. (2015). Zinc oxide nanostructures for NO_2 gas–sensor applications: a review. *Nano-Micro Letters.* 7 (2): 97–120.

23 Jung, W.S. (2002). Reaction intermediate (s) in the conversion of β-gallium oxide to gallium nitride under a flow of ammonia. *Materials Letters* 57 (1): 110–114.

24 Meneghesso, G., Verzellesi, G., Danesin, F. et al. (2008). Reliability of GaN high-electron-mobility transistors: state of the art and perspectives. *IEEE Transactions on Device and Materials Reliability* 8 (2): 332–343.

25 Guo, J.D., Pan, F.M., Feng, M.S. et al. (1996). Schottky contact and the thermal stability of Ni on n-type GaN. *Journal of Applied Physics* 80 (3): 1623–1627.

26 Hofstetter, M., Howgate, J., Schmid, M. et al. (2012). In vitro bio-functionality of gallium nitride sensors for radiation biophysics. *Biochemical and Biophysical Research Communications* 424 (2): 348–353.

27 Li, Q. and Wang, Q. (2016). Ferroelectric polymers and their energy-related applications. *Macromolecular Chemistry and Physics* 217 (11): 1228–1244.

28 Wan, C. and Bowen, C.R. (2017). Multiscale-structuring of polyvinylidene fluoride for energy harvesting: the impact of molecular-, micro-and macro-structure. *Journal of Materials Chemistry A* 5 (7): 3091–3128.

29 Bystrov, V.S., Paramonova, E.V., Bdikin, I.K. et al. (2013). Molecular modeling of the piezoelectric effect in the ferroelectric polymer poly(vinylidene fluoride) (PVDF). *Journal of Molecular Modeling* 19 (9): 3591–3602.

30 Nakamura, K. (2012). *Ultrasonic Transducers: Materials and Design for Sensors, Actuators and Medical Applications.* Elsevier.

31 Guo, H.F., Li, Z.S., Dong, S.W. et al. (2012). Piezoelectric PU/PVDF electrospun scaffolds for wound healing applications. *Colloids and Surfaces. B, Biointerfaces* 1 (96): 29–36.

32 Dang, Z.M., Fan, L.Z., Shen, Y., and Nan, C.W. (2003). Dielectric behavior of novel three-phase MWNTs/$BaTiO_3$/PVDF composites. *Materials Science and Engineering B* 103 (2): 140–144.

33 Kono, A., Shimizu, K., Nakano, H. et al. (2012). Positive-temperature-coefficient effect of electrical resistivity below melting point of poly (vinylidene fluoride) (PVDF) in Ni particle-dispersed PVDF composites. *Polymer* 53 (8): 1760–1764.

34 Sultana, A., Ghosh, S.K., Sencadas, V. et al. (2017). Human skin interactive self-powered wearable piezoelectric bio-e-skin by electrospun poly-l-lactic acid nanofibers for non-invasive physiological signal monitoring. *Journal of Materials Chemistry B* 5 (35): 7352–7359.

35 Yoshida, T., Imoto, K., Tahara, K. et al. (2010). Piezoelectricity of poly (l-lactic acid) composite film with stereocomplex of poly (L-lactide) and poly (D-lactide). *Japanese Journal of Applied Physics* 49 (9S): 09MC11.

36 Ochiai, T. and Fukada, E. (1998). Electromechanical properties of poly-l-lactic acid. *Japanese Journal of Applied Physics* 37 (6R): 3374.

37 Iitaka, Y. (1959). Crystal structure of β-glycine. *Nature* 183 (4658): 390–391.

38 Guerin, S., Stapleton, A., Chovan, D. et al. (2018). Control of piezoelectricity in amino acids by supramolecular packing. *Nature Materials* 17 (2): 180–186.

39 Zhou, Z., Qian, D., and Minary-Jolandan, M. (2016). Molecular mechanism of polarization and piezoelectric effect in super-twisted collagen. *ACS Biomaterials Science & Engineering* 2 (6): 929–936.

40 Minary-Jolandan, M. and Yu, M.F. (2009). Nanoscale characterization of isolated individual type I collagen fibrils: polarization and piezoelectricity. *Nanotechnology* 20 (8): 085706.

41 Silva, C.C., Lima, C.G., Pinheiro, A.G. et al. (2001). On the piezoelectricity of collagen–chitosan films. *Physical Chemistry Chemical Physics* 3 (18): 4154–4157.

42 Denning, D., Paukshto, M.V., Habelitz, S., and Rodriguez, B.J. (2014). Piezoelectric properties of aligned collagen membranes. *Journal of Biomedical Materials Research Part B: Applied Biomaterials* 102 (2): 284–292.

43 Joseph, J., Singh, S.G., and Vanjari, S.R. (2017). Leveraging innate piezoelectricity of ultra-smooth silk thin films for flexible and wearable sensor applications. *IEEE Sensors Journal* 17 (24): 8306–8313.

44 Yucel, T., Cebe, P., and Kaplan, D.L. (2011). Structural origins of silk piezoelectricity. *Advanced Functional Materials* 21 (4): 779–785.

45 Wang, Y., Kim, H.J., Vunjak-Novakovic, G., and Kaplan, D.L. (2006). Stem cell-based tissue engineering with silk biomaterials. *Biomaterials* 27 (36): 6064–6082.

46 Yang, Y., Ding, F., Wu, J. et al. (2007). Development and evaluation of silk fibroin-based nerve grafts used for peripheral nerve regeneration. *Biomaterials* 28 (36): 5526–5535.

47 Meinel, L., Betz, O., Fajardo, R. et al. (2006). Silk based biomaterials to heal critical sized femur defects. *Bone* 39 (4): 922–931.

48 Kholkin, A., Amdursky, N., Bdikin, I. et al. (2010). Strong piezoelectricity in bioinspired peptide nanotubes. *ACS Nano* 4 (2): 610–614.

49 Heerema, S.J. and Dekker, C. (2016). Graphene nanodevices for DNA sequencing. *Nature Nanotechnology* 11 (2): 127–136.

50 Xu, S., Zhan, J., Man, B. et al. (2017). Real-time reliable determination of binding kinetics of DNA hybridization using a multi-channel graphene biosensor. *Nature Communications* 8 (1): 1.

51 da Cunha, R.G., Zelenovskiy, P., Romanyuk, K. et al. (2015). Strong piezoelectricity in single-layer graphene deposited on SiO_2 grating substrates. *Nature Communications* 6 (1): 1–6.

52 You, Y.M., Liao, W.Q., Zhao, D. et al. (2017). An organic-inorganic perovskite ferroelectric with large piezoelectric response. *Science* 357 (6348): 306–309.

53 Dahiya, A.S., Morini, F., Boubenia, S. et al. (2018). Organic/inorganic hybrid stretchable piezoelectric nanogenerators for self-powered wearable electronics. *Advanced Materials Technologies.* 3 (2): 1700249.

54 Muratsugu, M., Ohta, F., Miya, Y. et al. (1993). Quartz crystal microbalance for the detection of microgram quantities of human serum albumin: relationship between the frequency change and the mass of protein adsorbed. *Analytical Chemistry* 65 (20): 2933–2937.

55 Deng, T., Li, J.S., Wang, H. et al. (2005). Piezoelectric immunoassay for complement C4 based on a Nafion-modified interface for antibody immobilization. *Journal of Immunological Methods* 299 (1–2): 1–8.

56 Funari, R., Terracciano, I., Della Ventura, B. et al. (2017). Label-free detection of gliadin in food by quartz crystal microbalance-based immunosensor. *Journal of Agricultural and Food Chemistry* 65 (6): 1281–1289.

57 Maraldo, D. and Mutharasan, R. (2007). 10-minute assay for detecting *Escherichia coli* O157: H7 in ground beef samples using piezoelectric-excited millimeter-size cantilever sensors. *Journal of Food Protection* 70 (7): 1670–1677.

58 Maraldo, D. and Mutharasan, R.A. (2007). Preparation-free method for detecting *Escherichia coli* O157: H7 in the presence of spinach, spring lettuce mix, and ground beef particulates. *Journal of Food Protection* 70 (11): 2651–2655.

59 Campbell, G.A., Uknalis, J., Tu, S.I., and Mutharasan, R. (2007). Detect of *Escherichia coli* O157: H7 in ground beef samples using piezoelectric excited millimeter-sized cantilever (PEMC) sensors. *Biosensors & Bioelectronics* 22 (7): 1296–1302.

60 Olsen, E.V., Sorokulova, I.B., Petrenko, V.A. et al. (2006). Affinity-selected filamentous bacteriophage as a probe for acoustic wave biodetectors of *Salmonella typhimurium*. *Biosensors & Bioelectronics* 21 (8): 1434–1442.

61 Pohanka, M. and Skládal, P. (2007). Piezoelectric immunosensor for the direct and rapid detection of *Francisella tularensis*. *Folia Microbiologica* 52 (4): 325–330.

62 Salam, F., Uludag, Y., and Tothill, I.E. (2013). Real-time and sensitive detection of *Salmonella typhimurium* using an automated quartz crystal microbalance (QCM) instrument with nanoparticles amplification. *Talanta* 15 (115): 761–767.

63 Guo, X., Lin, C.S., Chen, S.H. et al. (2012). A piezoelectric immunosensor for specific capture and enrichment of viable pathogens by quartz crystal microbalance sensor, followed by detection with antibody-functionalized gold nanoparticles. *Biosensors & Bioelectronics* 38 (1): 177–183.

64 Pohanka, M. (2018). Piezoelectric biosensor for the determination of tumor necrosis factor alpha. *Talanta* 1 (178): 970–973.

65 Naklua, W., Suedee, R., and Lieberzeit, P.A. (2016). Dopaminergic receptor–ligand binding assays based on molecularly imprinted polymers on quartz crystal microbalance sensors. *Biosensors & Bioelectronics* 15 (81): 117–124.

66 Dai, J., Zhang, Y., Pan, M. et al. (2014). Development and application of quartz crystal microbalance sensor based on novel molecularly imprinted sol–gel polymer for rapid detection of histamine in foods. *Journal of Agricultural and Food Chemistry* 62 (23): 5269–5274.

67 Ahmed, M.A., Abdelbar, N.M., and Mohamed, A.A. (2018). Molecular imprinted chitosan-TiO_2 nanocomposite for the selective removal of Rose Bengal from wastewater. *International Journal of Biological Macromolecules* 1 (107): 1046–1053.

68 Singh, A.K. and Singh, M. (2015). QCM sensing of melphalan via electropolymerized molecularly imprinted polythiophene films. *Biosensors & Bioelectronics* 74: 711–717.

69 Gupta, N., Shah, K., and Singh, M. (2016). An epitope-imprinted piezoelectric diagnostic tool for *Neisseria meningitidis* detection. *Journal of Molecular Recognition* 29 (12): 572–579.

70 Singh, A.K. and Singh, M. (2017). Electrochemical and piezoelectric monitoring of taurine via electropolymerized molecularly imprinted films. *Journal of Molecular Recognition* 30 (12): e2652.

71 Ebarvia, B.S., Ubando, I.E., and Sevilla, F.B. III (2015). Biomimetic piezoelectric quartz crystal sensor with chloramphenicol-imprinted polymer sensing layer. *Talanta* 144: 1260–1265.

72 Ebarvia, B.S., Binag, C.A., and Sevilla, F. III (2004 Mar). Biomimetic piezoelectric quartz sensor for caffeine based on a molecularly imprinted polymer. *Analytical and Bioanalytical Chemistry* 378 (5): 1331–1337.

73 Bartold, K., Pietrzyk-Le, A., Huynh, T.P. et al. (2017). Programmed transfer of sequence information into a molecularly imprinted polymer for hexakis (2, 2′-bithien-5-yl) DNA analogue formation toward single-nucleotide-polymorphism detection. *ACS Applied Materials & Interfaces* 9 (4): 3948–3958.

74 Sivashankar, S., Sapsanis, C., Agambayev, S. et al. (2016). Label-free detection of sex determining region y (sry) via capacitive biosensor. In: *2016 38th Annual International Conference of the IEEE Engineering in Medicine and Biology Society (EMBC)*, 4349–4352. IEEE.

75 Bui, V.N., Nguyen, T.T., Mai, C.T. et al. (2016). Procarcinogens–determination and evaluation by yeast-based biosensor transformed with plasmids incorporating RAD54 reporter construct and cytochrome P450 genes. *PLoS One* 11 (12): e0168721.

76 Karimi, A., Hayat, A., and Andreescu, S. (2017). Biomolecular detection at ssdna-conjugated nanoparticles by nano-impact electrochemistry. *Biosensors & Bioelectronics* 87: 501–507.

77 Kirimli, C.E., Shih, W.H., and Shih, W.Y. (2014). DNA hybridization detection with 100 zM sensitivity using piezoelectric plate sensors with an improved noise-reduction algorithm. *Analyst* 139 (11): 2754–2763.

78 Lian, Y., He, F., Wang, H., and Tong, F. (2015). A new aptamer/graphene interdigitated gold electrode piezoelectric sensor for rapid and specific detection of *Staphylococcus aureus*. *Biosensors & Bioelectronics* 65: 314–319.

79 Rijal, K. and Mutharasan, R. (2013). A method for DNA-based detection of *E. coli* O157: H7 in a proteinous background using piezoelectric-excited cantilever sensors. *The Analyst* 138 (10): 2943–2950.

80 Chen, S.H., Chuang, Y.C., Lu, Y.C. et al. (2009). A method of layer-by-layer gold nanoparticle hybridization in a quartz crystal microbalance DNA sensing system used to detect dengue virus. *Nanotechnology* 20 (21): 215501.

81 Pang, L., Li, J., Jiang, J. et al. (2006). DNA point mutation detection based on DNA ligase reaction and nano-Au amplification: a piezoelectric approach. *Analytical Biochemistry* 358 (1): 99–103.

82 Pesquero, N.C., Carvalho, F.C., Faria, R.C. et al. (2017). Artinm binding effinities and kinetic interaction with leukemia cells: a quartz crystal microbalance bioelectroanalysis on the cytotoxic effect. *Electroanalysis* 29 (6): 1554–1558.

83 Capobianco, J.A., Shih, W.Y., Adams, G.P., and Shih, W.H. (2011). Label-free Her2 detection and dissociation constant assessment in diluted human serum using a longitudinal extension mode of a piezoelectric microcantilever sensor. *Sensors and Actuators B: Chemical* 160 (1): 349–356.

84 Wang, X., Yu, H., Lu, D. et al. (2014). Label free detection of the breast cancer biomarker CA15. 3 using ZnO nanorods coated quartz crystal microbalance. *Sensors and Actuators B: Chemical* 195: 630–634.

85 Shan, W., Pan, Y., Fang, H. et al. (2014). An aptamer-based quartz crystal microbalance biosensor for sensitive and selective detection of leukemia cells using silver-enhanced gold nanoparticle label. *Talanta* 126: 130–135.

86 Kovar, D., Farka, Z., and Skládal, P. (2014). Detection of aerosolized biological agents using the piezoelectric immunosensor. *Analytical Chemistry* 86 (17): 8680–8686.

87 Bahk, Y.K., Kim, H.H., Park, D.S. et al. (2011). A new concept for efficient sensitivity amplification of a QCM based immunosensor for TNF-α by using modified magnetic particles under applied magnetic field. *Bulletin of the Korean Chemical Society* 32 (12): 4215–4220.

88 Ngo, V.K.T., Nguyen, D.G., Nguyen, H.P.U. et al. (2014). Quartz crystal microbalance (QCM) as biosensor for the detecting of *Escherichia coli* O157: H7. *Advances in Natural Sciences: Nanoscience and Nanotechnology* 5 (4): 045004.

8

Metal/Metal Oxide Nanoparticles-Based Biosensors for Detection of Infectious Diseases

Dipak Maity[1], Gajiram Murmu[2,3], Satya R. Sahoo[2,3], Ankur Tiwari[4], Siddharth Ajith[4], and Sumit Saha[2,3]

[1]*Department of School of Engineering and Technology, The Assam Kaziranga University, Jorhat, Assam, 785006, India*
[2]*Materials Chemistry Department, CSIR-Institute of Minerals & Materials Technology, Sachivalaya Marg, Acharya Vihar, Bhubaneswar, Odisha 751013, India*
[3]*Academy of Scientific and Innovative Research (AcSIR), CSIR- Human Resource Development Centre, Postal Staff College Area, Kamla Nehru Nagar, Ghaziabad, Uttar Pradesh, 201002, India*
[4]*Institute of Chemical Technology Mumbai-Indian Oil Campus, Department of Chemical Engineering, Samantapuri Mouza, Gajapati Nagar, Bhubaneswar, Odisha, 751013, India*

8.1 Introduction

As an integrated miniaturized device, a biosensor utilizes the biological element (antibody, enzyme, receptor protein, nucleic acid, whole-cell, or tissue section) as a sensing element coupled to a transducer to detect signals. Biosensors combine the selectivity of the biomolecule with the processing power of modern microelectronics and optoelectronics. They are thus an analytical tool with applications in medical diagnosis and many other fields [1]. Today's era of nanotechnological development has promoted the evolution and specialization of biosensors for different purposes. Nevertheless, one of the most relevant purposes of these devices is the detection of pathogens since bacterial and viral illnesses represent an essential aspect of human health [2, 3]. There are several detection techniques used for the detection of pathogens (like viruses and bacteria), namely reverse transcription–polymerase chain reaction (RT–PCR) [4], enzyme-linked immunosorbent assay (ELISA) [5], cell culture [6], etc. However, these techniques take a long time to get results and are laborious. Thus, a new approach has been developed based on nanotechnological advances where nanoparticles (NPs), especially metal/metal oxide nanoparticles (MMONs), have been used to create novel devices and technologies that show outstanding properties and performance [7, 8].

This chapter summarizes the metal and metal oxides nanoparticles as biosensors in detecting various infectious diseases. We have discussed various infectious diseases, their causing agents, and their diagnosis method and categorized them based

Point-of-Care Biosensors for Infectious Diseases, First Edition.
Edited by Sushma Dave and Jayashankar Das.

on their transmission routes. We addressed the MMONs used in biosensing and paid particular attention to gold, silver, platinum, copper, and zinc oxides. Subsequently, electrochemical, colorimetric, and fluorometric techniques were used to detect the disease. Lastly, a comparative study is presented for the biosensing techniques in detecting infectious diseases.

8.2 Biosensors

Infectious diseases are a massive threat to public health and the global economy. Quick and precise detection of pathogens is necessary to prevent the spread of such infections [9]. Regular techniques for detecting these microorganisms are tedious, expensive, and not pertinent for on-site observations [9]. Biosensors can provide a quick and reliable diagnostic for infectious diseases. NPs, because of their remarkable features (optical, chemical, and electrical features), have become essential participants in the field of biosensors. Various kinds of NP-derived biosensors for detecting infectious diseases have been discussed below.

8.2.1 Electrochemical Biosensors

Electrochemical recognition strategies are utilized due to their simple synthesis method, high sensitivity, and the scale-down ability for their applications [10, 11]. It has been found that nanomaterials have a wide range of applications in the preparation of electrochemical sensors ranging from analyte concentration (e.g. magnetic particle) to the catalyst of redox reactions enhancer of electrodes' surface conductivity [12–16]. Magnetic beads are regularly utilized in quick measures to capture the targets from crude samples before recognition [17]. The utilization of these does not just permit to pre-concentrate the analyte (for example, biomarkers) but also helps to improve the sensitivity of the sensors (as shown in Figure 8.1). It is a fact that the superparamagnetic properties of magnetic nanoparticles (MNPs) permit their rescattering in the absence of a magnetic field. The outer layer of MNPs can be effectively designed, which helps in providing an appropriate functional group for the immobilization of biomolecules; besides, it has been observed that immobilization of biomolecules decreases their biodegradation when in contact with complex biological or environmental systems [19, 20]. Electrochemical immunosensors have been viably utilized for microorganism detection [16, 19, 21] because of their high affectability and great applicability. Additionally, the sandwich assay empowers the development of a more explicit and sensitive immunosensor in contrast to direct recognition [19].

Synthesis of iron oxide (Fe_3O_4)/silicon dioxide (SiO_2)/gold nanoparticles (AuNPs) and their application in electrochemical immunosensing was reported by Fei et al. using the coprecipitation method of preparation; these, following covering with SiO_2 and pre-functionalization with 3-mercapto-propyltriethoxysilane (MPTES), were covered by AuNPs that filled in as securing element for *S. pullorum* and *S. gallinarum* antibodies. Fe_3O_4/SiO_2/AuNPs were used to capture these antibodies, and for its separation from the sample, an external magnetic field was applied. AuNPs/Fe_3O_4

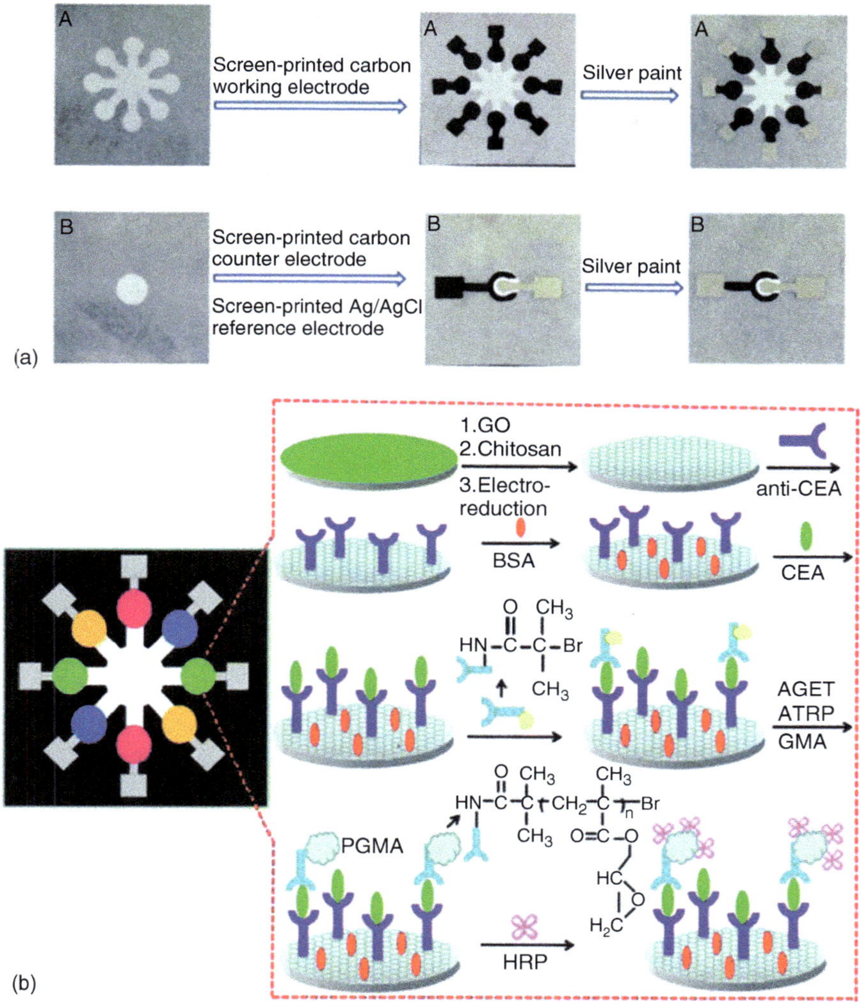

Figure 8.1 A paper-based microfluidic platform with electrochemical detection for multiplexed cancer biomarker detection. (a) Device fabrication procedures. (b) Schematic representation of the electrochemical immunoassay procedures using CEA as an example. Source: Reproduced from Wu et al. [18]. © 2014, Elsevier.

nanoparticles along the infectious bacteria were then re-dispersed in a buffer solution that contained horse radish peroxidase (HRP) labeled anti-*S. pullorum* and *S. gallinarum* to form a sandwich complex. The sandwich complex was dropped on a 4-channel screen-printed carbon cathode which was previously adjusted with electrodeposited AuNPs. This was imparted with a magnet and trailed by thionine and hydrogen peroxide expansion. It was seen that the antibody had been immobilized proficiently on the Fe_3O_4/SiO_2/AuNPs surface, and 93.95% of the signal of the electrode was kept up with following 30 days of storage, showing the capacity of Fe_3O_4/SiO_2/AuNPs to hold the bioactivity of the adsorbed antibody [19].

Luo et al. utilized a personal glucose meter (PGM) to foster a quick and sensitive immunosensor for *S. pullorum* and *S. gallinarum*. In the experiment, the microorganisms capture was carried out utilizing MNP-antibody, and this was trailed by introducing an enzymatic label (antibody-silica nanoparticles–glucose oxidase (GO*x*)). This resulted in forming a sandwich-like framework, following the magnetic capture and cleaning, which was scattered in a glucose solution. The immune NPs that were utilized in this work could retain their activity after storage at 4°C for a period of 90 days [22]. It was also noted that the reduction of glucose concentration came out to be proportional to the logarithm concentration of bacteria. AuNPs have been generally utilized in electrochemical sensors because of their electrical conductivity, biological compatibility, and high surface-to-volume proportion [23]. Additionally, gold nanomaterials can produce couple viably with other nanomaterials for the production of various Au-based nanocomposites biosensor like poly (diallyl dimethylammonium chloride)-functionalized graphene oxide (GO) and AuNps [24], chitosan/MWCNT/Polypyrrole/AuNPs [14], GO/AuNR/poly thionine [25] to foster new electrochemical sensors with further improving stability and selectivity.

8.2.2 Colorimetric Biosensors

The utilization of metal NPs in colorimetric discovery was first shown by Mirkin et al. in 1996 when they used AuNPs for colorimetric detection of target DNA [26]. Qualitative detection of analytes can be carried out through our naked eyes, and by using UV–vis spectrometry, quantitative investigation can be accomplished. These kinds of detection techniques provide advantages like low cost, simple operation, and ease of analysis. Interparticle plasmonic coupling between closely spaced metal NPs prompts a red-shifted termination range and thus an undeniable color change [27]. The change of color can be easily observed through naked eyes. For example, after the formation of NP aggregates of metal, for Au NPs, it was observed that the color of the solution changed from red to purple, and in the case of Ag NPs, a change occurred from yellow to brown. This detection method is very convenient as no hardware is required. Various plans have been produced to detect various species, including nucleic acids, tiny atoms, proteins, and pathogens [28, 29].

It was investigated by the Zourob group that enzyme-induced disaggregation of black magnetic nanobeads (MB) network occurred, because of peptidic cross-linking, for the optical detection of various pathogens. It can be described as follows: (i) In the presence of a magnetic field, MB networks were set up on gold substrates. (ii) During the addition of bacterial extract that contains proteases to the sensing surface, disaggregation of the MB network occurs due to the digestion of peptide, which exposes the Au particle located underneath. (iii) It was observed that the percentage of the observable area was shown to be proportional to the concentration of bacteria protease. The proposed sensors could achieve a Limit of Detection (LOD) of 7, 12, 49, 2.17×10^2 CFU ml^{-1} for *Methicillin-Resistant S. aureus, Escherichia coli O157: H7, Porphyromonasgingivalis, Listeria monocytogenes*, respectively [30–32].

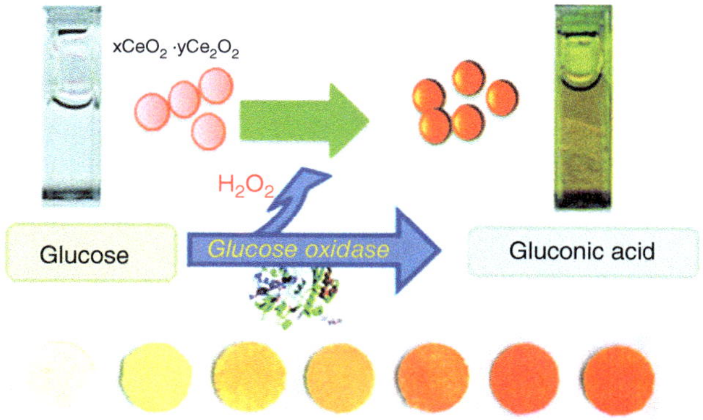

Figure 8.2 Schematic of the working principle of the colorimetric assays for detecting glucose with cerium oxide nanoparticles using a paper-based microfluidic device. Source: Reproduced with permission from Ornatska et al. [36]. © 2011, American Chemical Society.

Gold nanomaterials are among the most utilized material in colorimetric detection. AuNPs were additionally used to transduce the collaboration of antibiotics with the bacterial external film; this was displayed to empower the delicate location of microscopic organisms down to 10CFU ml^{-1}. All the more explicitly in their evidence of idea work, Singh et al. utilized a cationic antimicrobial (known as Colistin) as a bacteria recognition element and inducing element for AuNPs accumulation [33]. In another review, the nanocomposite of AuNPs and carbon nanotubes (CNT) was displayed to have more catalytic activity in contrast with AuNPs and CNT alone toward the oxidation of 3,3′,5,5′-tetramethylbenzidine (TMB) by H_2O_2. It was observed that the proposed nanocomposites were used in an immunosensor for the sensitive detection of flu A H_3N_2 infection down to 10 PFU ml^{-1} [34].

Apart from the change of color because of target-induced accumulation of metal NPs, diverse colorimetric behaviors can likewise be used to design various methodologies. Zhang et al. designed a plasmonic time–temperature indicator (TTI) to screen the microorganism's development and perishables in food by utilizing sharp color change when Ag shell developed on Au NRs [35]. The self-advancement of TTI and *E. coli* development cycle was synchronized at discretionary temperatures inside the typical temperature window after changing the concentration of reagents. This TTI can be used for the tracking of perishables as well as the growth of microbes according to a sharp change in color from red to green (shown in Figure 8.2).

8.2.3 Fluorescence Biosensors

Fluorescence-based sensors have drawn incredible attention due to their capability of presenting low detection limit, high sensitivity, quick response, basic instrumentation, and less cost. Various fluorescence nanomaterials, for example, metal nanoparticles, GO, silica nanobeads (dye-doped), quantum dots (QDs), carbon dots

(CDs), nanotubes (NTs), and upconversion nanoparticles, have been investigated for detection applications and broadly for biomedical diagnostics and bioimaging, for example, fluorophore-formed metal NPs and fluorescent gold nanoclusters (Au NCs) have been widely applied [28, 29, 37]. Compared with colorimetric methods, fluorescence-based techniques are more sensitive and offer better LOD, which can even reach the single-particle identification level. The remarkable properties of metal NPs, for example, high surface-to-volume proportion and excellent biocompatibility, make them ideal contenders for speeding up signal transduction and giving clear sign readout. Panda et al. planned a practical examination for fast quantification of bacterial cells by estimating the reduction of the fluorescence force of a positively charged highly fluorescent Au NP–polythiophene nanocomposite in the presence of bacterial cells [38].

It was reported that explicit separation and identification of *E. coli O157: H7* utilized an attractive magnetic immunoassay and QDs identification. The bacteria capture was carried in a twofold layer quartz channel utilizing the invulnerable attractive NP caught on the wall of the channel divider in the presence of a strong magnetic field [39]. Following marking with QD, the framed sandwich complex was gathered from the channel, and the fluorescence force of the sample was estimated. This showed a linear relation with the logarithm of the concentration of microbes in the value range from 8.9 to 8 900 00 CFU ml^{-1} [39]. It was noted that the fluorescence signal of the solution could be increased by enhancing the bacterial concentration. Also, water-soluble fluorescent mannose-secured Au nanodots with ultrasmall size and high quantum yield were utilized to foster an easy strategy for identifying *E. coli* [40]. The mannose-secured Au nanodots bind to the *E. coli* through the multivalent interactions between mannose receptors and the mannosylated Au nanodots on the microscopic organisms, which leads to a bright fluorescent cluster of cells.

One of the most generally used procedures for fluorescence detection is fluorescence resonance energy transfer (FRET)-initiated fluorescence quenching. The fluorescence of fluorophores can be viably quenched by metal NPs in the presence of the target, giving the turn-off/turn-on fluorescence signal change [28]. AuNPs have a strong fluorescence quenching capacity because of their enormous molar extinction coefficients and tunable extinction spectra, making them unique fluorescence quenchers for different applications [28]. When the fluorophores come into the closeness of AuNPs, the fluorescence will be extinguished by Au NPs. The peculiarity of Au NP-induced fluorescence extinguishing has been applied in various biosensing plans. These biosensors are mainly derived based on two approaches – target-initiated fluorescence quenching and target-instigated fluorescence recovery. In another review, Li et al. fostered an Au NP-based multichannel fluorescence biosensor to detect bacterial species in biofilms utilizing forms of Au NPs and fluorescent proteins [41]. The presence of biofilms caused interruption of the conjugates, bringing about a momentary three-color readout, where the ratiometric reaction of the readout was effectively applied to separate bacterial species and strains of six biofilms rapidly.

8.3 Types of Infectious Diseases

Infectious diseases have had significant impacts on human civilizations throughout the course of history. Infectious diseases were the cause of some of the significant events in the course of human existence, such as the death of Alexander the Great, failed invasion of Russia by Napoleon or the massive loss of life by Spanish flu, or the recent epidemic of COVID-19 the effects of which is still unfolding. Infectious diseases require large dense human populations to proliferate; this explains why their frequency and intensity have increased since humans started practicing agriculture and settling in communities. Now with increased globalization as travel time between nearly all parts of the world has collapsed, it has become ever more challenging to manage emerging infections (Figure 8.3a).

Infectious organisms such as bacteria, viruses, fungi, protozoa, parasites, or worms/helminths cause infectious diseases. A large fraction of the microorganisms causing infectious diseases is zoonotic. Zoonotic pathogens are those which initially did not affect humans but evolved to jump species and change target species, as suggested in the case of HIV, H1N1 flu, and possibly the recent novel covid-causing coronavirus. The point of concern regarding such pathogens is their ability to coevolve with their hosts. Hence, we must understand the transmission of infectious diseases and devise methods to overcome shortcomings in preventing transmission. Though recent advancements in diagnostic methods have made the detection of diseases easier, there is still an urgent requirement for rapid, affordable, mobile diagnostic devices with good selectivity and accuracy. Extensive studies have been conducted to study the various classes of infectious organisms. Though our understanding regarding them has helped to devise methods to prevent infections, yet by no means have we reached a stage to prevent their spread effectively, as was

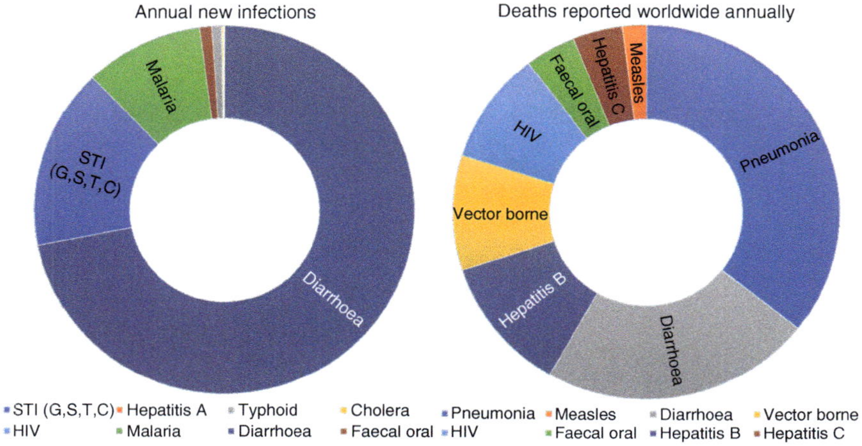

Figure 8.3 New infections which are recorded annually. STI (G, S, T, C) represents sexually transmitted infections (Gonorrhea, Syphilis, Trichomoniasis, Chlamydia) (a). Deaths occurring in persons infected by life-threatening infectious diseases (b).

evident during the subsequent waves of the COVID-19 pandemic. Studies have reported that many deaths that occur in persons infected by life-threatening infectious diseases are due to delays in diagnosis (Figure 8.3b) [42, 43].

Based on their transmission route, some of the major infectious diseases have been listed in Table 8.1. It is to be noted that infections can occur by more than one

Table 8.1 List of the transmission route of the disease, causing agent, and the conventionally used diagnosis methods.

Transmission route	Diseases	Causing agent	Diagnosis method
Air droplet	Tuberculosis	Bacteria (*Mycobacterium tuberculosis*)	Chest X-ray, Mantoux tuberculin test, blood test
	Pneumonia	Bacteria, fungi, or virus	X-ray, serological test, blood test, PCR
	Pandemic Influenza	Virus	Rapid antigen test, RT–PCR
	Trachoma	Bacteria (*Chlamydia trachomatis*)	Microscopy, cell culture, ELISA, RT–PCR
	Measles	Virus	Serological test, MRI, ELISA, RT–PCR
	Common cold	Virus (Rhinovirus, Adenovirus)	Symptoms based
Faecal-oral	Cholera	Bacteria(*Vibrio cholerae*)	Culture, PCR
	Typhoid	Bacteria(*Salmonella typhi*)	Microscopy, culture, Widal test, PCR
	Hepatitis A	Virus (Hepatitis A virus)	Serological test, RT–PCR
Sexual	Syphilis	Bacteria (Troponema pallidum)	Serological test, PCR
	Gonorrhea	Bacteria (*Neisseria gonorrhoeae*)	Culture, microscopy, NAAT
	HIV	Virus (Human immunodeficiency virus)	Serological, ELISA, PCR
	Zika virus	Flavivirus	RT–PCR, serological, ELISA
	Trichomoniasis	Parasite (*Trichomonas vaginalis*)	Microscopy, rapid antigen test, culture, PCR, ELISA, serological test
Vector-borne	Malaria	Protozoan (genus *Plasmodium*)	Microscopy, PCR, serological test (blood)
	Dengue	Virus (arbovirus)	Tourniquet test, antibody serology, PCR
	Diarrhea	Bacteria or viruses (*Escherichia coli*, *Salmonella* & *Rotaviruses*)	PCR, culture, microscopy

(Continued)

Table 8.1 (Continued)

Transmission route	Diseases	Causing agent	Diagnosis method
Blood	HIV	Virus (Human immunodeficiency virus)	Serological, ELISA, PCR
	Hepatitis B	Virus (family Hepadnaviridae)	ELISA, serological tests, PCR
	Hepatitis C	Virus (family Flaviviridae)	Serological test (antibody test), RT–PCR, immunosorbent assays

Abbreviation: PCR-Polymerase chain reaction, RT- Reverse transcriptase, ELISA- Enzyme-linked immunosorbent assay, NAAT- Nucleic acid amplification test. Source: Adapted from www.who.int/diseasecontrol_emergencies/publications/idhe_2009_london_inf_dis_transmission.pdf., www.cdc.gov/.

route but have been summarized under their most common transmission route for non-repetitions.

Route of transmission of infectious diseases is of utmost importance to understand transmissibility and the extent of an infectious disease's impact on the population. Their routes of transmission may be: through pathogens released along with air droplets on coughing or sneezing, bloodborne, through contact with body fluids, through sexual contact, injury or inoculation, and also through direct or indirect contact (www.cdc.gov, www.who.int/). The interaction between a pathogen and host is a complex dynamic process, and the extent of infection depends on many different factors of the host and the pathogen involved. Besides these, environmental factors are also a determining factor in exposure–outcome when a host comes in direct or indirect contact with the infecting agent [43]. The interaction between host and infecting agent occurs in stages, including exposure and initial infection, disease, recovery, or death. The aforementioned factors of the infecting agents include virulence, infectivity, and pathogenicity, while for the host, its susceptibility and immunity to the disease dictate the length of the infection and the final outcome (www.who.int/). Furthermore, the ability of the body to fight infections depends on many factors, including any underlying disease, age, or their nutritional status. The disease's stages and duration are unique to the infection and vary for any infection based on involved infecting agents, intrinsic hosts, and environmental factors.

The impact of noninfectious diseases can be easier to assess in terms of impact on public health. In contrast, infectious diseases generally require collective action, usually requiring coordination of governments and individuals, civil society, and other stakeholders for its elimination. This makes managing the spread of infectious diseases even more difficult. Robust Public health policy is essential for effective mitigation of the infectious disease outbreaks of which screening is an integral tool. Screening is the first step in the treatment cycle and determines whether a specific individual is infected or not (www.euro.who.int/en). It is necessary for an effective health strategy to have an easy, reliable, cost-effective, and fast screening tool/methods to contain outbreaks by minimizing transmissions. Biosensors can have a massive impact on conventional screening tools.

Conventional diagnosis methods face several limitations such as slow speeds, the requirement of instrumentations, contamination of samples when samples are transported from rural areas to labs, and the requirement of trained personnel for testing. Methods such as microscopy, culture methods, and polymerase chain reactions have biomolecules involved in the process, which has issues related to its stability. Newer methods such as ELISA and nucleic acid-based assays, though they give pretty accurate results, are suitable in laboratory settings with instrumentation facilities available and highly skilled persons conducting these tests [44]. Their usage as point-of-care (POC) devices for rapid assessment of infectious diseases has several constraints, which include lack of isolated environments, concerns regarding stability in the ambient conditions, the requirement of sophisticated instruments, higher costs also based on the external environment the selectivity and reproducibility of the tests may be impacted. These problems can be solved by using nanomaterial-based biosensors as nano biosensors can be utilized to prepare handheld POC devices making rapid, sensitive testing possible on-site. Due to NPs' inherent characteristic large surface area, nano-based biosensors have a synergic effect due to better mass transport of reactants enabling better analytical sensitivity. POC diagnostic techniques can also be efficiently utilized with minimal training in a setting that cannot support an entire laboratory.

8.4 Nanoparticles-Based Biosensors

Since preliminary biosensor studies, metal and metal oxides have been essential candidates for sensor applications. They can be synthesized in various nanostructures such as NPs, nanofibers (NFs), nanospheres (NSs), nanorods (NRs), nanotubes (NTs), nanowires (NWs), and nanosheets. Besides morphological versatility, MMONs also offer some advantages such as good biocompatibility, nontoxicity, chemical stability, high surface-to-volume ratio, excellent selectivity, high catalytic efficiency, and strong adsorption ability. Metal oxides nanostructures can be produced via relatively easy and cost-effective methods. Additionally, in recent years various metal oxides such as ZnO [45], Fe_3O_4 [46], Cu_2O [47], NiO [48], and TiO_2 [49] have been continuously produced as versatile and functional biosensors.

Nanomaterials are chemical substances or materials made up of particles or constituents in nanoscale dimensions and prepared or fabricated using nanotechnology and nanosciences. With the development of nanotechnology and nanosciences, many novel nanomaterials are being fabricated with novel properties that can be employed in the preparation of biosensors [50]. Recently, nanomaterials have proved to be pertinent in biosensing applications. By intelligent and innovative usage of these nanomaterial-based biosensors has enhanced the performance of biosensors in terms of both sensitivity and detection. These nanomaterial-based biosensors can detect biomolecules and have the potential to detection of disease-causing pathogens. Such nanomaterials and how they can recognize pathogens have been discussed in this section [51].

8.4.1 Recognition of Pathogens

Biosensors must have sufficient specificity and sensitivity to detect even low concentrations of biomolecules/pathogens. Thus, the sample should be concentrated and cleaned to make it contaminant-free by filtration or other separation processes. An ideal biosensor should discriminate between closely related pathogen and non-pathogen entities, detect a small number of targets, and be strongly stable over time. In nanomaterial-based biosensors, recognition elements (present on the surface of nanomaterial) interact with epitopes on the pathogen surface, and these interactions are monitored via a signal transduction method [52]. There are different types of nanomaterials used in biosensor applications, and these biosensors are a combination of nanomaterial, recognition elements, and signal transduction methods. The recognition elements are generally biomolecules that have an affinity for the epitopes. There are a variety of recognition elements like antibodies [53], aptamers [54], carbohydrates [55], and peptides [56] that can be employed in recognizing pathogens. Antibodies have been majorly used to detect bacteria, viruses, and other pathogens. There are well-established methods and nanomaterials [57, 58] that have highly selective and sensitive antibodies which can detect the pathogens effectively, and thus, it is primarily used in immunological recognition. Antimicrobial peptides also play an essential role in immune system response to pathogen infections [59].

Signal transduction methods are also the vital part of the biosensor-containing detector or transducers, which works by sensing the signal obtained from physicochemical changes due to interaction between the recognition elements and the analytes [60]. These transducers transform the signal into another one that can be evaluated and quantified [61, 62]. There are different signal transduction methods like optical, electrochemical, magnetic, and piezoelectric devices, out of which optical signal transduction dominates the biosensor field. In optical biosensors, the analysis is done by measuring photons by using optic fibers as transduction elements [63]. Several sensing mechanisms are employed in optical biosensors for analyte detection, such as absorption, colorimetry, fluorescence, luminescence, and surface plasmon resonance (SPR) [64]. These biosensors have lower noise and immunity to electromagnetic interference, which is advantageous over electrochemical and piezoelectric biosensors [65]. The electrochemical biosensor has also been applied for pathogen detection. In these biosensors, the analytes are analyzed through electrodes by measuring electrical signals resulting from interaction (redox reactions) between recognition elements and epitopes. Also, the analyses of different analytes are performed by different mechanisms like potentiometry, amperometry, and conductometry [66]. Moreover, electrochemical biosensor requires further improvement due to the nanomaterials development [67].

8.4.2 Metal/Metal Oxide Nanoparticles-Based Biosensors

Nanoengineered biosensors have altered the field of biomedical engineering by introducing smaller sensing structures for the highly selective and sensitive detection of biomolecules. Nanomaterials of varied forms, including single or hybrids/combinatorial nanostructures, can be designed with distinctive features that are significantly different from conventional bulk materials. The distinct molecular recognition interactions at the nano-realm level in combination with nanoscale components can further promote the high sensitivity of the biosensors. Nanostructured metal and metal oxides have received significant attention for biosensing applications owing to several characteristics such as ease of fabrication and controllable size/shape, biocompatibility, catalytic and optical properties, chemical stability, strong adsorption ability, and electron-transfer kinetics.

8.4.2.1 Gold Nanoparticles

Among the group of noble metal nanoparticles, gold nanoparticles (AuNPs) are extensively used in biosensing applications due to their unique optical and electronic properties, biocompatibility, simple and straightforward fabrication with modifications [68, 69]. GNPs have unique optical properties on their surface known as SPR. In this phenomenon, the electrons (in conduction band) start oscillating when the light of a particular wavelength is irradiated on it. When the particle is smaller than that of the incident wavelength, the oscillating electron cannot propagate along the surface. Thus, the electron density will be polarized on one side of the particle where the electrons oscillate due to SPR. This phenomenon was described by Mie theory [70, 71], and it is dependent on the size and shape of the NPs and the dielectric constant of the medium [72]. Therefore, any change in the medium or size and shape (from spherical to rod) will change the oscillation frequency, which will change the color of the NP. This medium and morphology dependence is very useful in biosensing applications. For example, Jena et al. developed a GNPs-based biosensor for sensing polyionic drugs, namely protamine and heparin, by reversible aggregation and de-aggregation of GNPs. This protamine drug induces aggregation of GNPs, which affects the SPR band, and therefore the color of the NPs shifts from red to blue [73].

Similarly, Wei et al. described aptamer-based colorimetric sensing of alpha-thrombin protein using unmodified GNPs-based probes [74]. GNPs based biosensors have been employed in food safety detection due to versatile shape and easy surface modifications. These GNPs based biosensors are rapidly developing, low-cost, portable, and on-site detection that can effectively detect contaminants and allergens [75]. Similarly, GNPs-based biosensors have also been developed for the sensing and measurement of not only enzyme activities [76] in biological specimens and DNA and immunogens [77].

8.4.2.2 Magnetic Nanoparticles

MNPs are those NPs that can be manipulated by using an external magnetic field. Materials like pure metal (Fe, Co, Ni, etc.) and oxides (like Fe_3O_4, γ-Fe_2O_3) are used to fabricate such MNPs. Among them, iron oxides are widely used for preparing MNPs. They show different magnetic behavior as compared to their bulk material. Due to their reduced size, the multi-domain structure is reduced to a single-domain structure, which increases coercivity value to the max, and therefore, these particles possess superparamagnetism [78]. In the presence of an external magnetic field, these particles have high saturation magnetization and magnetic susceptibility. In the absence of an external magnetic field, the magnetization can flip the direction randomly within a short time (Neel's relaxation time), and magnetization appears as an average zero. The temperature-dependent phenomenon disappears in the presence of an external magnetic field that aligns the magnetic moments. Due to their unique magnetic behavior, they are widely used in biosensing applications. MNPs are used to detect biomolecules and cells based on the magnetic resonance effect. This detection technique is termed as Diagnostic Magnetic Resonance. This detection technique detects a wide range of targets like DNA/mRNA, proteins, enzymes, drugs, pathogens, and tumor cells [79]. In addition to optical [80] or electrochemical [81] detection techniques combined with other often nanoscale labels, MNPs are a particularly sensitive method for high-resolution diagnostic magnetic resonance. A high-performance giant magnetoresistance, spin valve, or magnetic tunnel junction biosensor has been developed for this purpose [82, 83]. Magnetic labels are especially useful in biosensing applications since biological entities do not exhibit magnetic behavior or susceptibility, so that there will be no interference or background noise resulting from signal capture [84].

8.4.2.3 Quantum Dots

QDs are luminescent semiconducting nanocrystals extensively used for bioanalytics due to their unique photoluminescent properties. Several methods have been developed for fabricating water-soluble QDs to produce for use in biosensing and optical bar coding [85]. These synthesized QDs have significant advantages over traditional fluorescent dyes, including better stability, more vigorous fluorescent intensity, and different colors, which are adjusted by controlling the size of the dots. QDs are significantly more stable, more intense, and come in different colors, whose intensity can be adjusted by controlling the size. They also have significant advantages over traditional fluorescent dyes [86]. Thus, QDs offer a new platform in biosensing applications. QDs can also be used to study the interactions between proteins or determine how signal transduction proceeds in live cells by using FRET [87, 88]. Nowadays, QDs are available with inert or biocompatible coatings to provide functional groups for bioreceptor immobilization and anticipate possible toxicity issues [89]. Therefore, almost any biomolecule can be attached to these nanocrystals so long as photophysical recombination is not affected. Most colloidal QDs have been studied in terms of cadmium chalcogenides (S, Se, Te) [90–92] which provide a broad emission spectrum with a size-dependent absorption spectrum. This is because of the different bandgaps of semiconductor materials for different nanocrystal sizes (the bigger the particle, the smaller the bandgap), which produces different emission wavelengths from electron–hole recombination [93]. Deng et al. determined that green and orange CdTe QDs may be used as pH-sensitive fluorescent probes, which monitor proton (H^+) flux driven by ATP synthesis to detect viruses using antibody-antigen reactions in tandem [94].

8.4.2.4 Other Metal/ Metal Oxide Nanoparticles

Aside from the above-discussed NPs, many other metal and metal oxide nanoparticles could be used in biosensors [95, 96]. For example, Cheng et al. identified nano-TiO_2-based biosensors capable of detecting lactate dehydrogenase [97]. Similarly, ZnO nanoscale balls, which have a wide direct bandgap, have the ability to serve as excellent materials for enzyme immobilization and for the fabrication of efficient biosensors due to the ability to transfer electrons [95, 96] rapidly. Bioanalytical applications of nanoscale metal oxides include the immobilization of proteins and enzymes. Metal oxide-based semiconducting NWs or NTs play a significant role in electric, optical, electrochemical, and magnetic transducers [98]. The porous structures of metal/metal oxide nanoparticles provide enzymes with a protective microenvironment that preserves their activity and enzymatic stability. As a result, the active surface area of these NPs can be significantly increased for protein binding [99].

Among various metal oxide nanoparticles, TiO_2 has gained a particular interest because of its unique properties, such as high surface area and high catalytic efficiency, which can improve the interaction between biomolecules and electrode surfaces [100]. For example, high-purity TiO_2 nanoparticles were fabricated to study non-enzymatic glucose sensors [101]. The nanostructured ZnO has also received broad attention due to its exciting properties like remarkable photoluminescence, which makes them to be used as transducers for photoluminescence-based biosensors [102]. ZnO is a versatile biomimetic material for loading biomolecules [103]. The high isoelectric point ZnO nanoparticles can be used to adsorb enzymes with low isoelectric points to fabricate biosensors [104]. In addition, modified ZnO electrodes have also been used in sensing hemoglobin [104], microperoxidase [105], and Horseradish peroxidase (HRP) [106]. Nickel oxides are also promising material in biosensing, especially in enzymatic-based biosensors. Owing to its high biocompatibility and large surface area, it can be used to immobilize enzymes like catalase [107] and tyrosinase [108]. Despite being a poor catalyst in bulk, MnO_2 nanomaterials exhibit high catalytic activity. An enzyme field-effect transistor (ENFET) using MnO_2 nanoparticles as catalysts has been reported by immobilizing enzymes and MnO_2 at the gate surface [109]. As an example, Luo et al. developed a glucose-sensitive ENFET using a MnO_2 nanoparticle [110]. In addition to the metal oxides mentioned above, CuO [111], Bi_2O_3 [112], and CeO_2 [113] nanoparticles have been reported to be used in biosensor design. Furthermore, MgO [114], ruthenium oxide [115], and Ni-doped SnO_2 nanoparticles [116] were also modified on electrodes for HRP immobilization and H_2O_2 detection.

8.5 Metal/Metal Oxide Nanoparticles-Based Biosensors used for Infectious Diseases

Infectious diseases are transmissible diseases that are caused by microorganisms. These diseases can be transmitted through microorganisms such as bacteria, viruses, fungi, or parasites [117]. These diseases commonly have similar symptoms, like fever, diarrhea, fatigue, and muscle aches [118]. Whenever prone to a disease, they can also have very specific symptoms. These diseases are also treated differently depending on how severe they are and how much they affect your immune system [119]. A biosensor is a medium that detects environmental components or biocomponents using numerous techniques such as electrochemical optical. To detect infectious diseases, various nanomaterials are used because of their small size, increased surface area, and conductivity. We have discussed the following metal and metal oxides nanoparticle used in infectious diseases.

8.5.1 Gold Nanoparticles (AuNPs)

AuNPs have a pervasive history in chemistry and well-studied nanomaterial for over 150 years [120]. These materials have high chemical stability, biocompatibility, and high electrical conductivity [121]. AuNPs have the potential to be further functionalized and could be used in a different biomedical field [122]. Electrochemical biosensors employ AuNPs to be implemented into the electrochemically active areas of the materials and improve the electron transfer efficiency, which can help to achieve better performance with ultrahigh sensitivity and stability [123]. Furthermore, the presence of excellent electrical conductivity and electrochemical biosensors comprising AuNPs have been developed by many researchers to achieve minute target biomolecule detection with high sensitivity and selectivity. A magneto sandwich assay based on AuNPs for capturing of anti-hepatitis B surface antigen (α-HBsAg) IgG antibodies in humans [124], and chronoamperometric detection approaching the catalytic properties of AuNPs on the hydrogen evolution reaction (HER). The developed approach is an alternative to conventional ELISA & standard MEIA and is simple to use, low cost, and integrated into portable instruments. The reported assay takes the privilege of AuNPs tags as electrocatalytic labels and magnetic beads as a platform of immunoreactions. The detection of α-HBsAg IgG antibodies ranges from 3 to 10 mIU ml^{-1}.

Similarly, A label-free electrochemical immunosensor is constructed to detect Hepatitis C virus (HCV) core antigen with a sandwich immunosensor [125]. AuNPs–ZrO_2–chitosan nanocomposite was prepared by AuNPs directly synthesized on the surface of ZrO_2 NPs using chitosan as a reducing agent in a one-step procedure. The nanocomposite demonstrated excellent electrochemical and biocompatible behavior. Alongside, a nanocomposite is prepared from AuNPs, silica nanoparticles, and chitosan that is conjugated to a secondary antibody. The secondary antibody was immobilized on AuNPs–SiO_2–chitosan nanocomposite that enhances the signal responses of HCV immunosuppressor by accelerating the electron transfer. This resulted in a sandwich-type immunosensor that displays high sensitivity to HCV core antigen ranging from 2 to 512 ng ml^{-1} and a detection limit of 0.17 ng ml^{-1}.

Furthermore, an electrochemically constructed DNA biosensor that contains AuNPs deposited on Au electrode and mercaptobenzaldehyde for enhanced detection of short DNA related to Hepatitis B virus (HBV) [126]. The AuNPs were functionalized onto the Au electrode by the electrodeposition method. Later, the mercaptobenzaldehyde was self-assembled onto the electrode surface by an Au–S bond. The process offers a convenient and straightforward methodology for preparing self-assembled monolayer and covalently immobilized NH_2–HBV-ss-DNA. The oxidation peak current of ferricyanide with HBV-ss-DNA concentration presents a linear relationship ranging from 5.7×10^{-11} to 6.6×10^{-8} $mol\,l^{-1}$ and a detection limit of 7.6×10^{-12} $mol\,l^{-1}$ for target DNA. A successful differentiation between complementary HBV DNA three bases mismatched and noncomplementary ss-DNA displayed good selectivity for biosensors.

A FRET system containing gold nanorod (AuNRs) and fluorescein (FAM) for detection of HBV DNA sequences [127]. AuNRs wrapped with a thin layer of cetyltrimethylammonium bromide (CTAB) resulted in the AuNRs being positively charged. The FAM tag single-stranded DNA (FAM-ss-DNA) adsorbed to the positively charged AuNRs and formed the FAM-ss-DNA-CTAB-AuNRs ternary complex resulting in quenching of fluorescence intensity of FAM due to FRET occurring from FAM to AuNRs. When a complementary target DNA is added to the complex ternary solution, the fluorescence intensity further decreases. The fluorescence intensity declined linearly to the concentration of complementary DNA in the range from 0.045 to 6.0 $nmol\,l^{-1}$ with a detection limit as low as 15 $pmol\,l^{-1}$.

A biosensing technique was developed by a combination of the rolling circle amplification (RCA) method and quartz crystal microbalance (QCM) biosensor [128]. The powerful amplification ability RCA and high sensing ability of the QCM biosensor enable the detection of HBV with low consumption. Additionally, ligation, amplification, and detection occur inside the biosensor, minimizing contamination risk. RCA utilizes simple thermal control, avoids the complicated thermal cycling steps inside the biosensor, and amplifies the DNA to a detectable level. Based on the RCA reaction, it was observed that target DNA enabled 10-fold increase in the frequency shift with concentrations as low as 10^4 copies ml^{-1} of HBV genomic DNA sequence.

A simple, low-cost, ultrasensitive point of care (POCT)-based electrochemical detection technology was developed for detecting RNA of SARS-CoV-2 [129]. The sandwich-type electrochemical biosensor contains a p—sulfocalix [8] arene functionalized with graphene (SCX8-RGO) to enhance toluidine blue (TB) for the detection of the virus. The technology unaccompanied nucleic acid amplification and reverse transcription techniques by using a portable smartphone for the detection of RNA of SARS-CoV-2. The capture probe (CP) tagged with thiols deactivates on the surface of nanocomposite forms CP/Au@Fe_3O_4, which is sandwiched with labeled probe (LPs) namely Au@SCX8-TB-RGO-LP to form a sandwich structure of CP-target-LP. Thus, a biosensor was fabricated with high sensitivity due to the good conductivity of AuNP and RGO materials and TB-based supramolecular recognition of SCX8 on a screen-printed carbon electrode (SPCE) (shown in Figure 8.4). The LOD of the clinical specimen goes as low as 200 copies ml^{-1}.

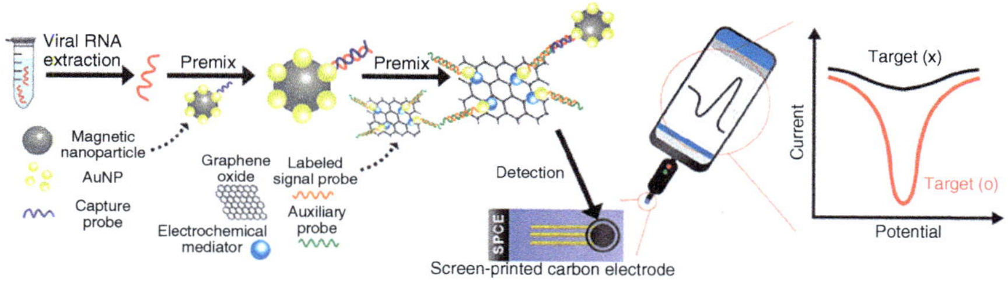

Figure 8.4 The capture probe (CP) captures the extracted viral RNA on the AuNPs and collected using magnetic nanoparticles. The screen-printed carbon electrode (SPCE) composed of graphene oxides, mediator and probes electrochemically detects the viral RNA. Source: Choi et al. [130] / Frontiers Media S.A / CC BY 4.0.

8.5.2 Silver Nanoparticles (AgNPs)

It represents unique optical properties that are influenced by shape, size, composition, uniformity, and deposition [131]. The SPR property of silver originates from the collective oscillation of electrons which can be applied to the photonic and sensory application. Thus, AgNPs can be modified based on the purpose into various forms and compositions [130]. A surface-enhanced Raman scattering (SERS) sensor was developed to detect HBV DNA. The sandwich NPs of AgNRs@MGITC@SiO_2 were coupled with Au nanoarray via linkage DNA target, and thus SERS signal is detected [132]. The coupling creates an intense, spatially distributed electromagnetic field, and thus the sensor can detect the HBV DNA with a LOD as low as 50 aM, and the linear range for biosensor is 0.5 fM–2 nM. The sensor detected and differentiated a single-base mismatch mutant of HBV DNA. The SERS intensities, when compared with the nanoarray and the film chips, conclude that the spatially increased plasmonic field coupled to a quasi-periodic array improves the SERS sensor's detection limit. The sensor further can be extended analytes such as small molecules, proteins, and other viruses. Similarly, SERS-based biosensor for direct immunoassay platform is demonstrated to detect influenza A virus [133]. PEGylated TBBT-labelled AuNPs constructed the biosensor covered with influenza an antibody as SERS probe and hydrophilic Au@Ag two-dimensional array as SERS substrate. The SERS signal was enhanced by four times taken immunoassay on nanoarray as SERS substrate rather than on Au film. The linear relation between the concentration of sample and SERS signal was in the range of 5–56 $TCID_{50}$ per ml, and the lowest detection limit of 6 $TCID_{50}$ per ml.

A simple, facile, versatile, low-cost, and label-free fluorescent molecular beacon (MB) was developed to analyze disease-related genes. The MB terminal part has a guanine-rich sequence (GRS) that enhances the fluorescence of AgNCs as the hairpin-shaped structure binds with the nanocluster [130]. When the target DNA

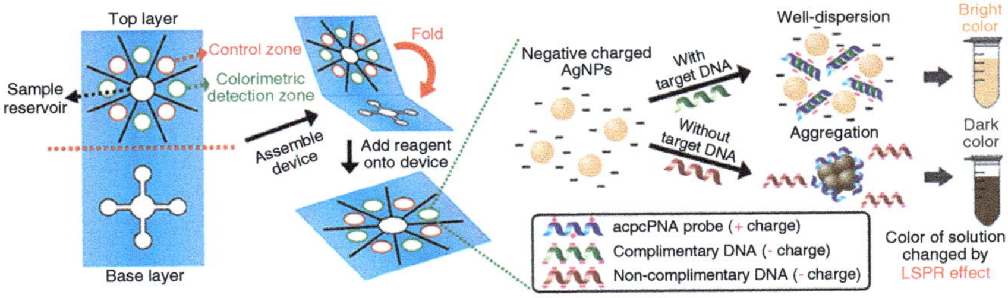

Figure 8.5 The colorimetric biosensor detects the influenza virus. The hybridization of probe and target DNA of influenza virus, causes the negatively charged AgNPs to be well dispersed, with a bright color. In absence of target DNA, the AgNPs gets aggregated with the probe DNA and cause the solution to darken. Source: Choi et al. [130] / Frontiers Media S.A / CC BY 4.0.

species is added, the hairpin-shaped loop opens and keeps the AgNCs away from the GRS, which decreases fluorescence. The developed biosensor can detect three disease-related DNA such as Human immunodeficiency virus (HIV), HBV, and human T-lymphotropic virus type-1 (HTLV-1) DNA with a LOD of 4.4, 6.8, and 8.5 nM respectively.

A colorimetric assay based on pyrrolidinyl peptide nucleic acid (acpcPNA) that induces aggregation of NPs for detection of DNA The colorimetric assay contains paper-based analytical devices (PAD) for the point-of-use technology developed for the simultaneous detection of bacterial and viral infectious diseases such as MERS-CoV, MTB, HPV [134]. The positively charged acpcPNA probe interacts with the negatively charged AgNPs leading to NP aggregation and a significant color change is observed (shown in Figure). The biosensor exhibits high selectivity against non-complementary target DNA, single-base mismatches, two base mismatches. The LOD for MERS-CoV, MTB, and HPV was found to be 1.53, 1.27, and 1.03 nM respectively (shown in Figure 8.5).

8.5.3 Platinum Nanoparticles (PtNPs)

Platinum is found to be an excellent catalyst due to its exceptional catalytic property for hydrogen redox reactions and high stability in acid electrolytes. Besides its rarity in nature and high cost is used as a small particle for both chemical and biological applications. Pt NPs, which are found to be in the range of 2–5 nm, possess excellent catalytic activity [135] that supports both electrochemical biosensors [136] and optical biosensors [137]. A nanohybrid of platinum nanoparticles–porous ZnO spheres-hemin (Pt–pZnO–hemin) was synthesized by combining hydrothermal method and ester bridging between porous ZnO and hemin to be used as alkaline phosphatase (ALP)-based immnunosensor for detection of influenza [138]. The electrochemically active 1-naphthol was generated by enzymatic hydrolysis of inactive 1-napthyl phosphate by ALP, then the nanohybrid catalytically oxidized 1-naphthol in the

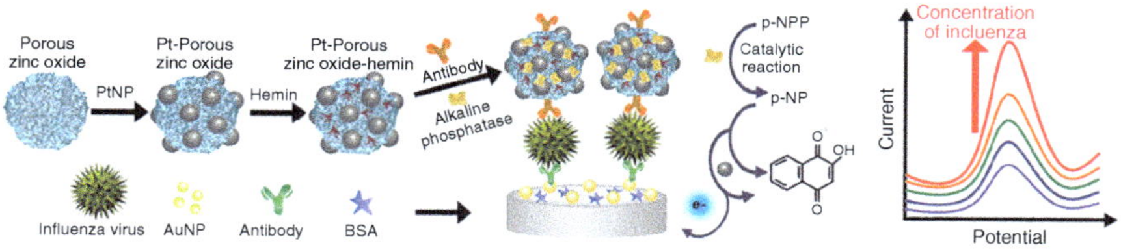

Figure 8.6 The sandwich-structured biosensor composed of Pt–porous zinc oxide–hemin and AuNPs captures the influenza virus between antibodies. The catalytic reaction of 1-napthyl phosphate generates 1-naphthol inside the Pt–pZnO–hemin structure, that is later catalytically oxidized to electrochemically detect the influenza virus. Source: Choi et al. [130] / Frontiers Media S.A / CC BY 4.0.

presence of H_2O_2 that resulted in an amplified electrochemical signal. The nanohybrid functionalized with an anti-influenza antibody becomes immobilized on the surface of the electrode, which displays an electrochemical signal when there is the presence of influenza (Figure 8.6). The virus was detected by the nanohybrid composite from a range of 0.001–60 ng ml^{-1} and a detection limit of 0.76 pg ml^{-1}.

A platinum nanoparticle-latex nanocomposite (Pt-P2VPs) was developed and used as a probe for the immunochromatographic test (ICT) strip [139]. The hybrid nanocomposite material was conjugated with monoclonal antibody (mAb) to detect influenza A (H1N1) antigen. The latex particle consists of poly(2-vinylpyridine) crosslinked with divinylbenzene (P2VPs) decorated with Pt NPs through emulsion polymerization and direct metallization. The hybrid nanocomposite particles are intensely colored even at low analyte concentrations. The lowest detectable concentration of influenza A antigen was found to be 2.5×10^{-2} HAU ml^{-1}. To demonstrate the performance of detecting H1N1 antigen by using Pt-P2VPs as a probe for ICT strips, the detection sensitivity was 16 times more when compared to conventional colloidal Au NPs. Further, the nanocomposite probe was functionalized with biotinylated protein for different biomedical applications.

8.5.4 Copper Nanoparticles (CuNPs)

Copper nanoparticles have opted for their potential to replace other expensive NPs. The small size and high surface-to-volume ratio allow the CuNPs to be able to interact closely with the virus antigen and detect them easily [140]. Alongside, nanotechnology provides new opportunities to investigate the role of CuNPs in viral detection. A significant disadvantage of using this particle is the ability to get oxidized at ambient conditions. Nonetheless, based on the well-established properties, the copper nanoparticle is used in various biosensors for pathogen detection [141]. Moreover, the chemical synthesis of CuNPs from various salts is cost-effective and straightforward. An ultrasensitive electrochemical biosensor based on a sandwich structure of copper nanoparticles and glucose oxidase for detecting single-stranded DNA (ss-DNA) of influenza virus [142]. Cupric hexacyanoferrate (Cu-HCF) NPs utilized

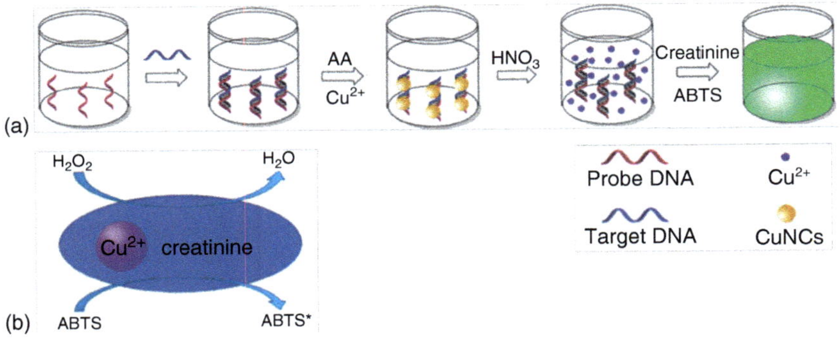

Figure 8.7 (a) The colorimetric detection of DNA based on DNA-templated copper nanoclusters. The target DNA hybridizes the capture probe to form double-stranded DNA (dsDNA). On reduction of $CuSO_4$ by ascorbic acid (AA) forms the DNA-templated CuNCs. And the addition of HNO_3, CuNCs oxidize to copper ions. After the addition of creatinine obtains a copper–creatinine complex which rapidly converts hydrogen peroxides (H_2O_2) and 2,2-azino-bis diammonium salts (ABTS) to oxidized ABTS which is green in color. (b) The principle of DNA detection using a peroxidase-like copper–creatinine complex. Source: Reproduced from Mao et al. [143] / with permission of Elsevier.

on the electrode surface by introducing an avidin–glucose oxidase (GOx-A) into the system to amplify the DNA hybridization. The performance of the biosensor was attributed to the excellent catalytic activity of GOx and the accumulation process of Cu-HCF NPs. The dynamic detection range was found between 1.0 fM and 10 pM with a detection limit of 1.0 fM. A colorimetric detection method was developed based on DNA-templated copper nanocluster (CuNCs) for detecting HBV [143]. The detection technique attributes detection of virus with the naked eye. Low-cost and unique physiochemical properties can be applied for biochemical analysis. This method detects three base pair mismatches in the target DNA, thus displays a potential application for distinguishing important genetic basis of diseases. The LOD observed here is 12×10^9 molecules. In conclusion, this method analyses the DNA with early diagnosis without any complicated pieces of equipment (Figure 8.7).

8.5.5 Zinc Oxide Nanoparticles (ZnONPs)

ZnO is an n-type semiconductor with a wide bandgap semiconductor under UV radiation. The presence of particular piezoelectric properties makes it a valuable senor-like property. It shows better binding ability with biological entities, making it a prerequisite for future biosensor application in medicine. It shows non-toxicity and compatibility with human skin. The piezotronics-assisted method for the label-free detection of DNA presented by the Schottky-contacted ZnO nanowire [144]. The ZnO nanowire DNA sensor measured the output current under different external stains and target cDNA concentration. In the presence of the piezotronics effect, the device's performance is enhanced significantly. Similarly, the result demonstrated improved output current and difference between two target cDNA

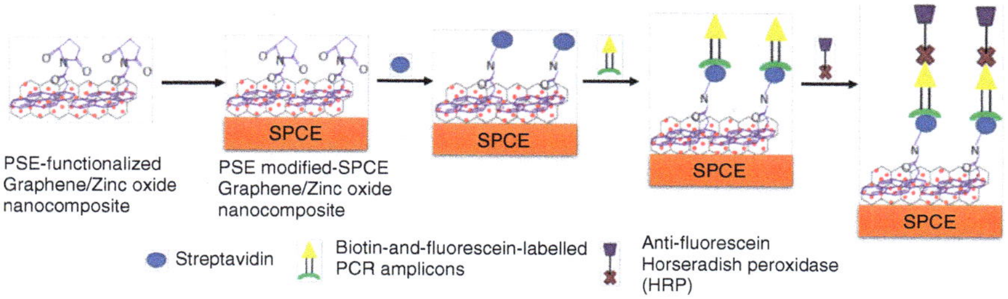

Figure 8.8 The development of Graphene/ZnO/PSE-modified nanocomposite for the enhanced electrochemical DNA biosensor on screen-printed carbon electrode (SPCE) for the detection of avian influenza virus (H5N1). Source: Reproduced from Low et al. [145] / with permission of Elsevier.

concentrations. The device detected the HIV one gene (HIV1). The concentration of target cDNA ranges from 0 to 10^{-10} M.

The Graphene/ZnO nanocomposite was synthesized using a low-temperature solvothermal method for sensing application [145]. The nanocomposite was further functionalized with pyrene succinimide ester (PSE), an organic bi-linker consisting of pyrene moiety and succinimidyl fragment. The bi-linker forms non-covalent bonds with the graphene while the ester group binds with the amine group of the biological compound. The modified G/ZnO nanocomposite DNA biosensor for detecting avian influenza virus (H5N1). The detection limit was found to be 7.4357 μm, with a range of 1–15 mM of H_2O_2 (Figure 8.8).

8.5.6 Miscellaneous Metal Oxide Nanoparticles

A fluorescent sensor based on virus-imprinted polymers (VIP) was made for highly selective detection of the Japanese encephalitis virus (JEV) [146]. The virus-imprinted method is imprinted through surface imprinting techniques with tetraethyl orthosilicate (TEOS) functioning as a building block and magnetic silicon microsphere ($Fe_3O_4@SiO_2$) acting as carrier materials. The VIP film selectively captures target JEV via recognition cavities and a high imprinting factor. The virus-imprinted magnetic microsphere enabled easy fluorescent detection, and the magnetic property of the polymer allowed magnetic separation. The fast magnetic separation, sensitive fluorescent detection, and specificity study demonstrated excellent selectivity for JEV.

The magnetic manganese ferrite nanoparticles ($MnFe_2O_4$) were synthesized and used in a microfluidic system that involves characterized micro-devices for the detection of influenza virus using fluorescence immunoassay (FIA) [147]. The produced magnetic $MnFe_2O_4$ NP are combined in a layer-by-layer (LBL) surface modification process to reduce background noise due to nonspecific adhesion of NPs. Further, the LOD on-chip diagnosis was found to be 0.007 HAU. The microfluidic system performs the immunoassay to detect the influenza infections within 20 minutes. Table 8.2 mentions the metal/metal oxides nanoparticles used as biosensors for infectious diseases.

Table 8.2. Metal/metal oxides nanoparticles biosensors for detecting infectious diseases.

Metal/metal oxides nps	Physiological properties	Detection techniques	Target (pathogen)/ analytes	Range of detection	Limit of detection	References
			Au nanoparticles			
AuNPs/ nCov-19Ab	21–30 nm	Electrochemical (DPV)	SARS-Cov-2; Protein	1 fM – 1 μm	90 fM (eCovsans); 120 fM (spiked saliva)	[148]
Au@RGOSCX8	Au precursor 338 nm	Electrochemical (DPV)	SARS-Cov-2; Nucleic acid	10^{-17} to 10^{-12} M (DPV)	200 copies/ml	[129]
AuNPs-DEP-MA	50 nm (Au NPs); spherical	Electrochemical (SWV)	MERS CoV; Protein	$1\,pg\,ml^{-1}$ to $10\,\mu g\,ml^{-1}$	$1.04\,pg\,ml^{-1}$	[149]
AuNP-PP-CdZnSeS/ZnSeS	57 nm; Spherical	Fluorescent	Influenza virus	10^{-14} to $10^{-9}\,g\,ml^{-1}$	$17.02\,fg\,ml^{-1}$	[150]
AuNPs-MSA	20 nm	Electrochemical (SPCE, 3-electrode system)	Hepatitis B virus (HBV); Protein	3 to $10\,mIU\,ml^{-1}$	$3\,mIU\,ml^{-1}$	[124]
AuNPs-ZrO_2-Cs	20 nm (AuNPs)	Electrochemical (3-electrode system with modified GCE)	Hepatitis C virus (HCV)	2 to $512\,ng\,ml^{-1}$	$0.17\,ng\,ml^{-1}$	[125]
AuNRs- CTAB-FAM-ssDNA	Gold nanorods	Fluorescence (FRET)	Hepatitis B virus (HBV); DNA	0.045 to $6.0\,nmol\,l^{-1}$	$15\,pmol\,l^{-1}$	[127]
Au-GNM-MB-NH_2-ssDNA	90 nm (Au NPs); spherical	Electrochemical (DPV)	Hepatitis B virus (HBV); DNA	5.7×10^{-11} to $6.6\times10^{-8}\,mol\,l^{-1}$	$7.6\times10^{-12}\,mol\,l^{-1}$	[126]
MB-AuNPs	AuNPs	Electrochemical (SWV)	Dengue virus (DENV)	10 nM to 1 pM	100 fM	[151]
AuNS@GO	Nanostars shaped Au	Electrochemical	SARS-CoV-2		$1.68\times10^{-22}\,\mu g\,ml^{-1}$	[152]

			Ag nanoparticles			
AgNrs-MGITC-SiO_2@AuNA	400 nm (length), 60 nm (diameter); nanorice	SERS	Hepatitis B virus (HBV); DNA	0.5 fM to 2 nM	50 aM	[132]
AgNCs		Fluorescence	HIV, HBV, HTLV-1; DNA	10 to 200 nM	4.4 nM, 6.8 nM, 8.5 nM	[153]
Ag@Au 2D aray	5 nm thick shell (Ag)	SERS (immunoassay)	Influenza A virus; Nucleoprotein	5 to 56 $TCID_{50}$/ml	6 $TCID_{50}$ per ml	[133]
AgNPs-GP-PAb		Electrochemical (LSV)	Avian Influenza Virus (AIV H7); Protein	1.6×10^{-3} to 16 $ng\,ml^{-1}$	1.6 $pg\,ml^{-1}$	[154]
AgNPs	Citrate stabilisedAgNPs	Colorimetric	MERS-CoV, MTB, HPV		1.53 nM (MERS-CoV), 1.27 nM (MTB), and 1.03 nM (HPV)	[134]
			Pt nanoparticles			
Pt-pZnO-hemin-Ab_2		Electrochemical (DPV, CV)	Influenza	0.001 to 60 $ng\,ml^{-1}$	0.76 $pg\,ml^{-1}$	[138]
Pt-P2VPs	70 nm (PtNPs), 368 nm (Pt-P2VPs, spherical)	Colorimetric (ICT strips)	Influenza (H1N1)	0 to 27.7 mm	2.5×10^{-2} HAU ml^{-1}	[139]
			Cu nanoparticles			
CuHCFNPs/GOx-A	50 nm (CuHCF)NPs	Electrochemical (DPV)	Influenza A virus; RNA	1.0 fM to 10 pM	103 copies/ml	[142]
CuNCs-dsDNA	6 nm (CuNCs)	Colorimetric	Hepatitis B Virus (HBV)	0 to 12 10^{13} molecule	12×10^{9}moleucle	[143]

(Continued)

Table 8.2. (Continued)

Metal/metal oxides nps	Physiological properties	Detection techniques	Target (pathogen)/ analytes	Range of detection	Limit of detection	References
			ZnO nanoparticles			
ZnO NW/FET/ DNA	130 μm (ZnO nanowire)	Electrochemical (Strain)	Human immunodeficiency virus 1 (HIV1)	0 to 10-10 M		[144]
ZnO/GP/PSE	200 nm; Spherical (ZnO)	Electrochemical (CV)	Avian Influenza H5 (H5N1)	1 to 15 mM	7.4537 μM	[145]
			Miscellaneous metal oxide nanoparticles			
Fe_3O_4@SiO_2	Microspheres of 300 nm	Fluorescence	Japanese encephalitis virus (JEV)	2.5–45 nM	0.32 nM	[146]
$MnFe_2O_4$	Magnetic nanoparticles with a diameter of 100 nm	Fluorescence Immunoassay (FIA)	Influenza infection		0.007 HAU	[147]

Ab: Antibody; DPV: differential pulse voltammetry; SARS-CoV-2: Severe Acute Respiratory Syndrome Coronavirus 2; SWV: Square Wave Voltammetry; MERS-CoV: Middle East respiratory syndrome coronavirus; SPCE: Screen-printed carbon electrode; GCE: Glassy carbon electrode; FRET: Fluorescence resonance energy transfer; SERS: Surface-enhanced Raman spectroscopy; HIV: Human immunodeficiency virus; HBV: Hepatitis B virus; HTLV-1: Human T-lymphotropic virus type 1; MTB: *Mycobacterium tuberculosis*; HPV: Human Papillomavirus; LSV: linear sweep voltammetry.

8.6 Comparative Studies of Biosensors for Infectious Diseases: Advantages & Limitations

Traditionally microscopy, culture, and serology have been usually used as methods for pathogen detection and usually involve expensive laboratory equipment operated by highly trained individuals. While these methods have been successful at diagnosis, their efficacy in preventing the spread of infectious diseases is not very good. In regions where such high-end laboratories are not available, the whole purpose of testing is defeated as by the time the result of the test is discovered; the disease may have spread from infected in their own local population. To prevent the spread of highly contagious infectious diseases, isolation and treatment of infected individuals are necessary. To achieve this, reliable biosensors can be used as POC devices with minimal instrumentation required.

As per World Health Organization standards, an ideal POC diagnostic device must-have attributes including affordability, sensitivity, specificity, user-friendliness, rapid/robust, equipment-free, and deliverability (ASSURED) [155]. Any biosensor consists of the following parts: a bioreceptor, transducer, and a final signal processing unit. The signal produced by the transducer determines the category of the different biosensors. The advantage of biosensor-based detection methods over traditional methods involving biomolecules is lower response time, better tuning opportunities with good selectivity, and stability for pathogen recognition [156]. The different biosensors and their advantages and limitations have been discussed briefly in the section.

8.6.1 Electrochemical Biosensors

Electrochemical biosensor generally utilizes semiconducting or conducting electrodes to detect a biochemical event in a sample. These can be used to detect nucleic acids, proteins, microorganisms, or other chemical specimens. In detecting diseases, biochemical events (for example, antigen–antibody interactions or induced events by using enzymes) can cause measurable alterations in the dielectric properties, conductivity, and potential in the tested sample. These signals form the basis for sensing changes in the electrochemical properties of the sample. The electrodes being used significantly impacts the effectiveness of the sensor, as signals being picked up are mainly due to reactions taking place near the electrode. The various electrodes that have been reportedly used include silicon, gold, platinum, graphene, oxides of alumina, and zirconia [157]. The electrochemical sensor can be further classified based on the signal being sensed, which include potentiometric (change in potential), amperometric (induced current), cyclic voltammetry, and impedimetric transducers (measuring both impedance and resistance) and conductometric (measuring conductance) [158].

Electrochemical sensors have significant advantages such as better sensitivity even at very low concentrations, fast response, reduced costs with lesser instrumentation requirements, and better portability that can enable the development of POC diagnostic devices. Besides, the testing is unaffected by sample characteristics such as turbidity or the presence of fluorescent compounds, as in the case of optical biosensors enabling better sensitivity. Though this diagnostic modality has numerous advantages, certain limitations and challenges impede its wide-scale adoption and commercialization. The factors plaguing further wide-scale adoption of devices based on this modality include inconsistent sensitivity when operating at different pH, ionic concentrations, and temperatures. The lack of cost-effective manufacturing to achieve required miniaturization and sensitivity is another area of concern as it impacts the cost of these POC devices [159].

8.6.2 Fluorescence-Based Biosensors

Fluorescent probes have been extensively used in different fields of biomedical applications. It is based on the emittance of photons on interaction with its target bioactive molecule accompanied by complex formation. The signal generated may be by different modes, including quenching, resonance-enhanced transfer, photoinduced electron transfer, and metal-enhanced SPR. Fluorescence-based biosensing produces better signal strength, low-detection limit rapid response, and simple instrumentations. It is pretty economical besides having a long history of its application in biosensing applications [160]. QDs, graphene nanoparticles, up-conversion NPs, and metallic NPs have been investigated for fluorescence-based biosensing [161].

The most challenging aspect of fluorescence-based biosensors is the development of stable, biochemically active fluorescent probes [162]. Polymer and enzyme-based fluorescence biosensors have also been investigated, and the results show that these can overcome some of the limitations of conventional fluorophores. Conventional chemical fluorophores have limitations, such as low solubility and unfavorable affinity to substrates. To overcome this limitation, protein, and enzyme-based fluorophores have also been developed, which have a better range of selectivity affinity and higher signal-to-noise ratio [163]. Besides, genetic engineering can modify protein-based fluorophores to obtain the required property of the fluorescent protein. The problems associated with using protein-based biosensors lie in the instability of protein and low quantum yield; this poses a challenge to the miniaturization of sensors as a POC device. Fluorescence-based biosensors suffer from limitations which include nonspecific binding and higher costs associated with the development of reagents [163].

8.6.3 Colorimetric Biosensors

We have seen that fluorescence biosensors and electrochemical methods have an advantage in terms of suitable sensitivity and rapid response time. However, the requirement of some instrumentation and miniaturization of such instruments for signal detection is one of the significant challenges in their development as POC diagnostic devices. This problem can be overcome by using colorimetric biosensors, which has output signal in the form of visible color change. The signals generated by colorimetric sensors require no instrumentation and can be detected by the

Table 8.3 Comparative features electrochemical and fluorescence biosensors and colorimetric biosensor.

Attribute	Electrochemical sensor	Fluorescence sensor	Colorimetric sensor
Affordability	Average	Good	Good
Instrumentation requirement	Yes (low powered)	Limited	None
Sensitivity	Varies (generally better than optical methods)	Good	Low
Selectivity	Good	Satisfactory (low in some cases)	Good
Resource requirement and ease of use	Good	Good	Best

naked eye [164]. This gives a significant advantage to the development of colorimetric systems for detecting infectious diseases as low skilled professionals can easily use it to detect infectious disease specifically in the low-middle income countries, which account for a significant number of deaths caused by infectious diseases (www.who.int/news-room/fact-sheets/detail/measles).

This diagnostic method provides faster detection times besides providing the utility of qualitative and quantitative analysis of the analyte based on the color and its gradient. Different methods for colorimetric sensors have been developed, including coupled with loop-mediated isothermal amplification method (LAMP), QDs-based, plasmonic nanoparticle-based sensors, enzyme substrate associated with paramagnetic particle and polymer-based colorimetric methods. Probes based on gold nanoparticles have been reported successful in lab tests [8]. The LAMP method, a relatively new method, has shown promising results in detecting viruses and can serve as a replacement to the conventional PCR method. The limitation of this method is its low sensitivity. This has been hindering the adoption of this method in clinics. Besides, there is also a problem with detecting infections caused by fast-mutating microorganisms, which is also a concern in fluorescence-based biosensors. The comparative features of the modalities such as colorimetric, electrochemical, and fluorescence biosensors are summarized in Table 8.3.

8.7 Conclusion and Future Prospects

The need for rapid, economic, and accurate diagnosis is undeniably crucial for public health management and effective government policies. The MMONs-based electrochemical, fluorescence, or colorimetric biosensors can provide the required turnaround time, sensitivity, and miniaturization as POC devices. The various types of biosensors using the above modalities can be further tuned to have sensitivities of the required range tailored to provide the best results in various usage conditions. The requirement of such devices is most extensively felt in low-middle-income countries where not enough well-equipped laboratory facilities are present. The biosensors must be developed keeping in mind the assured parameters of the WHO. This could significantly change the way infectious diseases are currently being screened and treated. The commercialization of biosensors has found limited success due to technical issues such as sensitivity, reproducibility, nonspecific binding to substrates, and lack of cost-effective miniaturized manufacturing methods.

As the world economy has faced severe setbacks and the extent of casualties caused by the recent COVID-19 pandemic, it is imperative to acknowledge our limitations and flaws in addressing and containing contagious infectious diseases. We can expect such outbreaks to happen in the future as well; therefore, an effective health policy for effective management of such outbreaks is necessary. The first action to controlling such an outbreak is screening and isolating the initially identified cases. The screening methods that we have been currently utilizing fit the criteria of good selectivity and sensitivity, but controlling the spread of contagious diseases requires economical POC devices that enable rapid on-site screening for isolation of the patient and prevent community transmission. The MMONs-based POC devices could provide the breakthrough for economical and tunable sensing devices with required sensitivity. However, MMONs based biosensors can be further tuned/perfected on the basis of requirements and thus may form the mainstay in effective management of infectious diseases.

Acknowledgment

Dipak Maity would like to thank the University of Petroleum and Energy Studies (UPES) for getting all the support. Gajiram Murmu acknowledges the Council of Scientific & Industrial Research (CSIR) and Satya Ranjan Sahoo would like to thank the University Grants Commission of India (UGC) for providing a Junior Research Fellowship. Sumit Saha wishes to thank Prof. Suddhasatwa Basu, Director, CSIR-Institute of Minerals & Materials Technology, Bhubaneswar, India, for in-house financial support (Grant number: CSIR-IMMT-OLP-112) and requisite permissions.

References

1 Solanki, P.R., Kaushik, A., Agrawal, V.V., and Malhotra, B.D. (2011). Nanostructured metal oxide-based biosensors. *NPG Asia Materials* 3 (1): 17–24.
2 Pandey, A., Gurbuz, Y., Ozguz, V. et al. (2017). Graphene-interfaced electrical biosensor for label-free and sensitive detection of foodborne pathogenic *E. coli* O157:H7. *Biosensors and Bioelectronics* 91: 225–231.
3 Cesewski, E. and Johnson, B.N. (2020). Electrochemical biosensors for pathogen detection. *Biosensors and Bioelectronics* 159: 112214.

4 Guliy, O.I., Zaitsev, B.D., Larionova, O.S., and Borodina, I.A. (2019). Virus detection methods and biosensor technologies. *Biophysics* 64 (6): 890–897.

5 Cheng, M.S., Ho, J.S., Tan, C.H. et al. (2012). Development of an electrochemical membrane-based nanobiosensor for ultrasensitive detection of dengue virus. *Analytica Chimica Acta* 725: 74–80.

6 Hematian, A., Sadeghifard, N., Mohebi, R. et al. (2016). Traditional and modern cell culture in virus diagnosis. *Osong Public Health and Research Perspectives* 7 (2): 77–82.

7 Rai, M., Gade, A., Gaikwad, S. et al. (2012). Biomedical applications of nanobiosensors: the state-of-the-art. *Journal of the Brazilian Chemical Society* 23 (1): 14–24.

8 Zhao, V.X.T., Wong, T.I., Zheng, X.T. et al. (2020). Colorimetric biosensors for point-of-care virus detections. *Materials Science for Energy Technologies* 3: 237–249.

9 Sheikhzadeh, E., Beni, V., and Zourob, M. (2021). Nanomaterial application in bio/sensors for the detection of infectious diseases. *Talanta* 230: 122026.

10 Housaindokht, M.R., Ashraf, N., and Sheikhzadeh, E. (2014). Design of a disposable solid state potentiometric sensor for codeine content control in pharmaceutical preparations. *Sensor Letters* 12 (9): 1341–1346.

11 Sheikhzadeh, E., Chamsaz, M., Turner, A.P.F. et al. (2016). Label-free impedimetric biosensor for *Salmonella Typhimurium* detection based on poly [pyrrole-*co*-3-carboxyl-pyrrole] copolymer supported aptamer. *Biosensors and Bioelectronics* 80: 194–200.

12 Singh, A., Choudhary, M., Singh, M.P. et al. (2015). DNA functionalized direct electro-deposited gold nanoaggregates for efficient detection of *Salmonella typhi*. *Bioelectrochemistry* 105: 7–15.

13 Xiang, C., Li, R., Adhikari, B. et al. (2015). Sensitive electrochemical detection of *Salmonella* with chitosan–gold nanoparticles composite film. *Talanta* 140: 122–127.

14 Güner, A., Çevik, E., Şenel, M., and Alpsoy, L. (2017). An electrochemical immunosensor for sensitive detection of *Escherichia coli* O157:H7 by using chitosan, MWCNT, polypyrrole with gold nanoparticles hybrid sensing platform. *Food Chemistry* 229: 358–365.

15 Lu, D., Pang, G., and Xie, J. (2017). A new phosphothreonine lyase electrochemical immunosensor for detecting *Salmonella* based on horseradish peroxidase/GNPs-thionine/chitosan. *Biomedical Microdevices* 19 (1): 12.

16 Zhu, F., Zhao, G., and Dou, W. (2018). A non-enzymatic electrochemical immunoassay for quantitative detection of *Escherichia coli* O157:H7 using Au@Pt and graphene. *Analytical Biochemistry* 559: 34–43.

17 Bonnet, R., Farre, C., Valera, L. et al. (2018). Highly labeled methylene blue-ds DNA silica nanoparticles for signal enhancement of immunoassays: application to the sensitive detection of bacteria in human platelet concentrates. *The Analyst* 143 (10): 2293–2303.

18 Wu, Y., Xue, P., Hui, K.M., and Kang, Y. (2014). A paper-based microfluidic electrochemical immunodevice integrated with amplification-by-polymerization

for the ultrasensitive multiplexed detection of cancer biomarkers. *Biosensors and Bioelectronics* 52: 180–187.

19 Fei, J., Dou, W., and Zhao, G. (2015). A sandwich electrochemical immunoassay for *Salmonella pullorum* and *Salmonella gallinarum* based on a AuNPs/SiO_2/Fe_3O_4 adsorbing antibody and 4 channel screen printed carbon electrode electrodeposited gold nanoparticles. *RSC Advances* 5 (91): 74548–74556.

20 Zhu, F., Zhao, G., and Dou, W. (2018). Electrochemical sandwich immunoassay for *Escherichia coli* O157:H7 based on the use of magnetic nanoparticles and graphene functionalized with electrocatalytically active Au@Pt core/shell nanoparticles. *Microchimica Acta* 185 (10): 455.

21 Hu, X., Dou, W., and Zhao, G. (2015). Electrochemical immunosensor for *Enterobacter sakazakii* detection based on electrochemically reduced graphene oxide–gold nanoparticle/ionic liquid modified electrode. *Journal of Electroanalytical Chemistry* 756: 43–48.

22 Luo, Y., Dou, W., and Zhao, G. (2017). Rapid electrochemical quantification of *Salmonella Pullorum* and *Salmonella Gallinarum* based on glucose oxidase and antibody-modified silica nanoparticles. *Analytical and Bioanalytical Chemistry* 409 (17): 4139–4147.

23 Xiao, T., Huang, J., Wang, D. et al. (2020). Au and Au-based nanomaterials: synthesis and recent progress in electrochemical sensor applications. *Talanta* 206: 120210.

24 Zhang, X., Jiang, Y., Huang, C. et al. (2017). Functionalized nanocomposites with the optimal graphene oxide/Au ratio for amplified immunoassay of *E. coli* to estimate quality deterioration in dairy product. *Biosensors and Bioelectronics* 89: 913–918.

25 Huang, H., Bai, W., Dong, C. et al. (2015). An ultrasensitive electrochemical DNA biosensor based on graphene/Au nanorod/polythionine for human papillomavirus DNA detection. *Biosensors and Bioelectronics* 68: 442–446.

26 Mirkin, C.A., Letsinger, R.L., Mucic, R.C., and Storhoff, J.J. (1996). A DNA-based method for rationally assembling nanoparticles into macroscopic materials. *Nature* 382 (6592): 607–609.

27 Halas, N.J., Lal, S., Chang, W.-S. et al. (2011). Plasmons in strongly coupled metallic nanostructures. *Chemical Reviews* 111 (6): 3913–3961.

28 Huang, X., Liu, Y., Yung, B. et al. (2017). Nanotechnology-enhanced no-wash biosensors for in vitro diagnostics of cancer. *ACS Nano* 11 (6): 5238–5292.

29 Ray, P.C., Khan, S.A., Singh, A.K. et al. (2012). Nanomaterials for targeted detection and photothermal killing of bacteria. *Chemical Society Reviews* 41 (8): 3193–3209.

30 Alhogail, S., Suaifan, G.A.R.Y., and Zourob, M. (2016). Rapid colorimetric sensing platform for the detection of *Listeria monocytogenes* foodborne pathogen. *Biosensors and Bioelectronics* 86: 1061–1066.

31 Alhogail, S., Suaifan, G.A.R.Y., Bizzarro, S. et al. (2018). On site visual detection of *Porphyromonas gingivalis* related periodontitis by using a magnetic-nanobead based assay for gingipains protease biomarkers. *Microchimica Acta* 185 (2): 149.

32 Suaifan, G.A.R.Y., Alhogail, S., and Zourob, M. (2017). Paper-based magnetic nanoparticle-peptide probe for rapid and quantitative colorimetric detection of *Escherichia coli* O157:H7. *Biosensors and Bioelectronics* 92: 702–708.

33 Singh, P., Gupta, R., Choudhary, M. et al. (2018). Drug and nanoparticle mediated rapid naked eye water test for pathogens detection. *Sensors and Actuators B: Chemical* 262: 603–610.

34 Ahmed, S.R., Kim, J., Suzuki, T. et al. (2016). Enhanced catalytic activity of gold nanoparticle-carbon nanotube hybrids for influenza virus detection. *Biosensors and Bioelectronics* 85: 503–508.

35 Zhang, C., Yin, A.-X., Jiang, R. et al. (2013). Time–Temperature indicator for perishable products based on kinetically programmable Ag overgrowth on Au nanorods. *ACS Nano* 7 (5): 4561–4568.

36 Ornatska, M., Sharpe, E., Andreescu, D., and Andreescu, S. (2011). Paper bioassay based on ceria nanoparticles as colorimetric probes. *Analytical Chemistry* 83 (11): 4273–4280.

37 Bhardwaj, N., Bhardwaj, S.K., Nayak, M.K. et al. (2017). Fluorescent nanobiosensors for the targeted detection of foodborne bacteria. *TrAC Trends in Analytical Chemistry* 97: 120–135.

38 Panda, B.R., Singh, A.K., Ramesh, A., and Chattopadhyay, A. (2008). Rapid estimation of bacteria by a fluorescent gold nanoparticle—polythiophene composite. *Langmuir* 24 (20): 11995–12000.

39 Xue, L., Zheng, L., Zhang, H. et al. (2018). An ultrasensitive fluorescent biosensor using high gradient magnetic separation and quantum dots for fast detection of foodborne pathogenic bacteria. *Sensors and Actuators B: Chemical* 265: 318–325.

40 Huang, C.-C., Chen, C.-T., Shiang, Y.-C. et al. (2009). Synthesis of fluorescent carbohydrate-protected Au nanodots for detection of concanavalin a and *Escherichia coli*. *Analytical Chemistry* 81 (3): 875–882.

41 Li, X., Kong, H., Mout, R. et al. (2014). Rapid identification of bacterial biofilms and biofilm wound models using a multichannel nanosensor. *ACS Nano* 8 (12): 12014–12019.

42 Kraef, C., Bentzon, A., Panteleev, A. et al. (2021). Delayed diagnosis of tuberculosis in persons living with HIV in Eastern Europe: associated factors and effect on mortality—a multicentre prospective cohort study. *BMC Infectious Diseases* 21 (1): 1038.

43 Tallury, P., Malhotra, A., Byrne, L.M., and Santra, S. (2010). Nanobioimaging and sensing of infectious diseases. *Advanced Drug Delivery Reviews* 62 (4): 424–437.

44 Shawky, S.M., Bald, D., and Azzazy, H.M.E. (2010). Direct detection of unamplified hepatitis C virus RNA using unmodified gold nanoparticles. *Clinical Biochemistry* 43 (13): 1163–1168.

45 Ahmad, M., Pan, C., Gan, L. et al. (2010). Highly sensitive amperometric cholesterol biosensor based on Pt-incorporated fullerene-like ZnO nanospheres. *The Journal of Physical Chemistry C* 114 (1): 243–250.

46 Masoomi-Godarzi, S., Khodadadi, A.A., Vesali-Naseh, M., and Mortazavi, Y. (2013). Highly stable and selective non-enzymatic glucose biosensor using carbon nanotubes decorated by Fe_3O_4 nanoparticles. *Journal of The Electrochemical Society* 161 (1): B19–B25.

47 Wang, L., Fu, J., Hou, H., and Song, Y. (2012). A facile strategy to prepare Cu_2O/Cu electrode as a sensitive enzyme-free glucose sensor. *International Journal of Electrochemical Science* 7 (12): 12587–12600.

48 Mu, Y., Jia, D., He, Y. et al. (2011). Nano nickel oxide modified non-enzymatic glucose sensors with enhanced sensitivity through an electrochemical process strategy at high potential. *Biosensors and Bioelectronics* 26 (6): 2948–2952.

49 Bao, S.-J., Li, C.M., Zang, J.-F. et al. (2008). New nanostructured TiO_2 for direct electrochemistry and glucose sensor applications. *Advanced Functional Materials* 18 (4): 591–599.

50 Holzinger, M., Le Goff, A., and Cosnier, S. (2014). Nanomaterials for biosensing applications: a review. *Frontiers in Chemistry* 2: 63.

51 Zhang, X., Guo, Q., and Cui, D. (2009). Recent advances in nanotechnology applied to biosensors. *Sensors* 9 (2): 1033–1053.

52 Vikesland, P.J. and Wigginton, K.R. (2010). Nanomaterial enabled biosensors for pathogen monitoring – a review. *Environmental Science and Technology* 44 (10): 3656–3669.

53 Campbell, G.A. and Mutharasan, R. (2008). Near real-time detection of *Cryptosporidium parvum* oocyst by IgM-functionalized piezoelectric-excited millimeter-sized cantilever biosensor. *Biosensors and Bioelectronics* 23 (7): 1039–1045.

54 Tombelli, S., Minunni, M., and Mascini, M. (2007). Aptamers-based assays for diagnostics, environmental and food analysis. *Biomolecular Engineering* 24 (2): 191–200.

55 Shen, Z., Huang, M., Xiao, C. et al. (2007). Nonlabeled quartz crystal microbalance biosensor for bacterial detection using carbohydrate and lectin recognitions. *Analytical Chemistry* 79 (6): 2312–2319.

56 Ngundi, M.M., Kulagina, N.V., Anderson, G.P., and Taitt, C.R. (2006). Nonantibody-based recognition: alternative molecules for detection of pathogens. *Expert Review of Proteomics* 3 (5): 511–524.

57 Ni, J., Lipert, R.J., Dawson, G.B., and Porter, M.D. (1999). Immunoassay readout method using extrinsic Raman labels adsorbed on immunogold colloids. *Analytical Chemistry* 71 (21): 4903–4908.

58 Kalele, S.A., Kundu, A.A., Gosavi, S.W. et al. (2006). Rapid detection of escherichia coli by using antibody-conjugated silver nanoshells. *Small* 2 (3): 335–338.

59 Boman, H.G. (2003). Antibacterial peptides: basic facts and emerging concepts. *Journal of Internal Medicine* 254 (3): 197–215.

60 Castillo-Henríquez, L., Brenes-Acuña, M., Castro-Rojas, A. et al. (2020). Biosensors for the detection of bacterial and viral clinical pathogens. *Sensors* 20 (23).

61 Shabaninejad, Z., Yousefi, F., Movahedpour, A. et al. (2019). Electrochemical-based biosensors for microRNA detection: nanotechnology comes into view. *Analytical Biochemistry* 581: 113349.

62 Zhai, Q. and Cheng, W. (2019). Soft and stretchable electrochemical biosensors. *Materials Today Nano* 7: 100041.

63 Konopsky, V.N. and Alieva, E.V. (2019). Imaging biosensor based on planar optical waveguide. *Optics and Laser Technology* 115: 171–175.

64 Yin, M.-j., Gu, B., An, Q.-F. et al. (2018). Recent development of fiber-optic chemical sensors and biosensors: mechanisms, materials, micro/nano-fabrications and applications. *Coordination Chemistry Reviews* 376: 348–392.

65 Chen, Y., Liu, J., Yang, Z. et al. (2019). Optical biosensors based on refractometric sensing schemes: a review. *Biosensors and Bioelectronics* 144: 111693.

66 Hammond Jules, L., Formisano, N., Estrela, P. et al. (2016). Electrochemical biosensors and nanobiosensors. *Essays in Biochemistry* 60 (1): 69–80.

67 Wang, H., Li, H., Huang, Y. et al. (2019). A label-free electrochemical biosensor for highly sensitive detection of gliotoxin based on DNA nanostructure/MXene nanocomplexes. *Biosensors and Bioelectronics* 142: 111531.

68 Li, Y., Schluesener, H.J., and Xu, S. (2010). Gold nanoparticle-based biosensors. *Gold Bulletin* 43 (1): 29–41.

69 Biju, V. (2014). Chemical modifications and bioconjugate reactions of nanomaterials for sensing, imaging, drug delivery and therapy. *Chemical Society Reviews* 43 (3): 744–764.

70 Mulvaney, P. (1996). Surface plasmon spectroscopy of nanosized metal particles. *Langmuir* 12 (3): 788–800.

71 Hao, E., Schatz, G.C., and Hupp, J.T. (2004). Synthesis and optical properties of anisotropic metal nanoparticles. *Journal of Fluorescence* 14 (4): 331–341.

72 Kelly, K.L., Coronado, E., Zhao, L.L., and Schatz, G.C. (2003). The optical properties of metal nanoparticles: the influence of size, shape, and dielectric environment. *The Journal of Physical Chemistry B* 107 (3): 668–677.

73 Jena, B.K. and Raj, C.R. (2008). Optical sensing of biomedically important polyionic drugs using nano-sized gold particles. *Biosensors and Bioelectronics* 23 (8): 1285–1290.

74 Wei, H., Li, B., Li, J. et al. (2007). Simple and sensitive aptamer-based colorimetric sensing of protein using unmodified gold nanoparticle probes. *Chemical Communications* 36: 3735–3737.

75 Hua, Z., Yu, T., Liu, D., and Xianyu, Y. (2021). Recent advances in gold nanoparticles-based biosensors for food safety detection. *Biosensors and Bioelectronics* 179: 113076.

76 Hutter, E. and Maysinger, D. (2013). Gold-nanoparticle-based biosensors for detection of enzyme activity. *Trends in Pharmacological Sciences* 34 (9): 497–507.

77 Pingarrón, J.M., Yáñez-Sedeño, P., and González-Cortés, A. (2008). Gold nanoparticle-based electrochemical biosensors. *Electrochimica Acta* 53 (19): 5848–5866.

78 Maity, D., Kandasamy, G., and Sudame, A. (2019). Superparamagnetic iron oxide nanoparticles for cancer theranostic applications. In: *Nanotheranostics:*

Applications and Limitations (ed. M. Rai and B. Jamil), 245–276. Cham: Springer International Publishing.

79 Haun, J.B., Yoon, T.J., Lee, H., and Weissleder, R. (2010). Magnetic nanoparticle biosensors. *Wiley Interdisciplinary Reviews. Nanomedicine and Nanobiotechnology* 2 (3): 291–304.

80 Bi, S., Yan, Y., Yang, X., and Zhang, S. (2009). Gold nanolabels for new enhanced chemiluminescence immunoassay of alpha-fetoprotein based on magnetic beads. *Chemistry - A European Journal* 15 (18): 4704–4709.

81 Mejri, M.B., Tlili, A., and Abdelghani, A. (2011). Magnetic nanoparticles immobilization and functionalization for biosensor applications. *International Journal of Electrochemistry* 2011: 421387.

82 Wang, S.X. and Li, G. (2008). Advances in giant magnetoresistance biosensors with magnetic nanoparticle tags: review and outlook. *IEEE Transactions on Magnetics* 44 (7): 1687–1702.

83 Konry, T., Bale, S.S., Bhushan, A. et al. (2012). Particles and microfluidics merged: perspectives of highly sensitive diagnostic detection. *Microchimica Acta* 176 (3): 251–269.

84 Tamanaha, C.R., Mulvaney, S.P., Rife, J.C., and Whitman, L.J. (2008). Magnetic labeling, detection, and system integration. *Biosensors and Bioelectronics* 24 (1): 1–13.

85 Han, M., Gao, X., Su, J.Z., and Nie, S. (2001). Quantum-dot-tagged microbeads for multiplexed optical coding of biomolecules. *Nature Biotechnology* 19 (7): 631–635.

86 Medintz, I.L., Uyeda, H.T., Goldman, E.R., and Mattoussi, H. (2005). Quantum dot bioconjugates for imaging, labelling and sensing. *Nature Materials* 4 (6): 435–446.

87 Jares-Erijman, E.A. and Jovin, T.M. (2003). FRET imaging. *Nature Biotechnology* 21 (11): 1387–1395.

88 Huang, X., Li, L., Qian, H. et al. (2006). A resonance energy transfer between chemiluminescent donors and luminescent quantum-dots as acceptors (CRET). *Angewandte Chemie International Edition* 45 (31): 5140–5143.

89 Biju, V., Itoh, T., and Ishikawa, M. (2010). Delivering quantum dots to cells: bioconjugated quantum dots for targeted and nonspecific extracellular and intracellular imaging. *Chemical Society Reviews* 39 (8): 3031–3056.

90 Murray, C.B., Norris, D.J., and Bawendi, M.G. (1993). Synthesis and characterization of nearly monodisperse CdE (E = sulfur, selenium, tellurium) semiconductor nanocrystallites. *Journal of the American Chemical Society* 115 (19): 8706–8715.

91 Park, J., Joo, J., Kwon, S.G. et al. (2007). Synthesis of monodisperse spherical nanocrystals. *Angewandte Chemie International Edition* 46 (25): 4630–4660.

92 Reiss, P., Protière, M., and Li, L. (2009). Core/shell semiconductor nanocrystals. *Small* 5 (2): 154–168.

93 Weller, H. (1993). Colloidal semiconductor q-particles: chemistry in the transition region between solid state and molecules. *Angewandte Chemie, International Edition in English* 32 (1): 41–53.

94 Deng, Z., Zhang, Y., Yue, J. et al. (2007). Green and orange CdTe quantum dots as effective pH-sensitive fluorescent probes for dual simultaneous and independent detection of viruses. *The Journal of Physical Chemistry B* 111 (41): 12024–12031.

95 Ansari, A.A., Kaushik, A., Solanki, P.R., and Malhotra, B.D. (2008). Sol–gel derived nanoporous cerium oxide film for application to cholesterol biosensor. *Electrochemistry Communications* 10 (9): 1246–1249.

96 Zhang, F., Ulrich, B., Reddy, R.K. et al. (2008). Fabrication of submicron IrO_2 nanowire array biosensor platform by conventional complementary metal–oxide–semiconductor process. *Japanese Journal of Applied Physics* 47 (2): 1147–1151.

97 Cheng, J., Di, J., Hong, J. et al. (2008). The promotion effect of titania nanoparticles on the direct electrochemistry of lactate dehydrogenase sol–gel modified gold electrode. *Talanta* 76 (5): 1065–1069.

98 Liu, A. (2008). Towards development of chemosensors and biosensors with metal-oxide-based nanowires or nanotubes. *Biosensors and Bioelectronics* 24 (2): 167–177.

99 Lu, X., Zhang, H., Ni, Y. et al. (2008). Porous nanosheet-based ZnO microspheres for the construction of direct electrochemical biosensors. *Biosensors and Bioelectronics* 24 (1): 93–98.

100 Bai, J. and Zhou, B. (2014). Titanium dioxide nanomaterials for sensor applications. *Chemical Reviews* 114 (19): 10131–10176.

101 Jeong, H., Yoo, J., Park, S. et al. (2021). Non-enzymatic glucose biosensor based on highly pure TiO_2 nanoparticles. *Biosensors* 11 (5).

102 Rodrigues, J., Pereira, S.O., Zanoni, J. et al. (2022). ZnO transducers for photoluminescence-based biosensors: a review. *Chemosensors* 10 (2): 39.

103 Zhao, Z., Lei, W., Zhang, X. et al. (2010). ZnO-based amperometric enzyme biosensors. *Sensors* 10 (2): 1216–1231.

104 Shi, X., Gu, W., Li, B. et al. (2014). Enzymatic biosensors based on the use of metal oxide nanoparticles. *Microchimica Acta* 181 (1): 1–22.

105 Zhu, X., Yuri, I., Gan, X. et al. (2007). Electrochemical study of the effect of nano-zinc oxide on microperoxidase and its application to more sensitive hydrogen peroxide biosensor preparation. *Biosensors and Bioelectronics* 22 (8): 1600–1604.

106 Negahdary, M., Asadi, A., Mehrtashfar, S. et al. (2012). A biosensor for determination of H_2O_2 by use of HRP enzyme and modified CPE with ZnO NPs. *International Journal of Electrochemical Science* 7 (6): 5185–5194.

107 Shamsipur, M., Asgari, M., Mousavi, M.F., and Davarkhah, R. (2012). A novel hydrogen peroxide sensor based on the direct electron transfer of catalase immobilized on nano-sized NiO/MWCNTs composite film. *Electroanalysis* 24 (2): 357–367.

108 Bayandori Moghaddam, A., Ganjali, M.R., Saboury, A.A. et al. (2008). Electrodeposition of nickel oxide nanoparticles on glassy carbon surfaces: application to the direct electron transfer of tyrosinase. *Journal of Applied Electrochemistry* 38 (9): 1233–1239.

109 Yin, L.-T., Chou, J.-C., Chung, W.-Y. et al. (2001). Glucose ENFET doped with MnO_2 powder. *Sensors and Actuators B: Chemical* 76 (1): 187–192.

110 Luo, X.-L., Xu, J.-J., Zhao, W., and Chen, H.-Y. (2004). A novel glucose ENFET based on the special reactivity of MnO_2 nanoparticles. *Biosensors and Bioelectronics* 19 (10): 1295–1300.

111 Li, Y., Wei, Y., Shi, G. et al. (2011). Facile synthesis of leaf-like CuO nanoparticles and their application on glucose biosensor. *Electroanalysis* 23 (2): 497–502.

112 Ding, S.-N., Shan, D., Xue, H.-G., and Cosnier, S. (2010). A promising biosensing-platform based on bismuth oxide polycrystalline-modified electrode: characterization and its application in development of amperometric glucose sensor. *Bioelectrochemistry* 79 (2): 218–222.

113 Saha, S., Arya, S.K., Singh, S.P. et al. (2009). Nanoporous cerium oxide thin film for glucose biosensor. *Biosensors and Bioelectronics* 24 (7): 2040–2045.

114 Zhao, J., Qin, L., Hao, Y. et al. (2012). Application of tubular tetrapod magnesium oxide in a biosensor for hydrogen peroxide. *Microchimica Acta* 178 (3): 439–445.

115 Periasamy, A.P., Ting, S.W., and Chen, S.-M. (2011). Amperometric and impedimetric H_2O_2 biosensor based on horseradish peroxidase covalently immobilized at ruthenium oxide nanoparticles modified electrode. *International Journal of Electrochemical Science* 6 (7): 2688–2709.

116 Lavanya, N., Radhakrishnan, S., and Sekar, C. (2012). Fabrication of hydrogen peroxide biosensor based on Ni doped SnO_2 nanoparticles. *Biosensors and Bioelectronics* 36 (1): 41–47.

117 Lim, H.J., Saha, T., Tey, B.T. et al. (2020). Quartz crystal microbalance-based biosensors as rapid diagnostic devices for infectious diseases. *Biosensors and Bioelectronics* 168: 112513.

118 Li, J.-Y., You, Z., Wang, Q. et al. (2020). The epidemic of 2019-novel-coronavirus (2019-nCoV) pneumonia and insights for emerging infectious diseases in the future. *Microbes and Infection* 22 (2): 80–85.

119 Selgelid, M.J. (2012). Infectious diseases. In: *Encyclopedia of Applied Ethics*, 2e (ed. R. Chadwick), 704–711. San Diego: Academic Press.

120 Giljohann, D.A., Seferos, D.S., Daniel, W.L. et al. (2010). Gold nanoparticles for biology and medicine. *Angewandte Chemie International Edition* 49 (19): 3280–3294.

121 Sardar, R., Funston, A.M., Mulvaney, P., and Murray, R.W. (2009). Gold nanoparticles: past, present, and future. *Langmuir* 25 (24): 13840–13851.

122 Choi, J.W., Kim, Y.J., Lee, J.M. et al. (2020). Droplet-based synthesis of homogeneous gold nanoparticles for enhancing HRP-based ELISA signals. *BioChip Journal* 14 (3): 298–307.

123 Jiang, P., Wang, Y., Zhao, L. et al. (2018). Applications of gold nanoparticles in non-optical biosensors. *Nanomaterials* 8 (12): 977.

124 de la Escosura-Muñiz, A., Maltez-da Costa, M., Sánchez-Espinel, C. et al. (2010). Gold nanoparticle-based electrochemical magnetoimmunosensor for rapid detection of anti-hepatitis B virus antibodies in human serum. *Biosensors and Bioelectronics* 26 (4): 1710–1714.

125 Ma, C., Xie, G., Zhang, W. et al. (2012). Label-free sandwich type of immunosensor for hepatitis C virus core antigen based on the use of gold nanoparticles on a nanostructured metal oxide surface. *Microchimica Acta* 178 (3): 331–340.

126 Mashhadizadeh, M.H. and Pourtaghavi, T.R. (2014). A highly sensitive and selective hepatitis B DNA biosensor using gold nanoparticle electrodeposition on an Au electrode and mercaptobenzaldehyde. *Analytical Methods* 6 (22): 8956–8964.

127 Lu, X., Dong, X., Zhang, K. et al. (2013). A gold nanorods-based fluorescent biosensor for the detection of hepatitis B virus DNA based on fluorescence resonance energy transfer. *The Analyst* 138 (2): 642–650.

128 Yao, C., Xiang, Y., Deng, K. et al. (2013). Sensitive and specific HBV genomic DNA detection using RCA-based QCM biosensor. *Sensors and Actuators, B: Chemical* 181: 382–387.

129 Zhao, H., Liu, F., Xie, W. et al. (2021). Ultrasensitive supersandwich-type electrochemical sensor for SARS-CoV-2 from the infected COVID-19 patients using a smartphone. *Sensors and Actuators, B: Chemical* 327: 128899.

130 Choi, H.K., Lee, M.-J., Lee, S.N. et al. (2021). Noble metal nanomaterial-based biosensors for electrochemical and optical detection of viruses causing respiratory illnesses. *Frontiers in Chemistry* 9, 672739.

131 Loiseau, A., Asila, V., Boitel-Aullen, G. et al. (2019). Silver-based plasmonic nanoparticles for and their use in biosensing. *Biosensors* 9 (2): 78.

132 Li, M., Cushing, S.K., Liang, H. et al. (2013). Plasmonic nanorice antenna on triangle nanoarray for surface-enhanced raman scattering detection of hepatitis B virus DNA. *Analytical Chemistry* 85 (4): 2072–2078.

133 Karn-orachai, K., Sakamoto, K., Laocharoensuk, R. et al. (2016). Extrinsic surface-enhanced Raman scattering detection of influenza a virus enhanced by two-dimensional gold@silver core–shell nanoparticle arrays. *RSC Advances* 6 (100): 97791–97799.

134 Teengam, P., Siangproh, W., Tuantranont, A. et al. (2017). Multiplex paper-based colorimetric DNA sensor using pyrrolidinyl peptide nucleic acid-induced AgNPs aggregation for detecting MERS-CoV, MTB, and HPV oligonucleotides. *Analytical Chemistry* 89 (10): 5428–5435.

135 Kinoshita, K. (1990). Particle size effects for oxygen reduction on highly dispersed platinum in acid electrolytes. *Journal of The Electrochemical Society.* 137 (3): 845–848.

136 Garlyyev, B., Kratzl, K., Rück, M. et al. (2019). Optimizing the size of platinum nanoparticles for enhanced mass activity in the electrochemical oxygen reduction reaction. *Angewandte Chemie International Edition* 58 (28): 9596–9600.

137 Cheng, T., Li, X., Huang, P. et al. (2019). Colorimetric and electrochemical (dual) thrombin assay based on the use of a platinum nanoparticle modified metal-organic framework (type Fe-MIL-88) acting as a peroxidase mimic. *Microchimica Acta* 186 (2): 94.

138 Yang, Z.-H., Zhuo, Y., Yuan, R., and Chai, Y.-Q. (2016). A nanohybrid of platinum nanoparticles-porous ZnO–hemin with electrocatalytic activity to construct an amplified immunosensor for detection of influenza. *Biosensors and Bioelectronics* 78: 321–327.

139 Matsumura, Y., Enomoto, Y., Takahashi, M., and Maenosono, S. (2018). Metal (Au, Pt) nanoparticle–latex nanocomposites as probes for immunochromatographic test strips with enhanced sensitivity. *ACS Applied Materials and Interfaces* 10 (38): 31977–31987.

140 Mokhtarzadeh, A., Eivazzadeh-Keihan, R., Pashazadeh, P. et al. (2017). Nanomaterial-based biosensors for detection of pathogenic virus. *TrAC Trends in Analytical Chemistry* 97: 445–457.

141 Magdassi, S., Grouchko, M., and Kamyshny, A. (2010). Copper nanoparticles for printed electronics: routes towards achieving oxidation stability. *Materials* 3 (9): 4626–4638.

142 Chen, X., Xie, H., Seow, Z.Y., and Gao, Z. (2010). An ultrasensitive DNA biosensor based on enzyme-catalyzed deposition of cupric hexacyanoferrate nanoparticles. *Biosensors and Bioelectronics* 25 (6): 1420–1426.

143 Mao, X., Liu, S., Yang, C. et al. (2016). Colorimetric detection of hepatitis B virus (HBV) DNA based on DNA-templated copper nanoclusters. *Analytica Chimica Acta* 909: 101–108.

144 Cao, X., Cao, X., Guo, H. et al. (2016). Piezotronic effect enhanced label-free detection of DNA using a Schottky-contacted ZnO nanowire biosensor. *ACS Nano* 10 (8): 8038–8044.

145 Low, S.S., Tan, M.T.T., Loh, H.-S. et al. (2016). Facile hydrothermal growth graphene/ZnO nanocomposite for development of enhanced biosensor. *Analytica Chimica Acta* 903: 131–141.

146 He, K., Chen, C., Liang, C. et al. (2016). Highly selective recognition and fluorescent detection of JEV via virus-imprinted magnetic silicon microspheres. *Sensors and Actuators, B: Chemical* 233: 607–614.

147 Hung, L.-Y., Chang, J.-C., Tsai, Y.-C. et al. (2014). Magnetic nanoparticle-based immunoassay for rapid detection of influenza infections by using an integrated microfluidic system. *Nanomedicine: Nanotechnology, Biology and Medicine* 10 (4): 819–829.

148 Mahari, S., Roberts, A., Shahdeo, D., and Gandhi, S. (2020). eCovSens-ultrasensitive novel in-house built printed circuit board based electrochemical device for rapid detection of nCovid-19 antigen, a spike protein domain 1 of SARS-CoV-2. *bioRxiv* 2020.04.24.059204.

149 Layqah, L.A. and Eissa, S. (2019). An electrochemical immunosensor for the corona virus associated with the Middle East respiratory syndrome using an array of gold nanoparticle-modified carbon electrodes. *Microchimica Acta* 186 (4): 224.

150 Nasrin, F., Chowdhury, A.D., Takemura, K. et al. (2020). Fluorometric virus detection platform using quantum dots-gold nanocomposites optimizing the linker length variation. *Analytica Chimica Acta* 1109: 148–157.

151 Lee, Y., Choi, J., Han, H.-K. et al. (2021). Fabrication of ultrasensitive electrochemical biosensor for dengue fever viral RNA Based on CRISPR/Cpf1 reaction. *Sensors and Actuators, B: Chemical* 326: 128677.

152 Hashemi, S.A., Golab Behbahan, N.G., Bahrani, S. et al. (2021). Ultra-sensitive viral glycoprotein detection NanoSystem toward accurate tracing SARS-CoV-2 in biological/non-biological media. *Biosensors and Bioelectronics* 171: 112731.

153 Cao, Q., Teng, Y., Yang, X. et al. (2015). A label-free fluorescent molecular beacon based on DNA-Ag nanoclusters for the construction of versatile Biosensors. *Biosensors and Bioelectronics* 74: 318–321.

154 Huang, J., Xie, Z., Xie, Z. et al. (2016). Silver nanoparticles coated graphene electrochemical sensor for the ultrasensitive analysis of avian influenza virus H7. *Analytica Chimica Acta* 913: 121–127.

155 Das, B., Franco, J.L., Logan, N. et al. (2021). Nanozymes in point-of-care diagnosis: an emerging futuristic approach for biosensing. *Nano-Micro Letters* 13 (1): 193.

156 Parihar, A., Ranjan, P., Sanghi, S.K. et al. (2020). Point-of-care biosensor-based diagnosis of COVID-19 holds promise to combat current and future pandemics. *ACS Applied Bio Materials* 3 (11): 7326–7343.

157 Grieshaber, D., MacKenzie, R., Vörös, J., and Reimhult, E. (2008). Electrochemical biosensors - sensor principles and architectures. *Sensors* 8 (3): 1400–1458.

158 Saylan, Y., Erdem, Ö., Ünal, S., and Denizli, A. (2019). An alternative medical diagnosis method: biosensors for virus detection. *Biosensors* 9 (2): 65.

159 Singh, R., Mukherjee, M.D., Sumana, G. et al. (2014). Biosensors for pathogen detection: a smart approach towards clinical diagnosis. *Sensors and Actuators B: Chemical* 197: 385–404.

160 Malekzad, H., Sahandi Zangabad, P., Mirshekari, H. et al. (2017). Noble metal nanoparticles in biosensors: recent studies and applications. *Nanotechnology Reviews* 6 (3): 301–329.

161 Li, J. and Zhu, J.-J. (2013). Quantum dots for fluorescent biosensing and bio-imaging applications. *The Analyst* 138 (9): 2506–2515.

162 Cheng, D., Peng, J., Lv, Y. et al. (2019). De novo design of chemical stability near-infrared molecular probes for high-fidelity hepatotoxicity evaluation in vivo. *Journal of the American Chemical Society* 141 (15): 6352–6361.

163 Strianese, M., Staiano, M., Ruggiero, G. et al. (2012). Fluorescence-based biosensors. In: *Spectroscopic Methods of Analysis: Methods and Protocols* (ed. W.M. Bujalowski), 193–216. Totowa, NJ: Humana Press.

164 Piriya, V.S.A., Joseph, P., Daniel, S.C.G.K. et al. (2017). Colorimetric sensors for rapid detection of various analytes. *Materials Science and Engineering C* 78: 1231–1245.

9

Biosensors for Point-of-Care Applications: Replacing Pathology Labs by Bedside Devices

Mayukh Sinha[1], *Sayak Banerjee*[1], *Sambit Majumdar*[1], *and Arindam Kushagra*[2]

[1]*Amity University Kolkata, Amity Institute of Biotechnology, Major Arterial Road, Action Area II, Newtown, Kolkata 700135, India*
[2]*Amity University Kolkata, Amity Institute of Nanotechnology, Major Arterial Road, Action Area II, Newtown, Kolkata 700135, India*

9.1 Introduction

Biosensors are analytical devices that have a combination of biological detecting elements such as sensor system and a transducer. It integrates a biological element such as enzymes or antibody with an electrical component. Electrical components are used for detecting, recording, and transmitting information for the existence of several chemical or biological materials in the environment. Biosensors be made up of:

- An analyte for the identification or detection of substances.
- Bioreceptors contain biomolecules which are used for the recognition of the target substrates.
- Transducer for the transformation of energy from one form to another.
- Electronics for the transduced signal to be processed and draw up for the display.
- Display for the generation of output in the readable form as numerical, graphical, and figure.

9.2 POCT Relevance in Healthcare

The bedside testing and monitoring play a significant role between the transition of diagnostics and treatment. It is quite evident that the physicians will run extensive tests before treatment, but to make life easier and to provide enough time to move the patient to healthcare facilities in case of any critical happenings, point-of-care treatment (POCT) proves it's significance. Normal laboratory tests take long time to produce the results and additional time is required to prepare the report of it, which might prove to be fatal for some patients. This delay in report production can be

Point-of-Care Biosensors for Infectious Diseases, First Edition.
Edited by Sushma Dave and Jayashankar Das.

resolved by POCT and this has a significant impact on the decision to be taken by the medical practitioner.

Portable and user-friendly POCT devices are available almost in every market. Nowadays, not only the POCT devices produce instant results but also near-accurate results. These features make POCT devices used by millions of users. POCT devices are not only used by patients but also by people having no chronic disease. This practice leads to higher health consciousness and maybe followed by medical consultations. Besides spreading awareness, POCT also reduces treatment costs (Figure 9.1).

Type 1a: Qualitative POCT Methods

- These methods are often regarded as test strips and these qualitative tests distinguish between the poistive and negative results.
- The visual display aids in reading the measuring signal and record it in a simple read-out device.

Type 1b: Unit wise POCT systems

- They are the most simple type of quantitative POCT methods in which the analytical reaction take place on the test strip.
- The readable calculated value is generated by the read-out device itself only.

Type 2: Benchtop POCT instruments

- These are usually complex analytical devices which are used in functional areas of central laboratories than keeping directly beside the bed.
- Benchtop instruments are a derived from automated analytical methods that are developed with improved speed of analysis, miniaturization and user-friendliness.

Type 3: Viscoelastic coagulation analyzers

- Being highly complex, these devices have a limited feasibility as POCT tools.
- Viscoelastic coagulation analyzers are operated in the central laboratory, and the clinical arena by trained personnel.

Type 4: Continuous POCT methods

- These methods were developed for the sole purpose of glucose monitoring only but eventually they have become accessible commercially.
- Continuous POCT can be operated using a slightly invasive methods in the subcutaneous tissue.

Type 5: Molecular biological POCT analyzers

- Presently this class of analyzers are under intensive development in the POCT field.
- Since, there is a high degree of complexity involved while carrying out th tests as well as interpreting the results, these methods are improbable to be found in the future.

Type 6: Direct-to-consumer testing (DTC)

- There are several generally available DTC, which a patient can select besides physician-ordered diagnostic tests conducted in clinical laboratories.
- This enables the patients to be knowledgeable and to answer questions about their own health and habits.

Figure 9.1 POCT device classes.

9.3 Self-Blood Glucose Monitoring

9.3.1 Introduction

POCT for monitoring of glucose, acquires enzymatic analysis method based on glucose dehydrogenase and glucose oxidase (GO). The reaction is detected electrochemically.

Previously, monitoring of glucose level was done by the physicians or by the qualified medical staff. With the improvement in medical technology day by day a self-glucose monitoring applications are available for patients. They do not need to visit any pathology center for their glucose monitoring, it can be done at home and reports are also known within a second.

There are many industries who produces these types of devices some of them like Dexcom Inc.; Johnson and Johnson; Abbott Laboratories, and many others. Nowadays these devices are available easily in all the nearby pharmacies.

9.3.2 Requirements for Self-Glucose Monitoring Device

- Glucose meter
- Lancets (needles)
- A lancing device
- Test strip

9.3.3 Types of Sensor-Based Monitoring System

To monitor the glucose levels, there are two types of sensor-based monitoring systems available:

- Continuous glucose monitoring (CGM)
- Flash glucose monitoring (FGM)

9.3.3.1 Continuous Glucose Monitoring (CGM)

Blood glucose meter measure the blood glucose from a drop of blood of a patient on a test strip but the CGM measures from interstitial fluid by placing the sensor under the skin [1, 2]. A CGM is a sensor that measures the glucose levels in blood for 24 hours a day automatically.

The benefits of CGM are that it gives reading with 500–1500 per 24 hours compared to self-monitoring blood glucose; looks after 24 hours of glucose changeability and postprandial blood glucose profiles [1, 2]. With the development of clinical studies metabolic control as measured by HbA1c, reduce in the periodicity and hypoglycaemia.

First CGM system is only promoted for a duration of 6–14 days. Later, new sensor needs to be inserted by replacing the old sensor and connect to parallel insulin pump. By various development it is expected to extend over 5–20 years [3]. However, its clinical advantages have enabled the patients of CGM system to keep growing from years after years not only to patients at home but also clinically [3].

9.3.3.2 Flash Glucose Monitoring (FGM)

Provide reading handheld reader is scanned over the sensor. The requirements for FGM are: sensor patch for worn on the upper arm for 14 days; touch screen reader device for data and information. These patients have a sensor embedded on the upper arm and a different touchscreen reader gadget [3]. At the point when the reader is swiped near the sensor, the sensor sends both a glucose levels and eight hours graph to the reader. This permits patients to get individual glucose readings (such as BGM) pattern data (such as CGM). Nonetheless, dissimilar to CGM, FGM doesn't have hypo or hyperglycemia cautions and will give a graph if it has been swiped in beyond eight hours [1].

CGM has been utilized by well-educated patients with type 1 diabetes whereas FGM-using patients are with type 2 diabetes who must perform few blood glucose measurements each day, perceiving how blood glucose ascends after a meal or lowers because of physical exercise.

9.4 Methods of Blood Glucose Monitoring

9.4.1 Enzymatic Assay Reaction

Glucose is oxidized by GO to gluconic acid in the existence of water and oxygen. The enzyme cofactor flavin adenine dinucleotide (FAD) diminishes to FADH [2]. Later on, FADH is oxidized by oxygen to form water. Consumption of oxygen are measured by electrochemical.

$$\beta\text{-D-Glucose} + H_2O + O_2 \xrightarrow{\text{Glucose Oxidase}} \text{D-gluconolactone} + H_2O$$

Oxidation of glucose-by-glucose oxidase

The assay response is profoundly distinct, the indicator response can be distributed to a variable's degree by substrates for example ascorbic acid or acetaminophen. The conclusion is on how much oxygen is in the sample.

In oxidation reaction glucose dehydrogenase, NAD, FAD, and PQQ are the first electron acceptor and reduced during the reaction to NADH, PQQH, and FADH. Glucose dehydrogenase system is more dominant in comparison to GO system as it lies in lower susceptibility to cross-react with redundancy therapeutics of the sample [2]. After the lysis of red blood cells and oxidization of glucose by glucose dehydrogenase, the subsequent NADH respond with tetrazolium salts which are diminished to formazan by diaphorase and estimated at 660/840 nm.

$$\beta\text{-D-Glucose} \xrightarrow{\text{Glucose Dehydrogenase}} \text{D-Gluconolactone} + H^+ + e^-$$

Oxidation of glucose by glucose dehydrogenase

9.4.2 Detection Method

Enzymatic reaction in blood can be detected photometrically or electrochemically.

By instituting sensor technology, a small volume of blood is required and measuring times also decrease and remove the direct contact of blood to the devices [2].

Amperometric technique for measuring has been applied to all sensor test strips. The benefit of this technique is the lower impact of the hematocrit (Hct) [2, 4]. If small amount of glucose is present in the sample, that can also be measured by this technique. The GMD (Glucose measuring device) is dependent on hematocrit.

9.4.3 Errors Occuring in Blood Glucose Monitoring

All devices have some errors like that blood glucose monitoring devices also have errors not because of analytical but also pre-analytical which are caused by not properly cleaning of skin or by not properly handling of devices. Some of the examples of causing errors are:

- Not properly drying hands before test which can dilute the blood sample.
- Not properly washing hands after contacting with fruits.
- Not storing of test strips at recommended temperature by the manufacturer.
- Using of expiry date test strip which can reduce the enzyme activity.
- Not closing of test strip container properly which causes damage to the enzyme because of moisture.

9.4.4 POCT for Blood Glucose Monitoring

The first utilization of POCT is found in Papyrus reports dating way back to 1500BC, in which Egyptian doctors utilizing ants to examine glycosuria in patients associated with having diabetes [5]. Jules Maumene was the first to invent a basic urine reagent "Strip" in 1850 made of sheep's wool having stannous chloride [6]. George Oliver was the first to invent a "Bedside urine testing" and advanced reagent papers for testing urine sugar [7, 8]. In 1965 Ames research group invented the first BG test strip, the Dextrostix, a paper reagent strips the utilized the GO reaction [9]. Boehringer Mannhein a German organization invented another BG test strip called ChemstripbG. Afterward Anton Clemens at Ames invented an instrument, the Ames Reflectance Meter, to create quantitative BG results from Dextrostix in 1970 [10].

Blood glucose monitoring are the accurate, rapid, and cost-effective techniques. They are remotely available and less time requirements for testing blood glucose make these devices as POCT [5]. It is currently the standard bedside glucose monitoring method in many establishments [11]. POCT BG meters require very little volumes of blood for determination (0.3 – 1 μl), rather than CLT which require 1 – 3 ml of blood. This is significant on the grounds that repeated venesection is costly and patients anemic [12].

9.5 Blood Gas Analysis

9.5.1 Introduction

The blood gas analysis (BGA) has been used in the intensive care units for decades and also on the grounds of anaesthesiology and as emergency medicine [13]. It helps in the detection of a patient's oxygenation status, ventilation, and acid–base balance in the body [14].

The BGA is conducted to analyze the following parameters:

- Partial oxygen pressure (pO_2), partial CO_2 pressure (pCO_2), and pH [15, 16].
- Blood oxygenation (O_2 saturation, sO_2), total hemoglobin (Hb) concentrations [16].
- Electrolyte's measurement such as Na^+, K^+, ionized Ca^{2+}, ionized Mg^{2+}, and Cl^- [17].
- Measurement of metabolites like glucose, lactate, creatinine, urea, bilirubin, and others [18].

9.5.2 Methodologies

To monitor the levels of the abovementioned entities, there are two different methods that are used [13, 19]. They are:

- Electrochemical method [20]
- Optical method [21]

9.5.3 Electrochemical Sensors

This type of sensor is equipped with pH glass electrode, a **Stow-Severing Haus** [15] pCO_2 electrode and a Clark pO_2 electrode [16, 20, 22]. They use gas-selective membranes which function as chemo-specific detection layers [20].

Electrochemical sensors are based on electrochemical reactions which occur between the surface of the electrode and blood [22]. Combination electrodes are used which consist of measuring and reference electrodes [20]. The effectivity of the CO_2 gas molecules diffusion process is being accelerated by the use of silicon membranes which in turn reduces the overall time of measurement [20]. Catalytically active platinum black (a powder form of platinum) is being fixed to the inner layer of an O_2 permeable Teflon membrane [22]. This provides an additional advantage in total analysis time reduction of less than 60 seconds at a sample volume of <400 ml which includes the CO-oximetry as well [16, 20].

9.5.4 Optical Sensors

Optical sensors emit spectroscopic signals or optodes. The determination of pO_2 is done by measuring the fluorescence of a dye [21]. The O_2 concentration in the blood sample modulates the fluorescence lifetime [19, 22]. According to **Stern–Volmer equation** [23], the intensity of the emitted fluorescence is inversely proportional to the pO_2 acting as a quencher [21]. The advantage is that there is nearly null consumption of O_2 during the whole process [22].

The pO_2 is measured by absorption in the infrared range as dissolved CO_2 absorbs light. According to the Lambert–Beer law (Eq. 2.1) [24], three various wavelengths are measured in order to determine the concentration of dissolved CO_2 and pCO_2.

$$A = \varepsilon bc$$

The pH in whole blood is determined by the means of spectroscopic absorption of specific light by a color indicator in the visible range by means of diode array photometer [21].

9.5.5 Measurement of the Blood Gas Parameters

The parameters to be measured are as follows:

- pH
- P_aCO_2
- P_aO_2
- Glucose and lactate metabolites
- Electrolytes
- Hemoglobin and bilirubin

Let us talk in detail about the techniques that are incorporated for the measurement of the above blood gas parameters.

9.5.6 pH

The reference electrode is the standard with which the chemical sample is being compared against. It consists of electrode jacket, electrolyte solution **(which has known pH)**, silver rod coated with silver chloride, and an outer membrane [15].

The pH electrode is a glass electrode and consists of 3D lattice with a central silicon atom and surrounded by four O_2 atoms [25]. There is incorporation of various metal oxides (Ca^{2+} and Na^{2+} ions) into the glass membrane for the sensitivity of the electrode [15]. The inner buffer solution has a constant and known ph. The presence of air bubble allows expansion of solution at 37 °C [15, 25].

Working: The metal oxide lose electron to the incorporated O_2 molecule and thus become cations. Charge is displaced across the membrane resulting in the flow of current across the glass [15]. One side of the glass is exposed to a buffer of known pH **(reference electrode)** and the other side is exposed to the sample/test solution **(blood)** [25]. The glass membrane is then partitioned with different H ions on either side establishing a potential difference via the concentration gradient [15]. The pH remains constant in spite of the change in H ions due to the action of the inner buffer solution.

9.5.7 $PaCO_2$

The Severinghaus electrode is the mainstay of CO_2 measurement for the arterial blood gas [15]. It is essentially a pH electrode but contains both pH and reference electrode in the same device [26]. This feature of the electrode makes the measurement time-consuming (1–2 minutes) due to the prolonged equilibration and calibration.

Working: The electrode is then presented to a thin film of sodium bicarbonate via a Silicon or a Teflon membrane which is permeable to CO_2 but impermeable to blood cells [15]. This equilibrates with the blood and forms the test solution [26]. There is buffer solution held in contact with the silicon membrane by nylon or glass wall. The bicarbonate buffer has a concentration of 0.001 $mol\,l^{-1}$ [27]. Before using, calibration occurs with two gases of known concentration 4% and 8% in either gaseous or liquid state [15, 26]. Reference electrode houses silver/silver chloride which is in direct contact with buffer solution. Silicon membrane is permeable to CO_2 but not to any other ions that can affect the change of pH [25]. The change in H ions will be read by Voltmeter, which gives a reading via calibration in units of $PaCO_2$ [15].

9.5.8 PaO_2

It is the measurement of how well the O_2 move from lungs to the blood. The Clark electrode is used for the task [16]. It is cheap, reliable, and small. It requires no external power supply [26]. The Clark electrode measures the tension of O_2 in the solution which is directly proportional or equated to O_2 content of the blood or gas mixture in ion oxide [16, 28]. The principle is that a certain number of O_2 molecules within a salt solution will produce a current. This is based on the principle of Ohm's law which states that voltage (V) is equal to the product of current (I) and resistance (R) [29];

$$V = IR$$

$$\text{Since R is constant;}$$

$$V \propto I$$

There are two electrodes within a single unit, both of which are found behind a single and semi-permeable membrane [29]. There is a polypropylene membrane which is O_2 permeable preventing bacteria, protein, and sediments reversing it [26]. The impurities interfere with the measurement and interpretation of the O_2 values, especially the bacteria which consumes excess O_2 and thus skews the values [16]. The increasing voltage between the platinum and silver electrode increases the rate of electrocatalysis (a type of catalysis which results in a rate change of an electrochemical reaction that is occurring on an electrode surface) [16, 26].

Working: The electrolyte solution contains phosphate buffer and it allows the O_2 to dissolve. The platinum cathode is used to measure the electrical activity which has been equated to O_2 tension, which is not actual P_aO_2 [16]. The basic principle is that the increase in the uptake of electron (O_2 tension) will result in increased electrical flow in the circuit [26]. Increasing the voltage between the electrodes results in a current (I) increase for any P_aO_2. At acertain point, the current

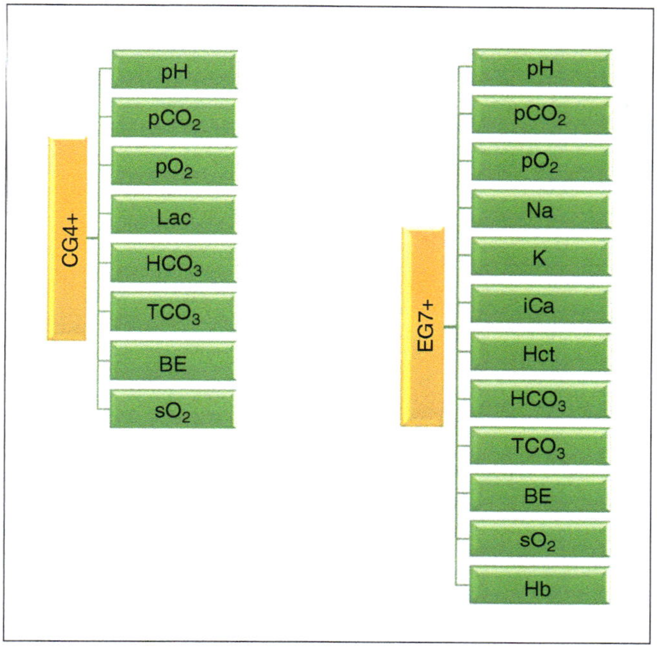

Figure 9.2 The two cartridges for the i-STAT BGA (blood gas analysis).

reaches a plateau of a certain potential difference (V) [16, 26]. At this point, the reaction is only limited by diffusion and depends on the properties of the permeable membrane. The following reaction shown in the figure (Figure 9.2) below takes place at the electrode.

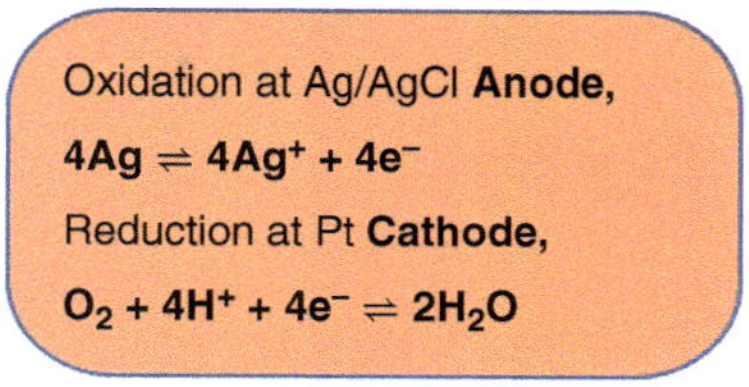

Reaction at Cathode and Anode

9.5.9 Glucose and Lactate Metabolites

The measurement of metabolites like glucose and lactate are vital via a rapid, reliable, and reproducible means. These values are determined by oxidation of the H_2O_2 [18]. There are two electrodes in the electrolyte solution [30]. It is important to know that there is a Ag/AgCl as cathode and a Pt as an anode along with this contrast with the Clark electrode [16]. Each electrode is then covered by a three-layered membrane which measures current, determined by the degree of concentration of glucose or lactate [18]. It is thus calibrated to read glucose or lactate [30].

Working: The current is being measured by an instrument known as an Ammeter. This is done by determining the degree of concentration of glucose or lactate [30]. The buffering in the electrolyte solution facilitates no change in the pH of the solution despite H^+ ions production [18].

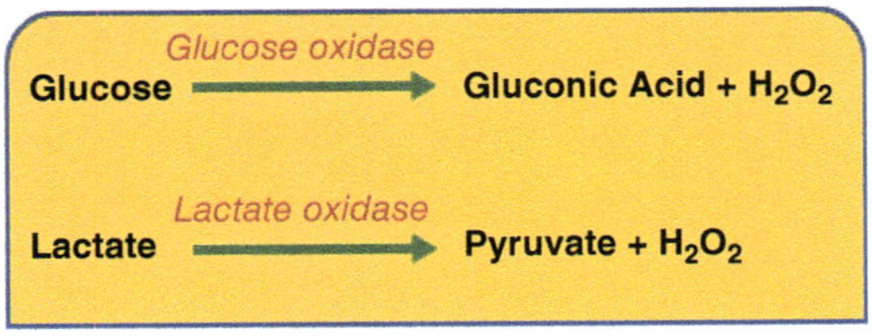

Electrochemical reaction of Glucose and Lactate[25][18]

The working can be explained with this electrochemical reaction as mentioned in the equation (Eq. 2.3) below:

The produced H_2O_2 then diffuses across the inner membrane towards the anode.

The following reactions take place at the electrode:

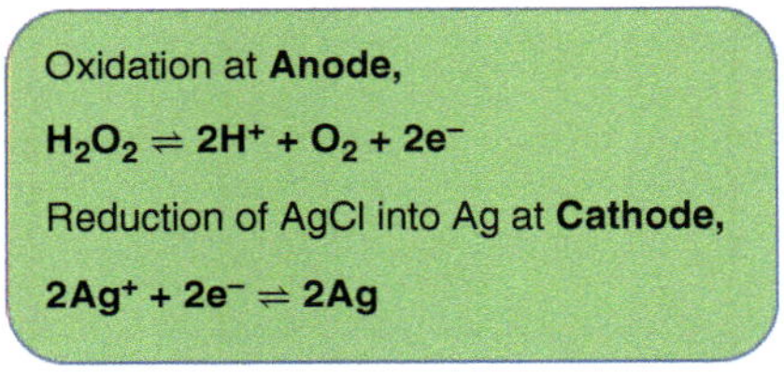

Equat Reaction at Cathode and Anode

9.5.9.1 Electrolytes

The blood gas analyzer also measures the presence of certain electrolytes such as **sodium (Na^+), potassium (K^+), calcium (Ca^{2+}), and chlorine (Cl^-)** present in the blood sample [17, 31].

Working: The measurement is based on using the standard electrode as discussed earlier. There is an ion-sensitive and selective poly vinyl chloride (PVC) membrane jacketing the Ag/AgCl electrode for **potassium(K^+), calcium(Ca^{2+}), and chlorine(Cl^-)** [17, 31]. The PVC membrane is being covered by cellophane paper so as to restrict the membrane from any contamination. Another electrode is the Sodium Na+ measuring electrode (Na^+ -ion exchanger or Na^+ -sensitive pin) which replaces the PVC [17].

9.5.9.2 Hemoglobin, Bilirubin

The blood sample provided in the blood gas analyzers undergoes spectrophotometry. The principle of spectrophotometry is the Beer–Lambert law [24] which has already been discussed earlier (Eq. 2.1) in the previous segment [32].

This method calculates the quantitative measurement of reflection or wavelength (λ) which facilitates the determination of fractional saturations of SaO_2

(Total hemoglobin (Hb), hemoglobin (Hb) derivatives), and total bilirubin concentration [17, 33]. The hemoglobin derivatives include the Oxy-Hb, de-oxy-Hb, Met-Hb, Carboxy-Hb, and Fetal Hb. Haematocrit, i.e. the ratio of the volume of RBCs in blood to that of the total blood volume) [17].

The setup consists of:

→ Lamp, the main light source.
→ Neon Lamp, provides the 585 nm spectral line.
→ Infrared filter, heat absorber.
→ Lens, facilitates the focusing of the light towards the cuvette that has been positioned within the hemolyzer.

Working: The hemolyzer consists of an ultrasonic resonator which emits 30 kHz ultrasound. Because of this, the erythrocytes are completely ruptured resulting in a homogenous sample (pigments are evenly distributed across the entire reference range) [33]. Then, the photodiodes (wavelength-specific) installed in the spectrophotometer will measure 128 types of wavelength (λ) in the range of 478–672 nm [17]. The property of the transmitted light is then analyzed by the principle of Beer–Lambert law [24].

9.5.10 POCT of Blood Gas Analysis

Blood gas and pH analyzers use electrodes to detect pH, partial pressure of CO_2 and O_2 in the blood [15, 16].

Chemistry analyzers use a dry reagent pad system in which a filter pad is impregnated with all reagents required for a particular reaction is placed on a thin plastic strip [20].

Electrolyte analyzers use the ion-selective electrode methodology in which measurements of the ion activity of the solution are made potentiometrically using an external reference electrode and an IAC containing an internal reference electrode [20]. Whole blood samples are placed in tubes on reaction cuvettes or on test strips and are loaded into the analyzers.

The most widely used **POCT** blood gas analyzer is **i-STAT** [34]. It has been developed by **Abbott Laboratories.** There are two types of cartridges (CG4+ and EG7+) that are used in i-STAT [34]. The parameters that are measured by the i-STAT are explained in the figure below (Figure 9.2) [34, 35]. Other used POCT devices for BGA are Epoc and OPTI CCA-TS2. These devices come under the device class of unit-use systems [35].

9.6 Urine Analysis

9.6.1 Introduction

Since decades, clinical diagnostics have become accessible, affordable as well as user-friendly due to the rapid development of POCT. This platform has enabled the patients to examine their own health status regularly in the comfort of their home so that diseases can be diagnosed at the earliest possible time point. Traditional methods such as qualitative or semiquantitative dipsticks and lateral flow immunoassays have been advanced to recent progresses in POCT such as microfluidics and paper strip-based assays where signals can be measured quantitatively using handheld devices.

Urine samples being complex contain a various component that indicate patient health. These classic POC test strips detect these parameters and can be used both by patients alone for self-testing and as screening tests in the central laboratory. Due to its high practicality and specificity, the introduction of test strips in the 20th century transformed urine diagnostics. This development of POCT devices for urine analysis with microfluidic technology can facilitate the detection of infections and frequent monitoring of chronic disease thereby decreasing the need of testing facilities [36]. In this article, we have reviewed upon the recent advancements in clinical urine analysis. Current technologies in sensing and imaging have been specifically discussed.

9.6.2 Methodologies

The conventional assay formats for POCT that have been dominating for decades include urine dipsticks and lateral flow assay. Owing to their miniaturized, superficial, and portable fabrication, these formats gained marketability.

9.6.2.1 Urine Dipsticks (Colorimetric Reagent Strip)

A patient's urine sample often show some pathological changes in urinalysis, and these changes are diagnosed using a urine test strip or dipstick. These test strips contain 10 different pads which change color upon the introduction of the urine sample. The results are then compared to a color-coded chart as per the specific concentration of the components or parameters [37] (Figure 9.3).

The colorimetric reagent strip works on ability of the components to change the color of definite acid–base indicators. The 5 mm wide plastic strip has chemicals infused in the respective pads that react with those components present in the urine sample thus imparting a distinctive color. Unlike paper strips, dipsticks facilitate the simultaneous determination of several compounds at once. The multiparameter strips included the analysis for the presence of leukocytes, nitrite, urobilinogen, protein, pH, hemoglobin, specific gravity, ketone, bilirubin as well as glucose. Thus, it has evolved to be a highly effective method for the rapid detection and screening of diseases with accurate results in a user-friendly manner [38]. These instruments are usually accompanied with a colorimetric chart with the help of which the colorimetric sensing can generate semiquantitative readouts (Figure 9.4).

9.6.2.2 Lateral Flow Immunoassay (Rapid Test)

Lateral flow immunoassays exists in diverse sizes, dimensions, and fabrications with either in a plastic housing or without one. They are reported to be designed

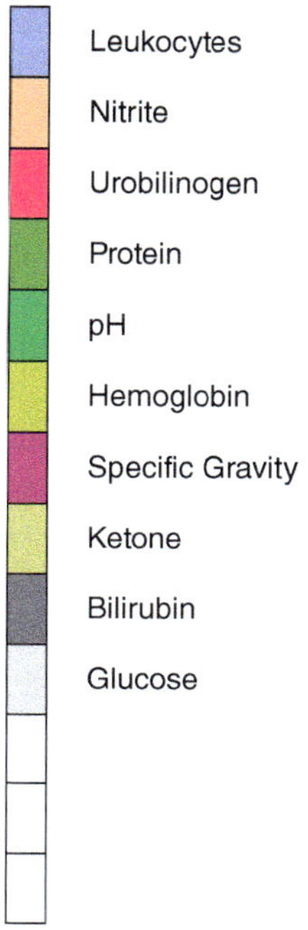

Figure 9.3 Urine Strip with different chemical Pads.

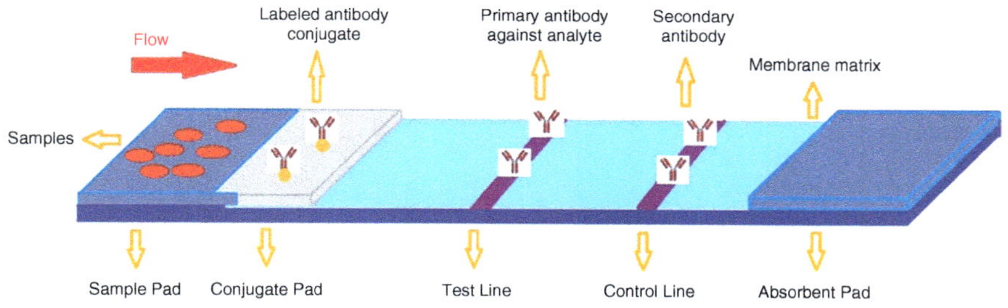

Figure 9.4 Schematic view of a typical lateral flow assay.

with very high sensitivity for the detection of infections; such as albumin is utilized as a biomarker for the detection of diabetes and human serum is detected in urine for renal diseases [39]. The assay works on the potential of antibodies to detect the presence of the analytes of interest, thus constituting the recognition part of the assay. Lateral immunoassays can be broadly divided into direct or sandwich assay formats and competitive assay formats.

The sideways flow of assay basically contains four segments overlapping one upon another, where the essential element is the sample pad which receives the urine sample and delivers it to the conjugate pad. As the sample reaches the conjugate pad, the analyte in the sample binds to the conjugated biomolecules which are held in the pad. This analyte–conjugate complex then reaches the third element, the nitrocellulose matrix, with preset lines containing immobilized antibodies which are fluorescently labeled. The first test line is composed of immobilized antibodies specific to the analyte and the second control one is composed of immobilized antibodies specific against free-labeled antibodies. The analyte and the analyte-specific antibody complex is caught in the test line of antibodies creating a colored spot visible to the naked eye. The rest unreacted labeled antibodies or the free conjugates get captured in the control line creating a visible spot too, hence yielding a positive result [40]. The remaining fluid is retained at the absorbent pad, i.e. the fourth element. This visual examination is colorimetric-based and the color intensity is ascertained with the help of a color chart. It can be used in screening as well as diagnosis via both qualitative as well as semiquantitative analysis [41].

9.6.3 Measurement of the Parameters in the Urine Sample

The parameters that are required to detect for the presence of any infection or disease for urine analysis are as follows:

- Protein
- Nitrite
- Leukocytes
- Bilirubin
- Urobilinogen,
- Specific gravity
- Hemoglobin
- Ketone
- pH
- Glucose

9.6.3.1 Protein

Although having general symptoms, proteinuria is a recurring disease, and it is detected in the infections of glomerulus and the urinary tract. When there is a positive detection of proteinuria, specific marker proteins are diagnosed differentially with help of classical test strips that follow the protein error test principle. This principle is based on the color change due to the interaction between the free protein amino groups in the urine and protons from tetrabromophenol indicator at constant pH. In case of urinary tract infections (UTI), the test strip might show false-positive results since the urine either remains contaminated or highly alkaline [42].

If there is a continuous detection of proteinuria or microalbinuria then an elevated albumin excretion and creatine clearance is observed by quantitative estimation. This measurement of albumin can be done either using immunochemical assay techniques or protein error strips [43]. In the immunoassay technique, the urine sample is passed through a test strip containing soluble red antibody–gold conjugate. Albumin, if present, binds to the conjugate, and the excess conjugate is retained in a capture zone such that only the albumin–gold conjugate flows to the detection zone imparting a light red color. On the other hand, the test strip based on indicator error principle shows a blue color due to the reaction with sulfophthalein indicator [44].

9.6.3.2 Nitrite

Gram-negative bacteria or nitrate-reducing bacteria, such as *E. coli*, *Klebsiella*, *Enterobacter*, and *Proteus* strains, cause bacteriuria, contributing to another major symptoms of UTI. However, this condition requires enough nitrate intake orally for its reduction to nitrite in the bladder. A rapid screening is done for the detection of nitrites, thereby the presence of any infection using a reactive test strip which follows the Griess reaction. Here, the nitrite reacts in an acidic medium with an aromatic amine (diazotization) producing a diazonium salt which consequently reacts with a chromogen to give a pink colorant [37].

1) Sulphanilamide + $NO2^-$ → Diazonium salt **[DIAZOTIZATION]**

2) Diazonium Salt + tetrahydrobenzoquinoline (chromatogen) → Pink Azodye

This too might give false negative results either due to large number of non-nitrite-producing bacteria or due to the presence of ascorbic acid which competes with the Griess reaction rendering low readouts.

9.6.3.3 Leukocytes

Often inflammation of the kidneys and the urinary tract due to bacterial infections give rise to leukocyturia. The majority of positive leukocytes excreted in the urine are granulocytes and during their lysis, an esterase enzyme is liberated which catalyzes the hydrolysis of an ester of indole carboxylic acid. This results in the release of indoxyl which in turn oxidizes in an acidic medium to impart a dark blue pigmentation, thus detecting the presence of leukocytes.

1) Reaction catalyzed by leukocyte esterase:

Indolecarboxylic acid ester → 3-hydroxy-5-phenylpyrrole (Indoxyl) + Acid

2) In acidic medium:

Indoxyl + Diazonium Salt → Dark Blue Azole dye

9.6.3.4 Bilirubin

Conjugated bilirubin is detected in urine when its concentration exceeds 20 $mg\,l^{-1}$ (bilirubinuria) causing disordered excretion of bilirubin, parenchymal liver damage, and cholestasis. The test strips employ the diazotization reaction for the detectionofbilirubinwherebilirubinreactswith2,4-dichloroanilineor2,6-dichlorobenzene-diazonium-tetrafluoroborate (diazonium salt) at an acidic pH. Azobilirubin is formed as a result of imparting different color gradations from pink to violet [45]. If the bilirubin sample is kept in the presence of light, it might get oxidized to biliverdin rendering false-negative results.

$$\text{Diazonium Salt} + \text{Bilirubin glucuronide} \rightarrow \text{VioletAzodye}\left(\text{Azobilirubin}\right)$$

9.6.3.5 Urobilinogen

When there is an increased production of bilirubin, often they breakdown via bacterial action in the intestine converting them to urobilinogen, thus causing urobilinogenuria. This gives rise to symptoms like hemolysis and primary liver diseases. Uroblinogen testing can be done using two different reaction test strips, one incorporating the diazotization reaction and other using the Ehrlich's reaction. In the diazotization reaction, urobilinogen reacts with 4-methoxybenzene-diazonium-tetrafluoroburate, and in the Ehrlich's reaction, the compound reacts with para-dimethyl-aminobenzaldehyde. Both the reactions yield a pinkish–red color complex in a strong acidic medium [46].

1) Diazotization reaction (acid medium):

Urobilinogen + 4-methoxibenzene-diazonium-tetrafluoroborate → Red Azodye

2) Ehrlich's reaction (acid medium):

Urobilinogen + p-dimethylaminobenzaldehide → Red Azodye

9.6.3.6 Specific Gravity

The test for specific gravity was mainly initially used as an indicator of diuresis, for a revised analysis in either highly diluted sample or a sample highly concentrated with specimens. The specific gravity varies substantially as it is directly proportional to the volume of the excreted fluid. Thus calculating specific gravity can account for the measurement of other parameters too. The chemical mechanism of the test is based on the change in dissociation constant (pKa) and involves two reagents in alkaline medium, a pH indicator, and a cation exchanger. Here, a polyelectrolyte gets ionized and an equal amount of cations are released rendering a color change on the pH indicator. This shift in indicator from blue to yellow determines the consequent drop in pH [37]. A rather high alkalinity or a high acidity could lead to false positive results.

1) Polyelectrolyte-H_n(polymethyl vinyl ether) + Cations^{n+} → Polyelectrolyte-Cations + nH^+

2) H^+ + Bromothymol blue → Bromothymol blue-H^+

9.6.3.7 Hemoglobin (Hb)

Several urological diseases are accompanied with hematuria in which due lysis (hemoglobinuria), intact erythrocytes, and free Hb are detected. The blood test uses hemoglobin's pseudo peroxidase to catalyze the reaction between peroxide ando-toluidine or 3,3′,5,5′-tetramethylbenzidine(chromogen). This chromogen acts as the indicator and is oxidized by cumene hydroperoxide imparting a color change in the reaction zone. The resultant color ranges from green to dark blue depending upon the amount of Hb [47].

$$\mathrm{H_2O_2 + Chromogen \rightarrow Oxidised\ chromogen(coloured) + H_2O}$$

9.6.3.8 Ketone

Ketones involve the intermediates in the metabolism of fatty acids viz. acetone, acetoacetate (AcAc) and beta-hydroxybutyrate (β-HBA). The presence of ketones is due to starvation when carbohydrate intake is absent and the proportions of each intermediate in urine differ from each other. The test trip follows the principle of Legal's test in which acetoacetate reacts with nitroferricyanide or sodium nitroprusside in alkali medium giving rise to a magenta-colored complex. A fresh urine sample should not be stored for too long as the acetate might get broken down by bacteria resulting in false-negative outcomes.

9.6.3.9 pH

Generally, pH of a fresh urine varies from 5.0–8.0 with the influence of diet and medicines. Meat-rich diets cause a slightly acidic pH whereas vegetable-rich diet likely cause a rather alkaline pH. Bromothymol blue and methyl red are the most used combination of indicators used in which the color change is detected showing the result.

9.6.3.10 Glucose

Diabetes mellitus is caused due to glucosuria or glycosuria, i.e. when the blood glucose level surpasses the renal threshold. Glucose is detected in urine using the test strip which follows the principle of the enzymatic oxidation of glucose followed by the secondary reaction which is carried out by peroxidase. In the presence of peroxidase enzyme, the peroxide produced from the glucose oxidation reaction is combined with a chromogen to form an oxidized colorant [48]. This reaction might indicate false-positive due to traces of oxidizing agents.

1) Catalysis in the presence of glucose oxidase:

Glucose + O_2 → D-glucono-δ-lactone + H_2O_2

2) Catalysis in the presence of peroxidase:

H_2O_2 + Chromogen → Oxidised chromogen (coloured) + H_2O

9.7 Conclusion

Despite having varied dimensions and design structure, the above mentioned POCT methods utilize the same amount of analytes, indicators, and receptors. Various researchers have approved these POCT methods pertaining to their ease of use, user-friendliness, easy manufacturing, and error-free performance. Personalized healthcare and personalized continuous monitoring have led to the rapid advancements of POCT methods. Additionally, the gradual increase in the continuous monitoring alternatives can be included under the catchphrase-mobile healthcare [49]. Over the last few years, it has been noticed that the implementation of smartphone-coupled analytical systems has expanded on a global scale, thus, embodying itself as a future technology for POCT applications. This current trend will loosen up the capacity at central laboratories for the intricate and time-consuming techniques like next-generation sequencing or mass spectrometric proteome analyses. In the developing mobile healthcare, the most promising one is the cellophane-based POC technologies for the output of colorimetric, electrochemical, lateral flow, and label-free assays for the detection of biomolecules, cells, microorganisms, nanoparticles, and other diagnostic applications [50]. It is not too late until POCT would be available for most of the people and this would allow them to monitor and take care of their own health.

References

1 Heinemann, L. and Freckmann, G. (2015). CGM versus FGM; or, continuous glucose monitoring is not flash glucose monitoring. *Journal of Diabetes Science and Technology* 9 (5): 947–950. https://doi.org/10.1177/1932296815603528.

2 Luppa, P. and Junker, R. (ed.) (2018). *Point-of-Care Testing: Principles and Clinical Applications*, 103. Springer.

3 Ashworth, L., Gibb, I., and Alberti, K.G. (1992). HemoCue: evaluation of a portable photometric system for determining glucose in whole blood. *Clinical Chemistry* 38 (8): 1479–1482.

4 Cardosi, M. and Liu, Z. (2012). Amperometric glucose sensors for whole blood measurement based on dehydrogenase enzymes. In: *Dehydrogenases* (ed. R.A. Canuto), 319. Rijeka, Croatia: InTech.

5 Rajendran, R. and Rayman, G. (2014). Point-of-care blood glucose testing for diabetes care in hospitalized patients: an evidence-based review. *Journal of Diabetes Science and Technology* 8 (6): 1081–1090.

6 Clarke, S.F. and Foster, J.R. (2012). A history of blood glucose meters and their role in self-monitoring of diabetes mellitus. *British Journal of Biomedical Science* 69 (2): 83–93.

7 Johnson, G. (1884). Clinical lecture on the various modes of testing for sugar in the urine. *British Medical Journal* 1 (1201): 1.

8 Kotwal, N. and Pandit, A. (2012). Variability of capillary blood glucose monitoring measured on home glucose monitoring devices. *Indian Journal of Endocrinology and Metabolism* 16 (Suppl 2): S248.

9 Free, A.H. and Free, H.M. (1984). Self testing, an emerging component of clinical chemistry. *Clinical Chemistry* 30 (6): 829–838.

10 Mendosa D. History of blood glucose meters: transcripts of interviews.

11 Atkin, S.H., Dasmahapatra, A., Jaker, M.A. et al. (1991). Fingerstick glucose determination in shock. *Annals of Internal Medicine* 114 (12): 1020–1024.

12 Perera, N.J., Stewart, P.M., Williams, P.F. et al. (2011). The danger of using inappropriate point-of-care glucose meters in patients on icodextrin dialysis. *Diabetic Medicine* 28 (10): 1272–1276.

13 Boemke, W., Krebs, M.O., and Rossaint, R. (2004). Blutgasanalyse. *Der Anästhesist* 53 (5): 471–494.

14 Davis, P. and Kenny, G. (2005). Further techniques of gas and vapour analysis. In: *Basic Physics and Measurement in Anaesthesia*, 219–226.

15 Severinghaus, J.W. and Astrup, P.B. (1986). History of blood gas analysis. III. carbon dioxide tension. *Journal of Clinical Monitoring* 2 (1): 60–73.

16 Suzuki, H., Hirakawa, T., Watanabe, I., and Kikuchi, Y. (2001). Determination of blood pO_2 using a micromachined Clark-type oxygen electrode. *Analytica Chimica Acta* 431 (2): 249–259.

17 Mirzazadeh, M., Morovat, A., James, T. et al. (2016). Point-of-care testing of electrolytes and calcium using blood gas analysers: it is time we trusted the results. *Emergency Medicine Journal* 33 (3): 181–186.

18 Jacobson, L.S. and Lobetti, R.G. (2005). Glucose, lactate, and pyruvate concentrations in dogs with babesiosis. *American Journal of Veterinary Research* 66 (2): 244–250.

19 Mikulcik, P. (2007). Rapidanalyse: Blutgase und mehr. In: *Siemens Medical Solutions*.

20 Wang, Y., Xu, H., Zhang, J., and Li, G. (2008). Electrochemical sensors for clinic analysis. *Sensors* 8 (4): 2043–2081.

21 Leiner, M.J. (1995). Optical sensors for in vitro blood gas analysis. *Sensors and Actuators B: Chemical* 29 (1–3): 169–173.

22 Omar, S. (2007). Electrochemistry (How a Blood Gas Machine Works). In: *University of Witwatersrand Part I Anaesthetics Refresher Course*.

23 Gehlen, M.H. (2020). The centenary of the Stern-Volmer equation of fluorescence quenching: from the single line plot to the SV quenching map. *Journal of Photochemistry and Photobiology C: Photochemistry Reviews* 42: 100338.

24 Swinehart, D.F. (1962). The beer-lambert law. *Journal of Chemical Education* 39 (7): 333.

25 Nolte D. (2018). Principles of Measurement Arterial blood gas, and electrodes.

26 Rabi, Y. and Ambalavanan, N. (2017). *Assisted Ventilation of the Neonate*, 6e. Elsevier.

27 Davis, P. and Kenny, G. (2005). Carbon dioxide measurement. In: *Basic Physics and Measurement in Anaesthesia*, 211–218.

28 Ortiz-Prado, E., Dunn, J.F., Vasconez, J. et al. (2019). Partial pressure of oxygen in the human body: a general review. *American Journal of Blood Research* 9 (1): 1.

29 Tenny KM, Keenaghan M. (2017). Ohms Law.

30 Leino, A. and Kurvinen, K. (2011). Interchangeability of blood gas, electrolyte and metabolite results measured with point-of-care, blood gas and core laboratory analyzers. *Clinical Chemistry and Laboratory Medicine* 49 (7): 1187–1191.

31 Dasgupta, A., Wahed, A., Dasgupta, A., and Wahed, A. (2014). Common poisonings including heavy metal poisoning. In: *Clinical Chemistry, Immunology and Laboratory Quality Control*, 337–351.

32 Prause, G., Ratzenhofer-Komenda, B., Offner, A. et al. (1997). Prehospital point of care testing of blood gases and electrolytes—an evaluation of IRMA. *Critical Care* 1 (2): 1–5.

33 Horecker, B.L. (1943). The absorption spectra of hemoglobin and its derivatives in the visible and near infra-red regions. *Journal of Biological Chemistry* 148 (1): 173–183.

34 Connelly, N.R., Magee, M., and Kiessling, B. (1996). The use of the iSTAT portable analyzer in patients undergoing cardiopulmonary bypass. *Journal of Clinical Monitoring* 12 (4): 311–315.

35 Hobin, H.P. (2020 Oct. 5; 2022 Mar. 26). i-STAT Analyser ABG machine. How its work. https://www.youtube.com/Watch?v=KcFTtGhYgbE&ab_channel=HobinHP.

36 Fogazzi, G.B. (2010). *The Urinary Sediment: An Integrated View*. Penerbit Buku Kompas.

37 Di Lorenzo, S. (2008). *Análisis de Orina y de los Liquidos Corporales*, 5. Argentina: Editorial médicapanamericana.

38 Yetisen, A.K., Akram, M.S., and Lowe, C.R. (2013). Paper-based microfluidic point-of-care diagnostic devices. *Lab on a Chip* 13 (12): 2210–2251.

39 Posthuma-Trumpie, G.A., Korf, J., and van Amerongen, A. (2009). Lateral flow (immuno) assay: its strengths, weaknesses, opportunities and threats. a literature survey. *Analytical and Bioanalytical Chemistry* 393 (2): 569–582.

40 O'Farrell, B. (2015). Lateral flow technology for field-based applications-basics and advanced developments. *Topics in Companion Animal Medicine* 30 (4): 139–147.

41 Koczula, K.M. and Gallotta, A. (2016). Lateral flow assays. *Essays in Biochemistry* 60 (1): 111–120.

42 Hofmann, W., Regenbogen, C., Edel, H., and Guder, W.G. (1994). Diagnostic strategies in urinalysis. *Kidney International. Supplement* 47.

43 Kutter, D. (1998). Chemistry: a chemical test strip to determine low concentrations of albumin and creatinine in urine. *Laboratory Medicine* 29 (12): 769–772.

44 Pugia, M.J., Lott, J.A., Clark, L.W. et al. (1997). Comparison of urine dipsticks with quantitative methods for microalbuminuria. *European Journal of Clinical Chemistry and Clinical Biochemistry* 35 (9): 693–700.

45 Jendrassik, L. (1938). Vereinfachtephotometrischemethodenzurbestimmung des blutbilirubins. *Biochemische Zeitschrift* 297: 81–89.

46 Laurine G. (1987). Análisis de orina: atlas color.

47 Kavoussi, L.R. and Wein, A.J. (ed.) (2007). *Campbell-Walsh Urology*. WB Saunders.

48 Heinemann, L., Asche, W., Withold, W., and Berger, M. (1994). Quantitative Beziehungzwischen der mitvierSchnelltestsbestimmten Ketonurie und der gleichzeitigvorliegenden Ketonämie. *Diabetes Stoffwechsel* 3: 339–342.

49 Luppa, P.B., Bietenbeck, A., Beaudoin, C., and Giannetti, A. (2016). Clinically relevant analytical techniques, organizational concepts for application and future perspectives of point-of-care testing. *Biotechnology Advances* 34 (3): 139–160.

50 Vashist, S.K., Mudanyali, O., Schneider, E.M. et al. (2014). Cellphone-based devices for bioanalytical sciences. *Analytical and Bioanalytical Chemistry* 406 (14): 3263–3277.

10

Strategic Synthesis of Diagnostic Novel Materials Against Infectious Diseases

Hardik Shyam Churi[1] *and Sushma Dave*[2]

[1]*SDSM College, Department of Biotechnology, 913, Govind, Fanasbhat, Kelwe Mahim, District Palghar, Mumbai, Maharashtra 401402, India*
[2]*JIET, Department of Applied Sciences, NH 62, Pali Road, Jodhpur, Rajasthan, India*

10.1 Introduction

Emerging infectious diseases pose a global threat to human beings. Infectious diseases are caused by infectious agents such as pathogenic microbes, fungi, and viruses. Pathogenic agents sometimes become lethal and may create pandemic situations. Rapid and ultrasensitive detection of these infectious agents helps to tackle such life-threatening conditions. Different routes of transmission spread infectious agents in humans. Due to the rise in population, industrial and other household wastes are directly dumped in dump yards, which increases the chances of contamination. Improper waste management systems decrease hygienic conditions. Plastic such as nonbiodegradable products clog the water bodies, which block the drainage system. In such conditions, waterborne pathogens can easily spread in nearby communities. Improper cleaning and hygiene maintenance seriously affect sanitization personnel which can spread such infections to the community. Pathogens can transmit through various routes such as contaminated air, water, and food, contaminated respiratory droplets, vectors carrying pathogens, physical contact, and accidental transfusion of contaminated body organs, and fluids.

Consumption of contaminated food and water increases the chances of infections. Cholera, hepatitis, typhoid, diarrhea, and so many stomach infections are spread through contaminated food and water. Lack of waste management techniques and training of the sanitization personnel are at high risk of various infections. Contaminated water is mixed with fresh water sources and then such pathogens enter into drinking water. Roadside food or street food suppliers most probably used such contaminated water for cleaning and washing utensils. Some patients disobey rules and regulations of masking and self-distancing due to which various infectious agents are spread through sneezing and coughing. Sneezing and

Point-of-Care Biosensors for Infectious Diseases, First Edition.
Edited by Sushma Dave and Jayashankar Das.

coughing produce various size droplets from which some are too small and easily respirable but contain thousands of infectious agents. These respirable droplets can travel for long-distance in the air that can easily spread the infection to nearby people. Large droplets easily get settled down on surfaces due to their large size and weight but may contain lakhs of infectious agents. These droplets contaminate surfaces that can easily come in contact with hands.

Various vectors can potentially spread diseases such as malaria, dengue, some viral fevers, schistosomiasis, and tularemia. Mosquitoes, flies, ticks, lice, aquatic snails, and sandflies are the main vectors that spread different infections. Due to a lack of knowledge and unhygienic health condition various sexually transmitted diseases (STD) are spread worldwide. Some infections are spread in communities just because of close contact with infected people. Nosocomial infections are spread from hospitals due to improper personnel training or carelessness of hospital staff. Sometimes lack of technology or accidentally contaminated body organs or fluids like blood is transmitted to a healthy person which can cause infections in patients.

These infectious agents are microscopic or some exist in nano size. Therefore the diagnosis of infectious diseases is difficult. For proper treatment, diagnosis of infectious agents is important. For the past few decades, diagnosis has been much easier due to novel technologies. Traditionally diagnosis of infectious agents was difficult and tedious, which required too much time and extended the beginning of treatment. The delay in treatment due to a time-taking diagnosis can cause the death of the patient. Sometimes due to the wrong diagnosis, doctors suggest the wrong medicines to the patient which creates serious side effects in the patient's body and sometimes causes death. To avoid such issues early accurate detection of infectious diseases is important which helps doctors to suggest proper medicines and give perfect treatment for specific infectious agents. Previously patients had to submit more blood for various medical tests. Improper training of medical workers can cause serious issues while blood sampling and in some cases accidentally infectious agents are transmitted through contaminated syringes.

The new advanced technologies help for early and accurate diagnosis of pathogens. Various novel technologies are available that use a noninvasive technique to avoid pain and transmission of parenteral route infections. Nanotechnology is an emerging field that helps to develop different types of tools and materials which can be beneficial for the diagnosis and treatment of various infectious diseases. Previously due to lack of technology thick needles were used for blood sampling but just because of nanotechnology thin size needles are used which is not that painful. Some biological components are produced by the patient's body which can be used as biomarkers for the detection of infectious agents. These biomarkers may contain proteins, enzymes, dead cells, chemicals produced by infectious agents, and microRNAs. The levels of these biomarkers are more abnormal in the patient's body than in a normal healthy person therefore detecting the levels of these biomarkers will be easy to detect infectious diseases. The abnormal amount of biological molecules can be detected by using biosensors. Biosensors play a crucial role in the detection of infectious diseases. Biological materials such as enzymes, antibodies, antigens, epitopes, and genetic material such as DNA-RNA, proteins, chemicals, cell debris, and other small components can be detected by using biosensors. New advanced technology helps to build noninvasive biosensors which are directly connected to smartphones and continuously monitor a patient's body. Nanotechnology can be used for the synthesis of drug carrier molecules which will peculiarly deliver the drug at the target site and will decrease the chances of side effects on the patient's body. Nowadays antibiotic resistivity increases in patients' bodies therefore nano size drug molecules can be delivered using nano carriers to tackle a resistivity issue. These nano size carriers require less amount so overdose toxicity in patients will decrease. Different combinations of nanotechnology-based methods show promising results in the medical and clinical fields for the diagnosis and treatment of infectious diseases. For an early and accurate diagnosis, on-site detection of infectious agents with high accuracy is important so that health professionals can suggest specific treatments and medications to the patient. Point-of-care (POC) technology has great potential for quick and ultrasensitive detection. Generally, POC tests are performed in front of the patient or nearby laboratories therefore results of the test can directly inform the patient as well as doctors. These POC devices mainly involve kit-based strips, biosensors, lateral flow devices, loading devices, and some polymerase chain reaction (PCR)-based devices. The reaction between ligands and biological material produces different responses. Various transducers are used in POC devices which convert these chemical responses into measurable signals which can be detected by various automated devices. The signals produced by reactions are colors, electrochemical, thermal radiation, magnetic response, and some optical responses. These readable signals are then forwarded to computers or other devices which help track or monitor the patient's health status.

This chapter mainly focuses on nanotechnology-based POC devices for various infectious diseases.

10.2 Detection Needs at the POC

POC tests peculiarly help in building the best management for infectious diseases. POC methods mainly involve mobile devices that are easily transportable and easy to handle. The POC test is conducted near the patient which provides quick and accurate on-site results, which is beneficial to providing proper treatment to the patient as soon as possible. Rapid on-site diagnosis reduces patient visits to multiple diagnosis centers; some noninvasive methods help to avoid parenteral route infection. The POC tools provide appropriate information to the medical officer who helps them to select specific treatments and medications for patients. An accurate diagnosis decreases the cytotoxicity in the patient's body. POC devices detect infectious agents or patients' body fluids that contain antibodies. Recent advanced technologies in POC use different constituents from body fluids to detect the causative agent [1, 2]. Nanotechnology helps to decrease the size of various devices required for diagnosis as well as decrease the cost so that anyone can access it. Traditionally detection tools required more blood for various clinical tests but now even a small amount of blood can be used for different diagnostic tests with the help of nanotechnology. A single blood test can be used for the detection of multiple causative agents. Various advanced nanomaterials are available in the market, potentially detecting various infectious diseases.

10.2.1 Nanomaterials for Malaria Parasites Detection

Malaria is a parasitic infection spread by mosquitoes. Female mosquitoes act as a vector that carries plasmodium parasites. When the female mosquito bites an infected person the plasmodium parasite enters into a female mosquito and then that carrier mosquito transfers that plasmodium into the healthy person through the bite. Different species of female mosquitoes are responsible for the transmission of malaria in healthy persons. Different types of mosquito species act as a carrier for the plasmodium parasites. Unhygienic conditions increase the growth of plasmodium parasites and mosquitoes across the globe [3]. Many underdeveloped and developing countries lack various hygienic and medical facilities due to which cleanliness and treatments of malaria are difficult. Millions of people have died in the past few decades because of malaria, therefore accurate diagnosis and treatment should be easily available for everyone [4]. The lack of diagnostic technology increases the difficulty of the detection of malaria. Recently malaria diagnosis is done by a simple microscopic technique and immunochromatographic rapid diagnostic tests (RDTs). For species detection, genetic information is needed from which nucleic acid amplification tests (NAATs) can be performed. These methods are the best choice for malaria diagnosis.

The best and easiest POC technology is the microscopic technique for primary screening of malaria. Two types of blood smear techniques are used for malaria detection. Doctors prefer thick and thin blood smears stained with Giemsa dye used for detection. A thick blood smear is prepared by placing a few blood drops on a glass slide, then lysed with water and stained with Giemsa dye. A thick blood smear may contain more parasites, which increase the sensitivity of the test. A thick blood smear allows for examining more blood than a thin smear. Preparation of a thick blood smear requires proper technicians and it is difficult to study morphological characters therefore thin blood smear technique is preferred. A thin blood smear helps to study the proper morphological characteristics of the parasite but the sensitivity is low compared to a thick blood smear. The technique is cheap and fast, and no special instruments are required [5, 6]. The microscopic technique required a well-trained technician and a microscope with reagents. In some regions lack of such things makes it difficult to diagnose and keep track of malarial parasite infection in individuals and communities. An alternative for microscopic technique is RDTs, which is a quick method for malarial parasite detection. This method has the advantage that it can detect parasitic infection in malaria-suspected patients and patients with negative microscopic tests. In negative microscopic tests, the malaria parasite remains in a dormant stage and is difficult to detect in blood smears. RDTs focus on the detection of proteins produced by the malaria parasite. An antigen produced by the parasite was detected by using fresh blood of a patient subjected to lateral flow immunochromatography with the help of immobilized antibodies [7]. This method simultaneously detects two or more antigens and has the potential to differentiate species-level infection.

Different types of proteins, enzymes, and biomolecules produce plasmodium parasites that can be specifically used as biomarkers for detection. Aldolase is commonly used as a biomarker for malarial parasitic detection but fails to distinguish malarial species. For species detection, some unique proteins are used as biomarkers. The *Plasmodium falciparum* produces Histidine Rich Protein 2 (HRP2), which is a biomarker for detection. *Plasmodium* lactate dehydrogenase also acts as a biomarker for the detection of *Plasmodiumvivax & Plasmodium falciparum* using specific pLDH antibodies [8]. Plasmodium parasites produce different enzymes that are important for their survival and growth like plasmodium glutamate dehydrogenase. This ubiquitous enzyme is used as a biomarker because the enzyme is absent in host blood and therefore peculiarly used for the detection of parasites [9]. For the metabolic activities of nucleic acid, specific enzymes are produced by the plasmodium parasite from which hypoxanthine guanine phosphoribosyl transferase is required for purines. This enzyme acts as a biomarker for plasmodium falciparum detection [10]. During the life cycle of a plasmodium parasite in a host cell, the parasite consumes blood cells from which heme molecules are released which are toxic to the parasite. To tackle this problem parasites convert that toxic heme molecule into nontoxic inert molecule hemozoin. Hemozoin is a crystal form of brown–red color pigment that serves as a biomarker for parasite detection [11]. Aptamers are chemical antibodies used to capture malaria parasites from the sample and then subjected to various techniques for further detection. Aptamers are also used to capture proteins and enzymes produced by plasmodium species. Infected red blood cells (iRBCs) contain hemozoin crystallites and can be separated by using magnetic resonance relaxometry (MRR) [12]. Gold nanoparticles (AuNPs) and oligonucleotides MSP10 are used for the detection of plasmodium DNA [13]. Various nanoparticles and nanomedicines are used for the treatment of plasmodium parasites and to decrease toxicity in patients.

Recent studies of malaria detection reveal that mutated malaria species lack HRP2 and HRP3 genes, which is difficult to diagnose using RDTs [14]. Therefore, accurate and specific detection of malaria parasites is possible through genetic information. Nucleic acid detection helps to overcome all the barriers to malaria diagnosis. The amount of nucleic acid obtained from a patient's blood sample is less therefore PCR is important for amplification. If the parasitemia is low then more PCR amplification is required to obtain an adequate amount of genetic material for diagnosis. PCR is more sensitive than microscopic and RDTs techniques can detect 0.5–5 parasites per μl. PCR in combination with the lateral flow technique shows promising results for the detection of parasites. This combined technique uses the colorimetric method to visualize the result. Researchers currently use various PCR-based tools for the detection and diagnosis of malaria infection.

Various colorimetric, quantum dots (QDs), and fluorescence aptamer-based biosensors are commercially available in that some chips are used as lab-on-a-chip methods. Microfluidic is a physics term that minimizes the time, and amount of required sample, waste, and instruments but increases the specificity and sensitivity of the test. These microfluidic devices are portable, easy to use, automated, and inexpensive diagnostic tools. These tools increase the attention of researchers and the diagnostic field for POC devices.

10.2.2 Nanomaterials for HIV

The whole world is suffering from various types of viral infections from which only a few have proper treatment but the rest are untreatable. Human Immunodeficiency Virus (HIV) attacks the human immune system and causes acquired immunodeficiency syndrome (AIDS) in approximately more than 30 million people across the world [15]. HIV is a STD and can also be caused by contaminated syringes, and sharp blades via parenteral routes. After many kinds of research, still, no proper vaccine therapy is available for HIV, only antiretroviral drugs are available that are prescribed by medical professionals to patients. Treatment of such life-threatening viral infection is possible in the early stages of infection because this virus peculiarly attacks the immune system, and in the late phase of infection, the immune system of the patient completely collapses. For better and more accurate treatment, diagnosis of HIV in an early stage of infection is important. Serological tests are traditionally employed for the detection of HIV infection. In this serological test, anti-HIV antibodies are used as biomarkers for the detection of HIV from a patient's body fluids such as blood or oral fluid [16]. Various techniques have the potential for rapid diagnosis of HIV with the help of fluorescent molecules.

AuNPs were used in bio barcode amplification assay (BCA) for the detection of the capsid protein of HIV (p24) that appears in patients' blood samples after two weeks of infection [17]. The latest modified version of BCA involves the use of antibodies on which magnetic nanoparticles (MNP) are coated. These MNP-coated antibodies are involved in the separation of the AuNPs-p24 antigen complex from the mixture [18]. The genetic material of HIV is RNA hence for completion of the life cycle reverse transcriptase (RT) is important because of that the amount of RT in the patient's blood increases. These RTs can be used as biomarkers. The ExaVir Load Assay is used for the detection of RT activity [19]. In this assay 5-bromodeoxyuridine, 5′triphosphate (BrdUTP) with an alkaline phosphatase-conjugated antibody is used for quantification of RT activity. A specific substrate is added for alkaline phosphatase which produces color that can be detected by colorimetric tools [20–23]. Various modified PCR versions are used for the detection of HIV. Some electrochemiluminescence biosensors used AuNPs and AgNPs (Silver Nanoparticles) to detect HIV. Oxide forms of metals create some important changes in biosensors which improve detection sensitivity. QDs, carbon nanotubes (CNT), and graphene molecules show promising results in detection. Different types of markers are generally used for the diagnosis of HIV. The $CD4^+$ T cells are destroyed by HIV therefore as HIV infection increases $CD4^+$ cells decrease. Some methods involve direct viral RNA detection by using NASBA [24]. For growth and other activities, transcription is important. HIV-1 Tat is a protein that appears quickly after infection and hence can be used as a biomarker for early detection. The HIV molecule itself has various protein molecules from which matrix p17 protein is used as a biomarker that can be detected with the help of surface-enhanced Raman spectrometry [25, 26]. Ultrasensitive and quick detection of HIV requires the detection of multiple biomarkers simultaneously to increase the detection rate to decrease the mortality rate in the whole world.

10.2.3 Nanomaterials for HBV

Millions of people are infected by the hepatitis B virus (HBV) across the world. For effective treatment, diagnosis of the virus is important. Hepatitis is a double-stranded DNA virus of the Hepadnaviridae family. The Hepatitis virus contains different proteins that can be used as biomarkers for detection. Various enzymes produced by HBV required for growth and activities can be beneficial for detection as biomarkers. Antigens of the virus or whole virus can be detected by using antibodies.

Different molecules of Hepatitis B (HB) virus can be used as target molecules such as surface proteins antigens (HBsAg), capsid protein antigens (HBcAg & HBeAg), and HBV genome (double-stranded DNA). Antibodies like anti-core protein antibodies (Anti-HBcAb), anti e protein antibodies (Anti-HBeAb), and anti-surface protein antibodies (Anti-HBsAb) are used for the detection of HB antigens in various commercially available POC tools [27]. The surface of the HB virus is made up of L, M, and P proteins. These surface proteins are responsible for the entry of the virus into the host cell therefore they serve as an important protein for vaccine production and detection [28, 29]. Traditionally PCR and immunoassay were used as diction tools for HBV. Nanotechnology creates a revolutionary change in the field of biomedical diagnostics. AuNPs and AgNPs are commonly used in some immunoassays but some modified versions involve MNP for antigen detection. An oligonucleotide sequence is used for immunoassays, one end of a sequence is incorporated on a glass slide and the free end is used for the recognition of viral nucleic acid, which will form a complex. AgNPs identify the complex and produce color that can be read out using spectroscopy [30]. Goat anti-HBsAb is coated on a polystyrene nanosphere (PSs) that reacts with target antigens to form a complex. This reaction complex reacts with hydrogen peroxide to reduce chloroauric acid and allows the synthesis of AuNPs. The color formation after the completion of the reaction can be observed by the naked eye [31]. Quantum dots are crystals that emit light of different wavelengths due to excitation by reaction, therefore preferentially used for POC tools in HBV detection. Streptavidin-bound MNP and horseradish peroxidase antibodies (HBsAb) are used in sandwich immunoassays. MNPs have the potential to detect the DNA of HBV using the lanthanide-DOTA complex [32]. HBV infection can create serious problems in a patient's body like liver failure, liver cancer, and liver cirrhosis [33]. In the case of hepatitis live cancer, alpha-fetoprotein acts as an important biomarker detected by using iron-oxide MNP [34]. Better and more efficient treatment should be given to the patient and for that specific POC, tools are required that detect the HB virus as early as possible. Various modified advanced versions of POC tools based on nanomaterials show promising results for the detection of HBV but still, more research is required. Mutations in the virus may sometimes change its structure but aptamer-based POC tools have the potential to create affordable POC procedures that will help with early and onsite HBV detection.

10.2.4 Nanomaterials for HPV

Human Papillomavirus (HPV) is a sexually transmitted infection (STI) transmitted through skin-to-skin contact and parenteral routes through contaminated syringes and blades. Most commonly, HPV causes infection in reproductive parts. HPV is a DNA virus of the *Papillomaviridae* family that contains more than 200 viruses. HPV does not cause cancer in the preliminary stage but attacks the immune system of the patient. HPV damages the immune system and cell-to-cell communication which leads to uncontrolled growth of cells. HPV is considered the main cause of cervical cancer in women. For efficient treatment of HPV, diagnosis is important, which can be possible with the help of nanotechnology-based POC tools and many advanced assays.

Currently, a gold standard technique for the diagnosis of HPV or cervical cancer is the direct invasive examination of a biopsied specimen of the cervix. Various other HPV diagnostic devices are commercially available but those are expensive and require a continuous electrical supply. Nanotechnology and microfluidic methods have the potential to develop affordable, easy to handle, POC tools for HPV biomarkers detection. Fluorescence-based streptavidin nanocrystals are used for HPV genotypes (HPV16, HPV18, and HPV45) detection. Nano-sized crystals are composed of polyethylene glycol (PEG) amine phospholipids labeled with fluorescein diacetate. The enzymatic reaction is further carried out by dimethyl sulfoxide (DMSO) and sodium hydroxide (NaOH) [35]. The capture probes specific for HPV are also used in combination with nanocrystals for hybridization. This enzymatic reaction produces fluorescence which can be detected using tools directly. Sandwich assay for HPV detection involves avidin-coated silica nanoparticles conjugated with biotinylated capture probes. Magnetic microparticles can detect HPV DNA strands hence used as capture molecules [36]. HPV-16 and HPV-18 are detected by using QDs that target DNA. A nano polymer biochip is used for the identification of HPV. Streptavidin cytochrome 5 is used to stain the DNA captured on a biochip which emits fluorescent color useful for the detection of HPV. Luminex Corporation patented a technique in which 46 probes were used to hybridize with colored microspheres for the detection of HPV DNA. The color complex was measured by flow cytometry or spectrophotometrically [37]. The DNA capture probe and fluorescent DNA probes for HPV detection are used in lateral flow DNA biosensors. These lateral flow DNA sensors are disposable and serve as a good POC tool for HPV detection [37, 38]. RT–PCR is also used for the detection of HPV that uses mRNA as a target. NASBA is the best method for the real-time detection of HPV-16 RNA molecules [37]. Various computer–mobile-based optical imaging methods are used for confirmation of cervical HPV infection. More research is required for affordable POC tools for detecting cervical cancer.

10.2.5 Nanomaterials for Dengue Virus

Dengue is a viral infection spread by an infected *Aedes* mosquito vector through a bite. Dengue virus remains in mosquitoes and the carrier mosquito-spread the virus to a healthy person. Approximately half of the population is living in the area which is considered a dengue risk zone. Each year millions of people are infected by the dengue virus therefore inexpensive affordable POC tools are important for the detection of the virus so that healthcare personnel can give accurate medications and treatment to patients without causing any cytotoxicity.

Dengue viruses can be detected by analyzing blood, serum, urine, and saliva from a patient's body by using biosensors. Various biosensors are commercially available that can be used for the detection of the dengue virus. The virus produces different types of proteins and enzymes for its growth and life cycle process. Two types of nonstructural (NS) proteins are produced by the virus on the fifth day after infection that can be used as a powerful biomarker for detection. NS proteins are detected by using optical biosensors [39, 40]. The surface Plasmon resonance (SPR) technique in combination with optical sensors is used for the detection of envelope (E) proteins with the help of IgM gold nano cellulose [40]. Four serotypes of dengue viruses (DENV) are closely related to each other and commonly cause dengue infection. Triangular silver nanoparticles (TAgNPs) have the potential to differentiate and detect serotypes. When TAgNPs attach to a specific serotype, RNA produces a color due to a reduction in pH. This biosensor will create a revolutionary change in biomedical POC tools and helps for the accurate identification of DENV in real time [41]. AuNPs bring the attention of researchers because the synthesis and handling of AuNPs are easy. Immunoassay-based AuNPs in combination with localized SPR produce two types of colors, red color indicates the dispersion of AuNPs and the bluish/purple color complex explains the aggregation of molecules. If patient samples contain target molecules then those molecules will attach to AuNPs that produce red color due to dispersion [42, 43]. Various serological tests are performed by using IgM and IgG antibodies for DENV detection. In dengue infection, IgM antibodies are produced after five days of symptoms that can be detected by using an enzyme-linked immunosorbent assay (MAC-ELISA) [44, 45]. Sometimes due to high doses of infection, patients can feel hemorrhagic shock after seven days, in such early days of infection RT–PCR is used for detection. IgG antibodies are produced after IgM antibodies can be detected by using IgG ELISA. The electrochemical impedance spectroscopy (EIS) technique in combination with AuNPs was used for the detection of 1–4 DENV [46]. CNT are recently used to detect the NS1 antigen of DENV and complete viral particles with the help of NIS [47]. Clustered regularly interspaced short palindromic repeats (CRISPR) Cas 13 tools have the potential to create accurate and ultrasensitive biosensors for the detection of DENV.

10.2.6 Nanomaterials for Ebola Virus

Ebola virus (EBOV) is a negative-strand RNA virus from the *filoviridae* family [48]. Five species of EBOV are considered as vulnerable. In 2014, some species of the

EBOV family caused outbreaks in the African continent. The genetic material of EBOV contains separated genes for seven proteins of which two are soluble. These soluble proteins damage the host immune system. EBOV is a zoonotic agent because close contact with infected animals transfers that EBOV to humans. EBOV infections cause fever in the first three days and then cause gastrointestinal issues with vomiting and diarrhea. All these complications create hemorrhagic fever and a complete collapse of the immune system.

Various laboratory diagnostics tools are available for the detection of EBOV but require some unique instruments. PCR is a rapid and sensitive technique used for nucleic acid detection from a blood sample but the instrument is required. Electron microscopy is a direct method for morphological detection of a virus but is less sensitive. For rapid, ultrasensitive, and on-site detection of viruses, affordable with a less complicated POC tool is important. The nanoplasmonic optofluidic biosensor was developed by researchers that have the potential to detect the whole virus [49]. The microarray-based assay uses nanoparticles for the detection of viral particles. When a viral particle from the sample attaches to a nanoparticle-antibody complex then according to the size of the genome the size of the nanoparticle–antibody changes that help for the detection of the target virus [50]. The fabricated monoclonal antibody shows promising results in the detection of the EBOV virus by using an electrochemical sensor [51, 52]. This electrochemical sensor has a high sensitivity for the detection of viruses and can detect pM levels. During EBOV infection IgG antibodies are formed, therefore IgG antibodies specific for EBOV can be used as a biomarker for detection with the help of the ELISA method [53]. Magnetic nanomaterials as a probe serve as a nano-enzyme strip that can be used for ultrasensitive detection of EBOV. This tool is easy and affordable and no particular instrument is required therefore it can be used for on-site detection in tribal areas. The main target of the nano-enzyme strip is glycoproteins of EBOV, after which the target and MNP reaction color is produced mainly due to peroxidase enzymes that are added to the strip as a substrate [54]. More research is important for early onsite POC detection tools to tackle such EBOV infections.

10.2.7 Nanomaterials for Mycobacterium Tuberculosis

Tuberculosis (TB) is one of the lethal diseases that kill thousands of people each year. TB is caused by mycobacterium TB that attacks the lungs but some species also attack other body parts termed extra pulmonary TB. In some cases, symptoms are observed quickly after the infection but in some cases, symptoms are observed after a year when their immune system fails to fight the infectious agent and becomes weak. Lack of diagnostic tools in some places various medical professionals prescribed the wrong medications or antibiotics to TB patients which increases drug resistance in patients across the world. An increase in drug resistance creates a serious issue for diagnosis and treatment because of mutant species that are now even becoming resistant to multidrug treatments. Therefore the accurate diagnosis of TB is important for better treatment. The development of an accurate, ultrasensitive, and reliable POC tool is necessary for the diagnosis.

A variety of nanomaterials serve as a diagnostic tool for the detection of TB. AuNPs have unique properties that can be used for detection. AuNPs attach with antibodies, antigens, or biomolecules that can be used as probes for the detection of the target molecule from the sample. AuNPs have high stability during antibody–antigen reactions which enhances the signals obtained by the reaction that further increases the sensitivity of the test. AuNPs-based probes are used for the detection of complementary DNA from the sample [55]. If the sample contains target complementary DNA then AuNPs probe solution when mixed with the sample turns pink in color while in absence of complementary DNA solution turns purple [55]. The modified version of this method involves the use of a specific sequence of TB IS6110. The IS6110 sequence coupled with the AuNPs probe shows a promising result for the detection of TB from simple tests which increases the sensitivity of the test [56]. SPR can be used with AuNPs that produce red color when Au particles collide. This optical sensor-based method can detect a minute amount of target molecules from the complex mixture [57]. Researchers developed an electrode made up of indium tin oxide (ITO) that serves as a detection tool for the direct detection of genomic DNA of TB [58]. The genomic DNA of TB can be directly isolated from a patient's body fluids containing an infectious agent and then used for detection. One more AuNPs detection tool is available, in which TB antigens and anti-TB rabbit antibodies are coated on AuNPs, and the nitrocellulose (NC) membrane serum sample of the patient is immobilized [59]. When coated AuNPs are spread on the NC membrane then AuNPs colloids and colors produce that red color that is visible to the naked eye. The silica nanoparticles are used in combination with fluorescence microscopy which helps in the detection of TB within in two hours [60]. MNP have the potential to detect bacterial calls with the help of IgG antibodies. Rifampicin mutation in TB can be detected by using streptavidin–biotin magnetic nano beads that specifically detect rpoB gene mutation [61]. Various QD nanocrystals are used for the detection of Tb from which a combination of probes with QDs can easily detect unamplified DNA and rRNA genes of TB [62]. Advanced and modified nanomaterials-based methods attract researchers to the development of automated POC biosensors which can help for ultrasensitive and fast diagnosis of wild and resistant TB species. Various biosensors use graphene oxides, glucose oxidase, and CNT as detection materials in biosensors.

10.2.8 Nanomaterials for Zika Virus

Zika virus belongs to the *Flaviviridae* family and requires mosquitoes as a vector to spread infection. Similar to malaria, the zika virus contains mosquito species that bite healthy people and cause infection. The main issue with the zika virus is that it can transmit to the fetus via an infected mother. This virus causes defects in the brain of the fetus. There is no specific treatment available for zika infection. Therefore proper diagnosis and species-level detection are important for the development of treatment [63].

RT–PCR is commonly used for diagnosis as well as for nucleic acid detection ELISA is preferred. For accurate, affordable, and ultrasensitive detection of viruses required POC tools that can involve the use of nanomaterial which enhances the quality of the test. Platinum nano probes have the potential to increase the sensitivity of paper microfluidic tools that use spectroscopy for the detection of the virus [63, 64]. The researcher used AuNPs on which biomolecules are immobilized specifically ssDNA is used which is beneficial for signal amplification of bioassay. Recent development in diagnostic POC tools uses reverse transcription loop-mediated amplification (RT-LAMP) technology for Zika virus detection. By using RT-LAMP researchers easily detect nucleic acid from a blood drop. Further modification in this technique uses fluorescence for rapid screening and also uses an automated device that passes singles to a smartphone. Zika is causing a pandemic and therefore more study is required for detection that will serve as data for vaccine and medicines development.

10.2.9 Nanomaterials for Biomarkers in Infectious Disease POCT

An infectious agent produces numerous molecules which are not normally produced by the host cell. These molecules serve as a marker for the detection of infectious agents. These molecules are required for various activities of the agent and are found in the patient's body fluids such as enzymes, proteins, antigens, lipids, carbohydrates, RNA, DNA, and metabolites. The amount of biomarkers present in the clinical sample helps to determine the infectious agent load. For early, accurate, and ultrasensitive diagnosis of infection, detection of biomarkers is important. Some of the body fluids are released by our body as waste or through excretory systems containing various biomarkers and can be used for the detection of infectious agents. Recent studies show that some noninvasive body fluids contain micro-RNAs (mi-RNAs) that act as a biomarker.

For the detection of biomarkers, various platforms are available but the main focus is on POC tools. Microfluidics technology has great potential in the development of affordable ultrasensitive POC tools for biomarker detection. Some microfluidic devices are used for real-time monitoring by using minute amounts of samples. Polymers are used to make POC tools more affordable to everyone across the globe. Various biomarkers for infectious diseases are already explained above. AuNPs have the potential to detect various biomarkers in combination with probes. AuNPs attached to the probe form complex with the target molecule producing color that can be used for real-time monitoring. Magnetic NPs and silver NPs serve as better options as they are less reactive to oxidation. Colorimetric assay of hepatitis detection involves the use of Nano rods in combination with AuNPs and shows promising results [65]. Streptavidin and biotin-like molecules are also used to enhance the sensitivity of the test. Some fluorescence-based biosensors are used for the detection of nucleic acids. QDs serve as the best tool for POC because it helps for ultrasensitive detection of target molecules by releasing color that can be readable through the naked eye. Polyclonal or monoclonal antibodies with paramagnetic NPs help for the recognition of antigens from clinical samples. The activity can be enhanced by adding enzymes and substrates which further increase the intensity of the signal. Various CRISPR detection tools are available that are used for the detection of pathogenic biomarkers. The aptamer is a chemical antibody that has the potential for the detection of proteins, carbohydrates, and numerous molecules produced by pathogens. Patients infected by *Mycobacterium tuberculosis* release t-cells against pathogens that produce gamma-interferons (INF-γ) that can be used as biomarkers for TB detection.

10.2.10 Nanomaterials for Pathogen Nucleic Acids

Nucleic acid is an important part of the pathogen that helps for growth and replication in the host. Mutation in genetic material increases the virulence of the pathogen and makes it difficult for the immune system of the host to fight against the pathogen. Pathogens can easily escape host defense mechanisms by altering surface coat protein structure or by entering a dormant stage in host cells. Numerous emerging pathogens are evolving into the world, increasing the chances of pandemics and potentially mutating at a high rate. Treatment of such highly mutating pathogens is difficult, therefore genetic-level detection of such pathogens is important and helps to control outbreaks. Electrophoresis, RT–PCR, and spectrophotometer are mainly used for nucleic acid detection [66] (Figure 10.1).

For the detection of nucleic acids nowadays biosensors are preferred that mainly involve the use of electrochemical and optical biosensors. AuNPs are used as immobilizing capture probes with oligonucleotide sequences in electrochemical

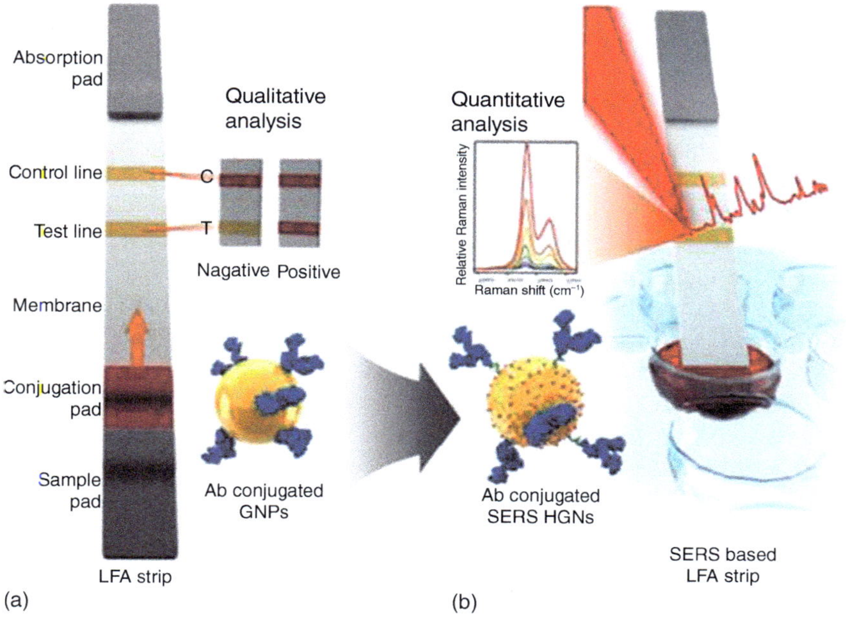

Figure 10.1 Illustration of the configuration of SERS-based lateral flow assay for detection of staphylococcal enterotoxin (Reproduced with permission from Application of a SERS-based lateral flow immunoassay strip for the rapid and sensitive detection of staphylococcal enterotoxin B). Source: Hwang et al. [67] / with permission of Royal Society of Chemistry.

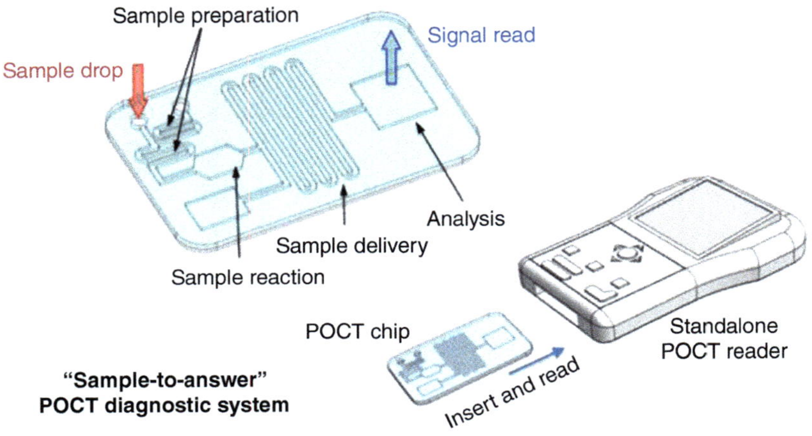

Figure 10.2 Schematic of an ideal microfluidic system with: "Sample-to-answer" characteristic for POCT reproduced with permission from point-of-care testing (POCT) diagnostic systems using microfluidic lab-on-a-chip technologies. Source: Jung et al. [73] / with permission of Elsevier.

biosensors that are peculiarly used for the detection of a nucleic acid of *Helicobacter pylori*. AuNPs have the potential to bind with disulfide bonds. Some AuNPs-based disposable electrochemical biosensors are used for the detection of respiratory pathogens such as *Pneumonia* species [68]. Carboxyl polymer-coated MNP were used for the detection of SARS-CoV-2 RNA [69]. QDs and other dye reagents are used to increase the sensitivity of the test. Liposomes are nano-sized used as dye carriers in disposable biosensors used for the detection of RNA of dengue virus [70]. Drinking water mainly contains some pathogenic strains of *Escherichia coli* that can be detected by using a similar liposome-based disposable biosensor. Labeled metal nanoparticles show promising results for DNA and RNA detection. A combination of Nanorods and AuNPs shows promising results for the detection of the hepatitis virus [71] abnormal conditions or infected patients, the body produces different mi-RNAs that are found in body fluids such as sweat, saliva, urine, plasma, serum, and semen that can be used as a biomarker to monitor infection. Approximately 92 mi-RNAs were found in the serum sample of the TB patient [72]. Nucleic acid lateral flow immunoassay is a promising tool for pathogen detection. Nuclei acid tests (NATs) are preferred for pathogen-specific detection of DNA or RNA. CRISPR Cas12a has great potential for ultrasensitive detection of target nucleic acid from minute samples [72] (Figures 10.2 and 10.3).

10.2.11 Nanomaterials for Antibodies and Proteins

Pathogens and viruses produce different proteins and enzymes necessary for the survival, protection, growth, and replication of the virus. Every pathogen produces unique proteins that are different from other pathogenic proteins, even species from the same family can produce different proteins and have different structures that

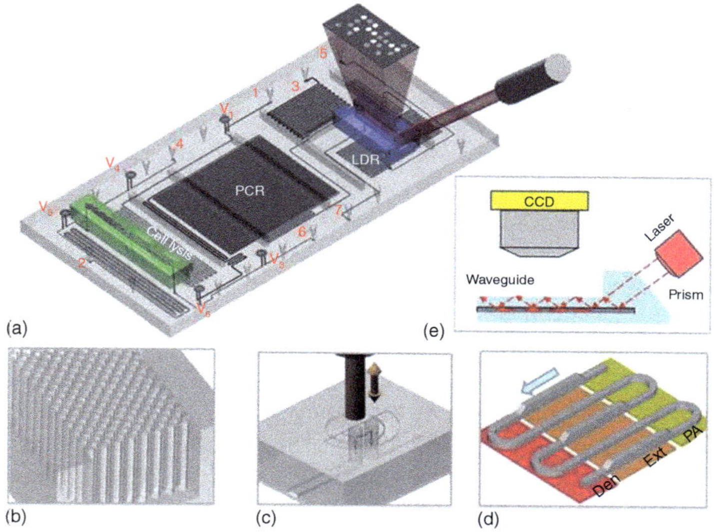

Figure 10.3 An integrated microfluidic chip for sensitive detection of DNA from M. tuberculosis with on-chip PCR. Source: Reproduced with permission from Wang et al. [74], John Wiley & Sons.

help for accurate detection of strain. For ultrasensitive detection of pathogenic strains, specific target detection molecules and rapid detection tools are important. The host body produces antibodies against pathogens as a defense mechanism that can be used for diagnosis. Traditionally various serological, PCR, and immunoelectrophoresis tests are performed for the detection of proteins and antibodies.

In response to SARS-CoV-2, the human body produces IgM antibodies that can be detected by using AuNPs lateral flow assay. Spike proteins of SARS-CoV-2 from nasal samples can be detected by using a graphene sheet coated with an antibody [75].

Aptamers are made up of ssRNA and ssDNA peculiarly attached to the target molecule and are therefore used for the detection of proteins and antibodies. Various aptamer-based biosensors are used for the detection of surface proteins of viruses with the help of colorimetric automated devices. For diagnosis of HIV specific and accurate detection tools are important but sometimes virus managed to escape the host immune system and in the early stage of HIV, the p24 antigen is used as a biomarker for HIV diagnosis. In the late stage of HIV, the anti-p24 antibody level increases while the p24 antigen level decreases to an undetectable level. An anti-p24 antibody is detected using different automated devices. Streptavidin-biotin labeled with QDs is used for the detection of biomarkers of HIV, HBV, syphilis, and malaria [76]. Nanowire electrodes are used for the detection of viral particles [77]. AuNPs conjugated with aptamers are used to detect small molecules, proteins, and particles of *E. coli* and *S. typhi* [78]. Various gold, silver, magnetic, and metal oxide-based nanomaterials are used for the detection of proteins, molecules, and antibodies. CRISPR is a new tool that detects specific target molecules.

10.3 Technology Advancements in Infectious Disease POCT

Numerous emerging diseases cause threats to humans. To fight against these life-threatening diseases diagnosis is important, which helps to provide accurate treatment to the patient. Traditionally detection tools required more samples and time but POC tools required a minute amount of samples and can detect pathogens in very few minutes. POC tools are ultrasensitive to overcome various challenges and help to detect the infectious agent as early as possible.

Lots of assay techniques are currently used in combination with nanomaterials which increases the speed of these tests. The serological test uses QDs and colorimetric reactions that help with the onsite detection of pathogens. Aptamer and CRISPR-based tests have the potential to detect pathogens accurately at species or strain levels. Microfluidic and electrochemical biosensors show promising results and automated devices produce signals which can be easily available on smartphones. Lateral flow assay and NAATs are now used in POCT. The main aim of the POCT tools is to minimize the cost of detection and minimize the risk of parenteral root infection due to accidents or carelessness of medical personnel. Numerous noninvasive POC techniques are currently used that detect biomarkers from patients' body fluids. Biomarkers such as antigens, antibodies, miRNAs, biomolecules, enzymes, and proteins are produced by an infectious agent and the host body. These POC tools are directly attached to automated devices and are beneficial to read out the reaction of the test. Some wearable devices contain biosensors that help for real-time continuous monitoring of patients made up of small chips, tattoos, watches, eye lenses, and spectacles. Disposable paper-based chips are available that are inexpensive and can be used in tribal areas for monitoring various diseases. Lateral flow assay in combination with disposable paper devices offers various advantages and by using nanomaterials it is easy to detect pathogens in a minute amount of blood sample without using special instruments.

10.4 Futuristic Developments

The world is filled with various pathogens from which most are unknown and some are life-threatening. Some infectious agents create pandemic or epidemic-like situations but still, the world doesn't have specific medicines or treatments for the diseases. Due to the increase in population, the temperature of the earth increases and creates global warming that decreases the availability of freshwater. Insufficient fresh water increases unhygienic conditions in poor countries which increase the risk of infections in the community. For such tribal and poor countries as well as for the whale world, POC tools should be developed. Aptamer and CRISPR are two new fields that have the potential to detect specific pathogens from a minute amount of samples [79]. Artificial intelligence (AI), neural links, and metaverse are useful for tracking the spread of infection in patients as well as across the globe [80].

Generally, analysis of minute analytes is difficult because the signals obtained from these analytes or reactions easily escape from detectors, or sometimes the data is too complex and difficult to analyze [81]. With Nanotechnology, new signal processing techniques, improved transducers, many advanced ML algorithms for processing data are emerging, and the intelligent biosensors will address very many challenges such as categorization based on the target analyte, anomaly detection and interferences in real samples, noise reduction by a better understanding of the signal from the noise and object identification and pattern recognition for on-site detection or diagnosis [82]. AI employing machine learning algorithms can be an effective tool to overcome the challenges faced by biosensors in processing big sensing data for complex samples and obtain reasonable analytical results which are overlapped due to noise and low resolution of the sensing data [83, 84]. The ability of ML methods to discover hidden relations between sample parameters as well as obtained signals when properly deployed is another advantage of intelligent sensors. Feiyun Cui et al in their review "Advancing Biosensors with Machine Learning" explained how machine learning can bridge biosensors with chemometrics [85]. Lussier et al. had given guidelines to execute ML algorithms in data analysis particularly to extract chemical information collected from experiments involving RAMAN and SERS techniques [86]. Jie Yang et al. provided a practical guide to employ deep learning algorithms to minimize signal artifacts for vibrational spectral analysis on preprocessing and feature selection techniques by taking the experiences as training data [87]. Ho et al. could use deep learning approaches to identify about 30 bacterial pathogens using Raman spectroscopy with more accuracy even with spectra having low signal-to-noise ratio [88]. Inevitably, all biosensors have issues in signal to noise, bioreceptors, data processing and which intern cause less reliable with anticipated accuracy and limit biosensors commercialization. Therefore, nanomaterials in combination with POC show promising results but more research will help for early on-site detection with the treatment of life-threatening diseases.

References

1 Huh, A.J. and Kwon, Y.J. (2011). "Nanoantibiotics": a new paradigm for treating infectious diseases using nanomaterials in the antibiotics resistant era. *Journal of Controlled Release* 156 (2): 128–145. ISSN 0168-3659, https://doi.org/10.1016/j.jconrel.2011.07.002.

2 Kozel, T.R. and Burnham-Marusich, A.R. (2017). Point-of-care testing for infectious diseases: past, present, and future. *Journal of Clinical Microbiology* 55 (8): 2313–2320. https://doi.org/10.1128/JCM.00476-17.

3 Nate, Z., Gill, A.A.S., Chauhan, R., and Karpoormath, R. (2022). Recent progress in electrochemical sensors for detection and quantification of malaria. *Analytical Biochemistry* 643: 114592. ISSN 0003-2697, https://doi.org/10.1016/j.ab.2022.114592.

4 Nair, C.B., Manjula, J., Subramani, P.A. et al. (2016). Differential diagnosis of malaria on truelab uno®, a portable, real-time, microPCR device for point-of-care applications. *PLoS One* 11 (1): e0146961. https://doi.org/10.1371/journal.pone.0146961.

5 Berzosa, P., de Lucio, A., Romay-Barja, M. et al. (2018). Comparison of three diagnostic methods (microscopy, RDT, and PCR) for the detection of malaria parasites in representative samples from Equatorial Guinea. *Malaria Journal* 17 (1): 333. https://doi.org/10.1186/s12936-018-2481-4. PMID: 30223852; PMCID: PMC6142353.

6 Coleman, R.E., Maneechai, N., Rachaphaew, N. et al. (2002). Comparison of field and expert laboratory microscopy for active surveillance for asymptomatic *Plasmodium falciparum* and *Plasmodium vivax* in western Thailand. *The American Journal of Tropical Medicine and Hygiene* 67 (2): 141–144. https://doi.org/10.4269/ajtmh.2002.67.141. PMID: 12389937.

7 Costa, A., Aguiar, A., Silva, A., and Pereira, D. (2021). Point-of-care strategies applied to malaria diagnosis. *Current Topics and Emerging Issues in Malaria Elimination* https://doi.org/10.5772/intechopen.96721.

8 Jatin, P., Chakma, B., Patra, S., and Goswami, P. (2014). Potential biomarkers and their applications for rapid and reliable detection of malaria. *BioMed Research International* 2014: 852645. https://doi.org/10.1155/2014/852645. Epub 2014 Apr 2. PMID: 24804253; PMCID: PMC3996934.

9 Srivastava, A., Philip, N., Hughes, K.R. et al. (2016). Stage-specific changes in *Plasmodium* metabolism required for differentiation and adaptation to different host and vector environments. *PLoS Pathogens* 12 (12): e1006094. https://doi.org/10.1371/journal.ppat.1006094.

10 Raman, J., Ashok, C.S., Subbayya, I.N. et al. (2005). *Plasmodium falciparum* hypoxanthine-guanine phosphoribosyltransferase. *The FEBS Journal* 272: 1900–1911. https://doi.org/10.1111/j.1742-4658.2005.04620.x.

11 Coronado, L.M., Nadovich, C.T., and Spadafora, C. (2014). Malarial hemozoin: from target to tool, biochimica et biophysica acta (BBA). *General Subjects* 1840 (6): 2032–2041. ISSN 0304-4165, https://doi.org/10.1016/j.bbagen.2014.02.009.

12 Kong, T.F., Ye, W., Peng, W.K. et al. (2015). Enhancing malaria diagnosis through microfluidic cell enrichment and magnetic resonance relaxometry detection. *Scientific Reports* 5: 11425. https://doi.org/10.1038/srep11425. PMID: 26081638; PMCID: PMC4469967.

13 Alnasser, Y., Ferradas, C., Clark, T. et al. (2016). Colorimetric detection of plasmodium vivax in urine using MSP10 oligonucleotides and gold nanoparticles. *PLoS Neglected Tropical Diseases* 10 (10): e0005029. https://doi.org/10.1371/journal.pntd.0005029.

14 Ling, X.X., Jin, J.J., Zhu, G.D. et al. (2019). Cost-effectiveness analysis of malaria rapid diagnostic tests: a systematic review. *Infectious Diseases of Poverty 8* (1): 104. https://doi.org/10.1186/s40249-019-0615-8.

15 Damhorst, G.L., Watkins, N.N., and Bashir, R. (2013). Micro- and nanotechnology for HIV/AIDS diagnostics in resource-limited settings. *IEEE Transactions on*

Biomedical Engineering 60 (3): 715–726. https://doi.org/10.1109/TBME.2013.2244894. PMID: 23512111.

16 Aberg, J.A., Kaplan, J.E., Libman, H. et al. (2009). Primary care guidelines for the management of persons infected with human immunodeficiency virus: 2009 update by the HIV Medicine Association of the Infectious Diseases Society of America. *Clinical Infectious Diseases* 49 (5): 651–681. https://doi.org/10.1086/605292.PMID: 19640227.

17 Tang, S. and Hewlett, I. (2010). Nanoparticle-based immunoassays for sensitive and early detection of HIV-1 capsid (p24) antigen. *The Journal of Infectious Diseases* 201 (Suppl 1): S59–S64. https://doi.org/10.1086/650386. PMID: 20225948; PMCID: PMC2943740.

18 Kim, E.-Y., Stanton, J., Korber, B. et al. (2008). Detection of HIV-1 p24 Gag in plasma by a nanoparticle-based bio-barcode-amplification method. *Nanomedicine (London, England)* 3: 293–303. https://doi.org/10.2217/17435889.3.3.293.

19 Ekstrand, D.H., Awad, R.J., Källander, C.F., and Gronowitz, J.S. (1996). A sensitive assay for the quantification of reverse transcriptase activity based on the use of carrier-bound template and non-radioactive-product detection, with special reference to human-immunodeficiency-virus isolation. *Biotechnology and Applied Biochemistry* 23 (2): 95–105. PMID: 8639277.

20 Borrego, P., Gonçalves, M.F., Gomes, P. et al. (2018). A serological point-of-care test for the detection of IgG antibodies against Ebola virus in human survivors. *ACS Nano* 12 (1): 63–73. https://doi.org/10.1021/acsnano.7b07021. Epub 2018 Jan 5. PMID: 29303554.

21 Greengrass, V., Lohman, B., Morris, L. et al. (2009). Assessment of the low-cost Cavidi Exa Vir Load assay for monitoring HIV viral load in pediatric and adult patients. *Journal of Acquired Immune Deficiency Syndromes* 52 (3): 387–390. https://doi.org/10.1097/QAI.0b013e3181b05f62. PMID: 19617845; PMCID: PMC2784031.

22 Mine, M., Bedi, K., Maruta, T. et al. (2009). Quantitation of human immunodeficiency virus type 1 viral load in plasma using reverse transcriptase activity assay at a district hospital laboratory in Botswana: a decentralization pilot study. *Journal of Virological Methods* 159 (1): 93–97. https://doi.org/10.1016/j.jviromet.2009.03.008. Epub 2009 Mar 20. PMID: 19442851.

23 Steegen, K., Luchters, S., De Cabooter, N. et al. (2007). Evaluation of two commercially available alternatives for HIV-1 viral load testing in resource-limited settings. *Journal of Virological Methods* 146 (1–2): 178–187. https://doi.org/10.1016/j.jviromet.2007.06.019. Epub 2007 Aug 7. PMID: 17686534.

24 Farzin, L., Shamsipur, M., Samandari, L., and Sheibani, S. (2020). HIV biosensors for early diagnosis of infection: the intertwine of nanotechnology with sensing strategies. *Talanta* 206: 120201. https://doi.org/10.1016/j.talanta.2019.120201. Epub 2019 Jul 31. PMID: 31514868.

25 Carneiro, L., Moraes, M., Ribeiro, S., and Brolo, A. (2019). Detection of the peptide P17-1 (HIV) by surface enhance Raman scattering (SERS). *Química Nova* 42: 1038–1043. https://doi.org/10.21577/0100-4042.20170430.

26 De Francesco, M.A., Caruso, A., Fallacara, F. et al. (1998). HIV p 17 enhances lymphocyte proliferation and HIV-1 replication after binding to a human serum factor. *AIDS* 12 (3): 245–252. https://doi.org/10.1097/00002030-199803000-00001. PMID: 9517986.

27 Xi, Z., Huang, R., Li, Z. et al. (2015). *ACS Applied Materials & Interfaces* 7: 11215–11223.

28 Toita, R., Kawano, T., Kang, J.H., and Murata, M. (2015). Applications of human hepatitis B virus preS domain in bio- and nanotechnology. *World Journal of Gastroenterology* 21 (24): 7400–7411. https://doi.org/10.3748/wjg.v21.i24.7400. PMID: 26139986; PMCID: PMC4481435.

29 Tseng, T.C., Liu, C.J., Chen, C.L. et al. (2015). Higher lifetime chance of spontaneous surface antigen loss in hepatitis B carriers with genotype C infection. *Alimentary Pharmacology & Therapeutics* 41 (10): 949–960. https://doi.org/10.1111/apt.13170. Epub 2015 Mar 23. PMID: 25809540.

30 Wang, S., Ai, Z., Zhang, Z. et al. (2020). Simultaneous and automated detection of influenza A virus hemagglutinin H7 and H9 based on magnetism and size mediated microfluidic chip. *Sensors and Actuators B: Chemical* 308: 127675. ISSN 0925-4005, https://doi.org/10.1016/j.snb.2020.127675.

31 Chen, R., Yongqin, H., Chen, M. et al. (2021). Naked-eye detection of hepatitis B surface antigen using gold nanoparticles aggregation and catalase-functionalized polystyrene nanospheres. ACS Omega 6 (14): 9828–9833. https://doi.org/10.1021/acsomega.1c00507.

32 Yang, L., Xing, G., Wang, L. et al. (2015). Acute kidney injury in China: a cross-sectional survey. *Lancet* 386 (10002): 1465–1471. https://doi.org/10.1016/S0140-6736(15)00344-X. PMID: 26466051.

33 Shevtsov, M., Zhao, L., Protzer, U., and van de Klundert, M.A.A. (2017). Applicability of metal nanoparticles in the detection and monitoring of hepatitis B virus infection. *Viruses* 9 (7): 193. https://doi.org/10.3390/v9070193. PMID: 28753992; PMCID: PMC5537685.

34 Cao, L., Huang, C., Cui Zhou, D. et al. (2021). Proteogenomic characterization of pancreatic ductal adenocarcinoma. *Cell* 184 (19): 5031–5052.e26. https://doi.org/10.1016/j.cell.2021.08.023. PMID: 34534465; PMCID: PMC8654574.

35 Riccò, R., Meneghello, A., and Enrichi, F. (2011). Signal enhancement in DNA microarray using dye doped silica nanoparticles: application to human papilloma virus (HPV) detection. *Biosensors and Bioelectronics* 26 (5): 2761–2765. ISSN 0956-5663, https://doi.org/https://doi.org/10.1016/j.bios.2010.10.024.

36 Xiang, D.S., Zeng, G.P., and He, Z.K. (2011). Magnetic microparticle-based multiplexed DNA detection with biobarcoded quantum dot probes. *Biosensors & Bioelectronics* 26 (11): 4405–4410. https://doi.org/10.1016/j.bios.2011.04.051. Epub 2011 May 5. PMID: 21602038.

37 Tasoglu, S., Tekin, H., Inci, F. et al. (2015). Advances in nanotechnology and microfluidics for human papillomavirus diagnostics. *Proceedings of the IEEE* 103: 161–178. https://doi.org/10.1109/JPROC.2014.2384836.

38 Xu, Y., Liu, Y., Wu, Y. et al. (2014). Fluorescent probe-based lateral flow assay for multiplex nucleic acid detection. *Analytical Chemistry* 86 (12): 5611–5614. https://doi.org/10.1021/ac5010458. Epub 2014 Jun 6. PMID: 24892496.

39 Nawaz, M.H., Hayat, A., Catanante, G. et al. (2018). Development of a portable and disposable NS1 based electrochemical immunosensor for early diagnosis of dengue virus. *Analytica Chimica Acta* 1026: 1–7. https://doi.org/10.1016/j.aca.2018.04.032. Epub 2018 Apr 18. PMID: 29852984.

40 Ramdzan, N.S.M., Fen, Y.W., Omar, N.A.S. et al. (2019). Optical and surface plasmon resonance sensing properties for chitosan/carboxyl-functionalized graphene quantum dots thin film. *Optik* 178: 802–812. ISSN 0030-4026, https://doi.org/10.1016/j.ijleo.2018.10.071.

41 Vinayagam, S., Rajaiah, P., Mukherjee, A., and Natarajan, C. (2018). DNA-triangular silver nanoparticles nanoprobe for the detection of dengue virus distinguishing serotype. *Spectrochimica Acta Part A: Molecular and Biomolecular Spectroscopy* 202: 346–351. ISSN 1386-1425, https://doi.org/10.1016/j.saa.2018.05.047.

42 Cordeiro, M., Ferreira Carlos, F., Pedrosa, P. et al. (2016). Gold nanoparticles for diagnostics: advances towards points of care. *Diagnostics (Basel).* 6 (4): 43. https://doi.org/10.3390/diagnostics6040043. PMID: 27879660; PMCID: PMC5192518.

43 Fan, W., Lu, N., Huang, P. et al. (2017). Glucose-responsive sequential generation of hydrogen peroxide and nitric oxide for synergistic cancer starving-like/gas therapy. *Angewandte Chemie (International Ed. in English)* 56 (5): 1229–1233. https://doi.org/10.1002/anie.201610682. Epub 2016 Dec 9. PMID: 27936311.

44 Innis, B.L., Nisalak, A., Nimmannitya, S. et al. (1989). An enzyme-linked immunosorbent assay to characterize dengue infections where dengue and Japanese encephalitis co-circulate. *The American Journal of Tropical Medicine and Hygiene* 40 (4): 418–427. https://doi.org/10.4269/ajtmh.1989.40.418. PMID: 2540664.

45 Kabir, M.A., Zilouchian, H., Younas, M.A., and Asghar, W. (2021). Dengue detection: advances in diagnostic tools from conventional technology to point of care. *Biosensors* 11: 206. https://doi.org/10.3390/bios11070206.

46 Luna, D.M.N., Avelino, K.Y.P.S., Cordeiro, M.T. et al. (2015). Electrochemical immuno sensor for dengue virus serotypes based on 4-mercaptobenzoic acid modified gold nanoparticles on self-assembled cysteine monolayers. *Sensors and Actuators B: Chemical* 220: 565–572. ISSN 0925-4005, https://doi.org/10.1016/j.snb.2015.05.067.

47 Palomar, Q., Gondran, C., Holzinger, M. et al. (2017). Controlled carbon nanotube layers for impedimetric immuno sensors: high performance label free detection and quantification of anti-cholera toxin antibody. *BiosensBioelectron* (97): 177–183. https://doi.org/10.1016/j.bios.2017.05.052. Epub 2017 Jun 1. PMID: 28599177.

48 Kaushik, A., Tiwari, S., Dev Jayant, R. et al. (2016). Towards detection and diagnosis of Ebola virus disease at point-of-care. *Biosensors & Bioelectronics* 75:

254–272. https://doi.org/10.1016/j.bios.2015.08.040. Epub 2015 Aug 20. PMID: 26319169; PMCID: PMC4601610.

49 Yanik, A.A., Huang, M., Kamohara, O. et al. (2010). An optofluidic nanoplasmonic biosensor for direct detection of live viruses from biological media. *Nano Letters* 10 (12): 4962–4969. https://doi.org/10.1021/nl103025u. Epub 2010 Nov 5. PMID: 21053965; PMCID: PMC3123676.

50 Daaboul, G.G., Lopez, C.A., Chinnala, J. et al. (2014). Digital sensing and sizing of vesicular stomatitis virus pseudotypes in complex media: a model for Ebola and Marburg detection. *ACS Nano* 8 (6): 6047–6055. https://doi.org/10.1021/nn501312q. Epub 2014 Jun 4. PMID: 24840765; PMCID: PMC4466106.

51 Cruz, A.F., Norena, N., Kaushik, A., and Bhansali, S. (2014). A low-cost miniaturized potentiostat for point-of-care diagnosis. *Biosensors & Bioelectronics* 62: 249–254. https://doi.org/10.1016/j.bios.2014.06.053. Epub 2014 Jun 30. PMID: 25016332.

52 Vasudeva, A., Kaushikb, A., Tomizawab, Y. et al. (2013). An LTCC-based microfluidic system for label-free, electrochemical detection of cortisol. *Sensors and Actuators B: Chemical* 182: 139–146. https://doi.org/10.1016/j.snb.2013.02.096.

53 Brangel, P., Sobarzo, A., Parolo, C. et al. (2018). A serological point-of-care test for the detection of IgG antibodies against Ebola virus in human survivors. *ACS Nano* 12 (1): 63–73.

54 Duan, D., Fan, K., Zhang, D. et al. (2015). Nanozyme-strip for rapid local diagnosis of Ebola. *Biosensors & Bioelectronics* 74: 134–141. https://doi.org/10.1016/j.bios.2015.05.025. Epub 2015 May 11. PMID: 26134291.

55 Baptista, P., Pereira, E., Eaton, P. et al. (2008). Gold nanoparticles for the development of clinical diagnosis methods. *Analytical and Bioanalytical Chemistry* 391 (3): 943–950. https://doi.org/10.1007/s00216-007-1768-z. Epub 2007 Dec 21. PMID: 18157524.

56 Tsai, T.T., Huang, C.Y., Chen, C.A. et al. (2017). Diagnosis of tuberculosis using colorimetric gold nanoparticles on a paper-based analytical device. *ACS Sensors* 2 (9): 1345–1354. https://doi.org/10.1021/acssensors.7b00450. Epub 2017 Sep 13. PMID: 28901134.

57 Khan, I., Saeed, K., and Khan, I. (2019). Nanoparticles: properties, applications, and toxicities. *Arabian Journal of Chemistry* 12: 908–931.

58 Zhu, L., Song, L., Liu, W. et al. (2017). Gold nanoparticles-modified indium tin oxide microelectrode for in-channel amperometric detection in dual-channel microchip capillary electrophoresis. *Analytical Methods* 29: 4319–4326. https://doi.org/10.1039/C7AY01008G.

59 Stephen, B.J., Singh, S.V., Datta, M. et al. (2015). Nanotechnological approaches for the detection of mycobacteria with special references to *Mycobacterium avium* subspecies paratuberculosis (MAP). *Asian Journal of Animal and Veterinary* 10: 518–526. https://doi.org/10.3923/ajava.2015.518.526.

60 Qin, D., He, X., Wang, K. et al. (2007). Fluorescent nanoparticle-based indirect immunofluorescence microscopy for detection of mycobacterium tuberculosis. *Journal of Biomedicine & Biotechnology* 2007: 89364. https://doi.org/10.1155/2007/89364.

61 Engström, A., de la Torre, T.Z.G., Strømme, M. et al. (2013). Detection of rifampicin resistance in Mycobacterium tuberculosis by padlock probes and magnetic nanobead-based readout. *PLoS One* 8: e62015. https://doi.org/10.1371/journal.pone.0062015.

62 Gupta, A.K., Singh, A., and Singh, S. (2020). Diagnosis of tuberculosis: nanodiagnostics approaches. In: *Nano Bio Medicine* (ed. S. Saxena and S. Khurana). Singapore: Springer https://doi.org/10.1007/978-981-32-9898-9_11.

63 Draz, M.S., Venkataramani, M., Lakshminarayanan, H. et al. (2018). Nanoparticle-enhanced electrical detection of Zika virus on paper microchips. *Nanoscale* 10 (25): 11841–11849. https://doi.org/10.1039/c8nr01646a. PMID: 29881853; PMCID: PMC6033661.

64 Mauk, M.G., Song, J., Bau, H.H., and Liu, C. (2017). Point-of-care molecular test for Zika infection. *Clinical laboratory international* 41: 25–27. PMID: 28819345; PMCID: PMC5556939.

65 Wang, Y., Yu, L., Kong, X., and Sun, L. (2017). Application of nanodiagnostics in point-of-care tests for infectious diseases. *International Journal of Nanomedicine* 12: 4789–4803. https://doi.org/10.2147/IJN.S137338. PMID: 28740385; PMCID: PMC5503494.

66 Tram, D.T., Wang, H., Sugiarto, S. et al. (2016 Dec). Advances in nanomaterials and their applications in point of care (POC) devices for the diagnosis of infectious diseases. *Biotechnology Advances* 34 (8): 1275–1288. https://doi.org/10.1016/j.biotechadv.2016.09.003. Epub 2016 Sep 26. PMID: 27686397; PMCID: PMC7127209.

67 Hwang, J., Lee, S., and Choo, J. (2016). Application of a SERS-based lateral flow immunoassay strip for rapid and sensitive detection of staphylococcal enterotoxin B. *Nanoscale* 8: https://doi.org/10.1039/C5NR07243C.

68 Bébéar, C.M., de Barbeyrac, B., Pereyre, S. et al. (2008). Activity of moxifloxacin against the urogenital mycoplasmas *Ureaplasma spp., Mycoplasma hominis and Mycoplasma genitalium and Chlamydia trachomatis. Clinical Microbiology and Infection* 14 (8): 801–805. https://doi.org/10.1111/j.1469-0691.2008.02027.x. PMID: 18727805.

69 Zhen, Z., Haodong, C., Wenxing, S., Xiaoling, R., Wenhua, Z., Xuefeng, Y. (2020). A simple magnetic nanoparticles-based viral RNA extraction method for efficient detection of SARS-CoV-2. https://doi.org/10.1101/2020.02.22.961268.

70 Baeumner, A.J., Schlesinger, N.A., Slutzki, N.S. et al. (2002). Biosensor for dengue virus detection: sensitive, rapid, and serotype specific. *Analytical Chemistry* 74 (6): 1442–1448. https://doi.org/10.1021/ac015675e. PMID: 11922316.

71 Chen, Q., Yuan, L., Wan, J. et al. (2014). Colorimetric detection of hepatitis E virus based on reverse transcription loop mediated isothermal amplification (RT-LAMP) assay. *Journal of Virological Methods* 197: 29–33. https://doi.org/10.1016/j.jviromet.2013.11.006. Epub 2013 Nov 21. PMID: 24269797.

72 Chen, H., Liu, K., Li, Z., and Wang, P. (2019). Point of care testing for infectious diseases. *Clinica Chimica Acta* 493: 138–147. https://doi.org/10.1016/j.cca.2019.03.008. Epub 2019 Mar 8. PMID: 30853460; PMCID: PMC6462423.

73 Jung, W., Han, J., Choi, J. et al. (2014). Point-of-care testing (POCT) diagnostic systems using microfluidic lab-on-a-chip technologies. *Microelectronic Engineering* 132: 46–57. https://doi.org/10.1016/j.mee.2014.09.024.

74 Wang, H., Chen, H.-W., Hupert, M.L. et al. (2012). Fully integrated thermoplastic genosensor for the highly sensitive detection and identification of multi-drug-resistant tuberculosis. *Angewandte Chemie, International Edition* 51: 4349–4353. https://doi.org/10.1002/anie.201200732.

75 Seo, G., Lee, G., Kim, M.J. et al. (2020). Rapid detection of COVID-19 causative virus (SARS-CoV-2) in human nasopharyngeal swab specimens using field-effect transistor-based biosensor. *ACS Nano* 14 (4): 5135–5142. https://doi.org/10.1021/acsnano.0c02823. Epub 2020 Apr 20. Erratum in: ACS Nano. 2020 Sep 22;14(9):12257-12258. PMID: 32293168; PMCID: PMC7172500.

76 Klostranec, J.M., Xiang, Q., Farcas, G.A. et al. (2007). Convergence of quantum dot barcodes with microfluidics and signal processing for multiplexed high-throughput infectious disease diagnostics. *Nano Letters* 7 (9): 2812–2818. https://doi.org/10.1021/nl071415m. Epub 2007 Aug 18. PMID: 17705551.

77 Patolsky, F., Zheng, G., Hayden, O. et al. (2004). Electrical detection of single viruses. *Proc Natl Acad Sci USA* 101 (39): 14017–14022. https://doi.org/10.1073/pnas.0406159101. Epub 2004 Sep 13. PMID: 15365183; PMCID: PMC521090.

78 Kim, J., Mohamed, M.A.A., Zagorovsky, K., and Chan, W.C.W. (2017). State of diagnosing infectious pathogens using colloidal nanomaterials. *Biomaterials* 146: 97–114. https://doi.org/10.1016/j.biomaterials.2017.08.013. Epub 2017 Aug 17. PMID: 28898761; PMCID: PMC7124370.

79 Rohan, P., Hardik, C., and Sushma, D. (2022). Point-of-care electrochemical biosensors using CRISPR/Cas for RNA analysis. *Biosensors for Emerging and Re-Emerging Infectious Diseases* 317–333. https://doi.org/10.1016/B978-0-323-88464-8.00003-8.

80 Sushma, D. and Kirubavathy, S.J. (2022). Biosensors based on metal-organic framework (MOF): paving the way to point-of-care diagnosis. *Electrochemical Applications of Metal-Organic Frameworks* 255–267. https://doi.org/10.1016/B978-0-323-90784-2.00004-6.

81 Radhakrishnan, S., Kim, B.-S., and Dave, S. (2022). Surface modification with nanomaterials for electrochemical biosensing application. *Advanced Nanomaterials for Point of Care Diagnosis and Therapy* 101–120. https://doi.org/10.1016/B978-0-323-85725-3.00002-7.

82 Dave, S., Dave, A., Radhakrishnan, S. et al. (2022). Biosensors for healthcare: an artificial intelligence approach. *Biosensors for Emerging and Re-Emerging Infectious Diseases* 365–383. https://doi.org/10.1016/B978-0-323-88464-8.00008-7.

83 Sahoo, M., Gadnayak, A., Nayak, A. et al. (2022). Advantages of silicon nanowire-based biosensors as wireless technology for infectious disease diagnosis. In: *Biosensors for Emerging and Re-Emerging Infectious Diseases*, 407–417. Academic Press.

84 Swayamprabha, S., Nayak, A., Gadnayak, A. et al. (2022). Quantum dots enabled point-of-care diagnostics: a new dimension to the nanodiagnosis. In: *Advanced Nanomaterials for Point of Care Diagnosis and Therapy*, 43–52. Elsevier.

85 Cui, F., Yue, Y., Zhang, Y. et al. (2020). Advancing biosensors with machine learning. *ACS Sensors* 5 (11): 3346–3364. https://doi.org/10.1021/acssensors.0c01424.

86 Lussier, F., Thibault, V., Charron, B. et al. (2020). Deep learning and artificial intelligence methods for raman and surface-enhanced raman scattering. *TrAC Trends in Analytical Chemistry* 124: 115796. https://doi.org/10.1016/j.trac.2019.115796.

87 Yang, J., Jinfan, X., Zhang, X. et al. (2019). Deep learning for vibrational spectral analysis: recent progress and a practical guide. *Analytica Chimica Acta* 1081: 6–17. https://doi.org/10.1016/j.aca.2019.06.012.

88 Ho, C.-S., Jean, N., Hogan, C.A. et al. (2019). Rapid identification of pathogenic bacteria using Raman spectroscopy and deep learning. *Nature Communications* 10 (1, 1): –8. https://doi.org/10.1038/s41467-019-12898-9.

11

Development of a Diagnostic Kit for Point-of-Care Biosensors: Fundamentals and Applications

Vijay Vaishampayan[1], Prabir Kulabhushan[2], Ishita Dasgupta[3], Ashish Kapoor[4], and Sarang P. Gumfekar[1]

[1]*Indian Institute of Technology, Department of Chemical Engineering, Nangal Road, Rupnagar, Punjab 140001, India*
[2]*University of Oxford, Kavli Institute of NanoScience and Discovery, Department of Physics, New Biochemistry Building, Dorothy Hodgkin Rd, Oxford OX13QU, UK*
[3]*Pondicherry University, Nanoscience and Technology Department, R Venkat Raman Nagar, Kalapet, Puducherry 605014, India*
[4]*Harcourt Butler Technical University, Department of Chemical Engineering, Kanpur, Uttar Pradesh 208002, India*

11.1 Introduction

The fabrication, development, and use of biosensors have become a center of attraction for researchers over the globe in recent decades due to their wide variety of applications, including disease monitoring, food quality analysis, environmental applications, and drug delivery. The significant challenges concerned with biosensor technology are the efficient biorecognition steps, transducer performance, reproducibility, sensitivity, etc. [1]. The device that assists in detecting changes in chemical, physical, or biological parameters is commonly known as sensor. The sensing devices can be helpful in the point-of-care (POC) analysis in resource-limited settings where sophisticated health monitoring facilities are not available. These devices offer tremendous advantages such as accuracy, portability, user-friendliness, frugality, and rapid detection. [2–5] Frugality in the POC biosensors offered a significant advantage in the developing countries where the large population cannot afford the sophisticated analytical testing due to its huge cost. The POC tests can help monitor a particular biomarker or analyte solution continuously. They can significantly reduce the turnaround time compared to conventional laboratory-based methods, which may help manage crucial cases effectively [6, 7]. The POC can be a game changer in resource-limited settings as it utilizes fewer resources. Also, a highly skilled workforce may not need to use such devices. Numerous people are using POC devices for monitoring blood pressure [8], blood glucose [9], and oxygen monitoring devices [10] to reduce nurses' workloads and other first-line healthcare workers. Detection of

Point-of-Care Biosensors for Infectious Diseases, First Edition.
Edited by Sushma Dave and Jayashankar Das.

blood glucose level and cardiological analysis are the major sectors that have contributed to most diagnostic market [9]. Easy to use and rapid diagnosis stimulated the penetration of POC devices in developing countries. Infectious diseases such as dengue, malaria, and influenza have collectively piled on top of threatening diseases such as tuberculosis (TB), urinary tract infections, and cardiovascular diseases. Therefore, sophisticated high-end laboratory testing methods must be updated to POC tests to cater to remote areas' needs effectively [10, 11].

The biosensor is the most critical and essential component of any point-of-care diagnostics (POCD) and is directly responsible for the bioanalytical performance of an assay. Various types of biosensors, such as electrochemical [12, 13] surface plasmon resonance (SPR) [14, 15] and white light reflectance spectroscopy (WLRS) [16–18], have been developed and used for improved diagnosis. Complementary technologies such as lab-on-a-chip (LOC) microfluidic devices allow automation, integration, and signal readout for continuous advancements in POC devices [19–21]. There is now overwhelming evidence that POC testing can significantly strengthen the diagnostic precision of clinicians in a wide variety of areas.

11.2 Evolution of Biosensor

In 1906, M. Cramer displayed that acid concentration in liquids was proportional to the electric potential that arose between parts of the fluid located on opposite sides of a glass membrane [22]. L. C. Clark Jr. designed the first biosensor for oxygen detection in 1956. Before this development, various breakthroughs has accomplished by other scientists [23, 24].

The evolution of biosensors can be categorized into three generations based on the incorporation method of the biorecognition element. The first-generation biosensors analyzed the content of analyte solutions and products of the bioreceptor reactions, which diffuse to the transducer surface, thus, producing electrical signals. These sensors were referred to as mediator-less amperometric biosensors. Individual components and co-reactants, including mediators, nanomaterials, and auxiliary enzymes, were integrated into the biological components of the biosensors in the second-generation biosensors. These incorporation processes enhanced the analytical efficiency of the devices. These sensors were known as mediator amperometric biosensors [1, 25]. In developing third-generation biosensors, bioreceptors became the crucial part of the base sensing element. The sensors deployed enzymes and mediators on the same electrode rather than freely diffusing mediators in the electrolyte. Due to the transfer of electrons, a direct interaction was established between the enzymes and electrode material without any intermediate stages. Robustness, low designing cost, feasibility in the fabrication, and repeatability of the results were some advantages of the third-generation biosensors [1, 26]. Figure 11.1 represents the evolution of biosensors over the several years. In recent times, Girbi et al. develop a model for nerve-on-chip type biosensor for analyzing nerve impulse conditions [27]. Also, during the several waves of the deadly pandemic, many researchers developed microfluidic platforms for biosensing applications [28–31].

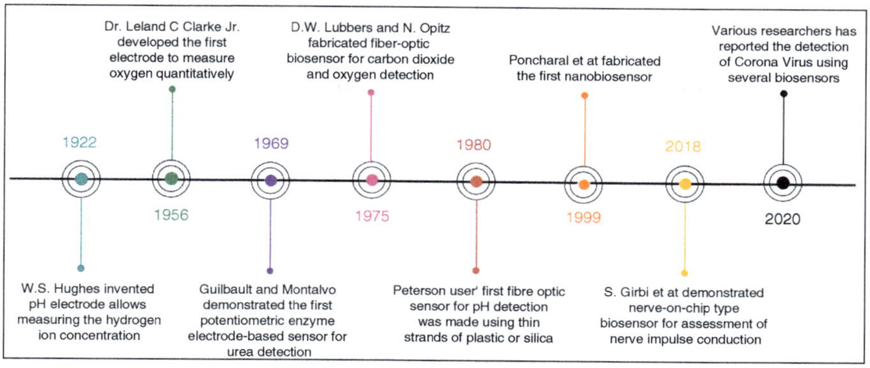

Figure 11.1 Evolution of biosensors over the period.

Thus, we know that biosensor devices include a natural element and a physicochemical detector. The primary function of these devices is to detect various analytes and biomolecules. So, the applications of biosensors are in a wide range. These devices are applicable in the medical, food industry, the marine sector as they offer good sensitivity and stability compared with the usual techniques. Recently, biosensors were developed based on the variety of immobilized bioreceptors that help create a selective analysis of chemical, biological, and biomedical substances. One of the recent advancements in biosensors was marker-free detection. The sensor surface itself considered the role of the marker and allowed for the detection of biological species for rapid detection. A biosensor can significantly impact detecting and measuring the COVID-19 virus in the air. Hence, the method enabled authorities to measure how a virus could stay in the air when mass spread. The biosensors can also be helpful for the detection of other infectious diseases. The healthcare facilities could receive the data using biosensors for the real-time monitoring of the diseases [32–34].

11.3 Biosensors for Point-of-Care Sensing

11.3.1 Fundamentals of Biosensor

The first biosensor was invented in 1962 to measure the glucose in biological samples by employing the electrochemical detection of oxygen and hydrogen peroxide and using immobilized glucose oxidase electrode. Since then, there is tremendous application of biosensors in myriad field of biomedical sciences. Biosensor refers to a powerful and analytical device containing a biological recognition element having application in diagnosis, biomedicine, environmental monitoring, and food safety. It is possible to detect a wide variety of analytes using biosensors that employ molecules with high affinity, such as antibody aptamers and enzymes, along with appropriate signal transduction. With advancement of nanotechnology, electrochemistry, and bioelectronics, there are several modifications of biosensors that have been improvised to enhance the sensitivity, specificity, and other parameters. Furthermore, the introduction of microfluidics and LOC methodologies advances the biosensor to miniaturization or portable device format with qualitative or semiquantitative read-out for nonspecialists and helps in ease of interpretation. The biosensors used for the detection of target analytes consist of three parts such as the bioreceptor or recognition element, the transducer, and the signal processing unit. The bioreceptors/recognition probe plays an important role by specific binding to the target and the transducer generate the measurable signal [35]. Conventionally, most of the biosensors employs poly/monoclonal antibodies as the recognition element [36]. The use of antibodies in biosensors have several limitations such as stability, batch-to-batch variation, cost, and time required for its production [37]. These issues subsequently affect the final performance of the biosensor during the bioanalysis. To overcome this, several other receptors such as aptamers, enzymes, whole cells, and antibody fragments are recently employed in various detection platforms.

Currently, the sensing strategy revolves around the label based and label-free technology of biosensors. The label-based biosensor consists of specific tags/labels to detect the target analytes in the presence of surrounding matrices. Usually, the labels are fluorescence, chemiluminescence, or radiolabeled for the detection of target molecules. In addition, the whole microbe-based sensor and bioluminescence also contributed largely to label-based biosensors [38]. Generally, the label such as fluorescent probes are attached with the biological recognition probe and after binding of the target molecule, the signals are generated and quantified. However, the fluorescent-based methods are commonly used due to their high sensitivity to detect the target molecule. Recently, several nanomaterials such as quantum dots, organic polymer nanoparticles, and up-conversion nanoparticles (UCNPs) have been used to avoid the fluorescent quenching. Contrarily, the label-free methods permit the detection of target molecules without any labeling of the ligand or the receptor and can be performed with minimal assay development. This technique overcomes the limitations of label-based technique such as hindrance of active binding epitope and the disadvantages associated with the fluorescent or radioactive probe. Briefly, this technique employs the biophysical properties such as refractive index SPR, localized surface plasmon resonance (LSPR), and molecular weight (micro cantilever) to monitor the presence of the target analytes. Figure 11.2 describes the various components of the biosensors. Recent advancement in the bioengineering, electronics paves the way for developing label-free biosensors and their application in the field of medicines and environmental sciences.

11.3.2 Bioreceptors in Biosensor

The biorecognition element or recognition probe plays a vital role for the successful design of efficient biosensor. Based upon the recognition probe, the biosensors are broadly classified into catalytic biosensors or non-catalytic biosensors or affinity biosensors. The catalytic biosensor includes the enzymes, whole cell, and microorganism as their recognition probe whereas the affinity-based biosensor consists of antibodies, aptamers, cell receptors, specific peptides, nanobodies. In this case, the

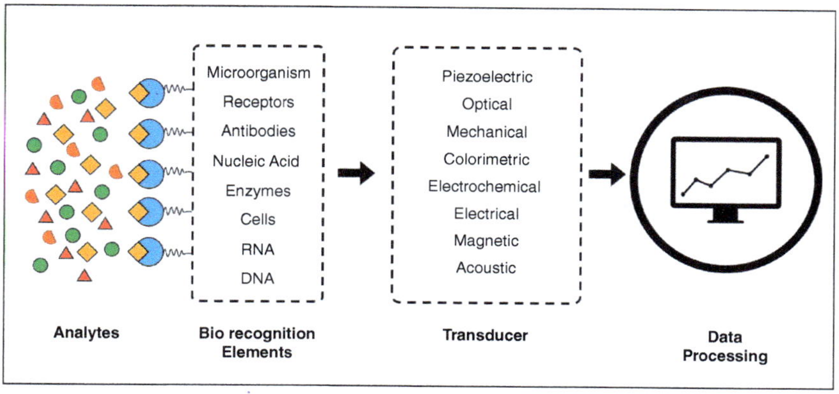

Figure 11.2 Components of biosensors.

target analytes bind to the probe irreversibly, which leads to generating a quantifiable signal. Enzymes are most ubiquitous to catalysis and their high specificity toward their target molecules. The application of enzymes in biosensor as recognition probe is based upon the catalytic reaction or the binding abilities of the target molecule. However, the enzyme structure is highly sensitive toward the surrounding matrices, therefore, it limits the stability and adaptability during the diagnosis. The biosensor generally employs two strategies such as (i) enzyme inhibition or activation by the enzyme (ii) measurement of the enzyme characteristics. The most common enzyme-based biosensors are glucose and urea biosensors [24, 39].

Antibodies are the major workhorse in the development of various biosensors due to their strong interaction with the antigen molecule. The antibodies generally consist of two heavy chains and two light chains connected together by disulfide bonds. The biosensor or the antibody-based sensor known as immunosensor, which are most well studied amongst various other sensors. The immunosensor commonly employs the poly or monoclonal antibodies for the detection of target molecule. The antigen–antibody complex is sensitively assessed through label measurement [40]. Castro et al. describes an electrochemical immunosensor for the detection of biomarker CA125 for the early diagnosis and monitoring of ovarian cancer. The detection module consists of a carbon electrode modified with poly(3-hydroxyphenylacetic acid) and anti-CA125 antibodies as probe for evaluate the biomolecular interactions (anti-CA125:CA125). The linear dynamic range of detection includes in the range of 5–80 $U\,ml^{-1}$ with a limit of detection of 1.45 $U\,ml^{-1}$ and also capable to detect the negative and positive sample. The immunosensor also able to detect the CA-125 in real biological samples using employing differential pulse voltammetry (DPV), SPR, and atomic force microscopy (AFM). Similarly, several other antibody-based biosensor have also been demonstrated in literature [41, 42]. The antibodies are most commonly employed for the development of biosensor, there is certain limitation concerning their stability, batch-to-batch variation, and prone to harsh condition. Recent development in search for other recognition probes indicated that aptamers are the most suitable candidate for the replacement of antibodies in biosensors.

Aptamers are single-stranded DNA or RNA which bind to the target analytes with high affinity as comparable with monoclonal antibodies due to their 2D or 3D structures. The aptamers are structurally and functionally stable over wide range of temperatures and pH. Furthermore, these are chemically synthesized and known as chemical antibodies having affinity constant of nanomolar to picomolar range. Aptamers are selected using an *in vitro* selection procedure known as systematic evolution of ligands by exponential enrichment (SELEX), from diverse oligonucleotide libraries. There are several modifications in the SELEX procedure such as cell-SELEX, capillary electrophoresis-based SELEX, microfluidics-SELEX, FACS-based SELEX, microtiter plate-SELEX have been carried out to get high-affinity aptamers with optimal binding properties [43]. Considering the superior binding properties aptamer, known as aptasensor are nowadays employed for the detection of various target molecules. Grabwoska et al. developed a multi-analyte platform for cardiac biomarkers using the aptamer and electrochemical platform. The biomarker used here is the brain natriuretic peptide (BNP-32) and cardiac troponin I (cTnI) are the vital cardiac biomarker. The aptamers were immobilized on the commercial gold-based screen printed by using Cu(I)-based "click" chemistry. The sensor demonstrated a linear dynamic range of $1\,\text{pg}\,\text{ml}^{-1}$–$1\,\mu\text{g}\,\text{ml}^{-1}$ in serum for BNP-32 and the linearity was observed from $1\,\text{pg}\,\text{ml}^{-1}$ to $10\,\text{ng}\,\text{ml}^{-1}$ for cTnI, which is demanded for early-stage diagnosis of heart failure [44]. Bacteriophages, known as phages has been exploited to design biosensor for their specific binding to the target bacteria. Several functionalization procedures have been performed in the phages to detect *Staphylococcus aureus* using lytic phages (detection limit $\sim 10^4\,\text{CFU}\,\text{ml}^{-1}$) by SPR [45] and *Salmonella* using magnetoelastic sensor in suspension and fat-free milk (detection limit $\sim 10^3\,\text{CFU}\,\text{ml}^{-1}$) [46]. Furthermore, the adsorption of phages is also reported on the gold and silica surface, which is later used in biosensor. Similarly, physical adsorption of the phages on sugar and amino acid-modified gold surfaces [47] as well as surface-modified silica particles [48] for pathogen capture have also been reported in the literature. In addition, phage-displayed peptides are also employed in the development of biosensors. These peptides are short and highly stable as aptamers, hence known as peptide aptamers. There are several reports suggesting the application of phage-displayed peptides in biosensors for the detection of various target molecules [49, 50].

11.3.3 Transducer in Biosensor

Transducers can be generically classified as electrochemical, optical, thermal, electrical, or gravimetric based on their underlying operating principles.

11.3.3.1 Electrochemical Biosensor

The electrochemical characteristics of the analyte and the transducer form the basis of the electrochemical biosensors. This type of biosensor demonstrated to be high sensitivity, selectivity for the detection of target analytes. The electrochemical reaction between bioreceptor and the transducer produces observable electrochemical signals. There are several types of electrochemical biosensors such as potentiometric, amperometric, impedimetric, conductometric, and voltammetric [51].

11.3.3.2 Potentiometric Biosensor

These biosensors measure the potential or charge accumulated due to the interaction between the target and the recognition probe relative to the reference electrode. The biochemical signals are transformed into the potential signal and generally, ion-selective and field-effect transistors are employed [52].

11.3.3.3 Amperometric Biosensor

This type of sensor measures the current produced due to the electrochemical oxidation or reduction of electroactive species and the current is proportional to the concentrations of the analytes. These biosensors are fast, sensitive, and provides a linear response as compared to other electrochemical biosensors [53].

11.3.3.4 Impedimetric Biosensors

It measures the electrical impedance produced at the electrode/electrolyte interface upon application of low amplitude AC voltage at the sensor electrode and then the in/out-of-phase current response is measured as a function of frequency using an impedance analyzer [51].

11.3.3.5 Voltammetric Biosensors

These biosensors measure the interaction of target analytes with the recognition probe measuring the current during controlled variation of applied potentials.

11.3.3.6 Optical Biosensor

The optical biosensors consist of biorecognition elements such as antibodies and aptamers with an optical transducer system. The transduction process involves the changes in the transmission, reflection, refraction, amplitude of light, or change in light polarization in response to the interaction between the target analytes and the recognition probe. The most commonly used optical biosensors are (i) fluorescence-based (ii) chemiluminescence (iii) SPR-based optical biosensors, and (iv) optical fiber-based optical biosensors [54]. However, in label based optical detection system, the optical signals are generated but the fluorescence, luminescent methods. Furthermore, several next-generation optical biosensors such as SPR, evanescent wave (EW) fluorescence, and optical waveguide interferometry are employed for the detection of target molecules [55] (Figure 11.3).

11.3.3.7 Gravimetric Biosensor

The principle of gravimetric biosensor involves the minute changes in the mass of binding materials such as proteins or antibodies on the surface producing a quantifiable signal The gravimetric biosensors include the piezoelectric, quartz crystal microbalance (QCM), and magnetoelastic-based biosensor (MES) employs the principle of mass change for the detection of target analytes. These transducers are used to identify pathogens and antigens by means of binding interactions. Su et al. developed an piezoelectric biosensor using lead titanatezirconate (PZT) ceramic resonator as transducer for the detection of cancer biomarkers. The detection of cancer biomarkers such as prostat-specific antigen (PSA) and α-fetoprotein (AFP) with

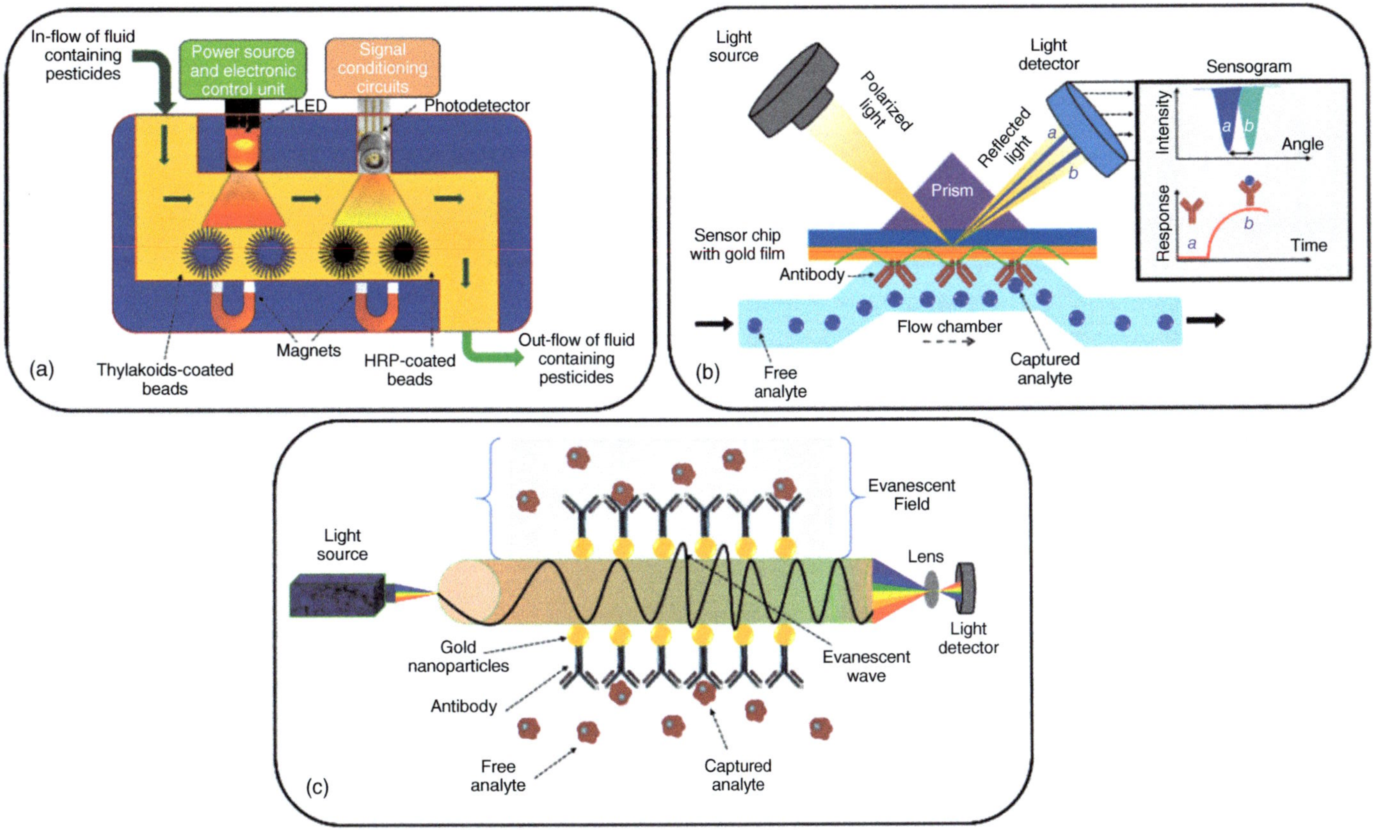

Figure 11.3 The schematic representation of schematic diagrams of (a) chemiluminescence biosensors, (b) surface plasmon resonance (SPR) biosensors, and (c) evanescent wave-based optical fiber biosensors. Source: The images are reproduced from Naresh and Lee [1], MDPI, CC BY 4.0.

high sensitivity (0.25 ng ml^{-1}) within 30 minutes with a sample volume of 1 μl which is highly beneficial for the clinical sample measurement. The sensor includes the two ceramic resonators connected in parallel, where one resonator is employed as sensing unit and other as control unit [56]. Similarly, Lim et al. developed an QCM biosensor for the detection miR-21 with a detection limit of 0.87 pM in the linear range from 0.1 pM to 10 μm with a correlation coefficient of 0.988 [57]. Furthermore, the sensor demonstrated that, highly effective measurement of miR-21 in serum samples, which can be an excellent alternative for clinical diagnosis.

11.3.3.8 Acoustic Biosensors

These biosensors are based on the change of physical properties of an acoustic wave in response that can be correlated to the presence of analytes. There are two classes of mass-balance acoustic transducers: bulk-acoustic wave (BAW) and surface-acoustic wave (SAW) devices. BAW devices can transmit an acoustic wave from one crystal surface to another whereas SAW devices can transmit an acoustic wave along a single crystal face from one location to another [58].

11.3.4 Materials Used to Fabricate Biosensors

In recent years, continuous monitoring of biological processes emphasized greatly on biosensors. Their applications range from chemical analysis, environmental monitoring to the drug, agriculture, and biological industries. Nowadays, numerous authors use electrochemical biosensing for different applications, making it a trending method. Various materials can be helpful in the fabrication of biosensors. It includes electrodes, outer casing materials, and materials for immobilization of species on the sensing element. Recently, many authors have utilized carbon and graphene oxide-based materials for making conductive inks. These prepared inks could be highly useful for making screen-printed electrodes [59–61]. These electrodes sometimes provide flexibility if fabricated on flexible materials. Metal oxide frameworks attracted various researchers over the globe in recent years due to their advantages, such as high surface area, modifiable surface, and large porosity. [62]

Polymers are the crucial material for biosensors due to their user-friendly processing, and their properties can be tailored easily. Some polymers cannot react with any analytes, making it a suitable choice for fabrication. Organic conductive polymers have provoked an enormous interest in preparing electrodes and substrates to immobilize biological species [62]. Various authors have tried biosensing using quantum dots, nanomaterials, and other optical techniques [40].

11.3.5 Biosensors for Infectious Diseases

11.3.6 Biosensor for the Detection of Dengue

Dengue is one of the most contagious viral diseases, caused by dengue viruse (DENV) infecting 390 million people yearly all over the world, mostly in subtropical and tropical areas [63]. Every year, approximately, 22 000 people die due to the dengue virus infection globally and affects more than 100 countries including the most developed country such as United States of America [64]. In the coming years, the spread of dengue is expected to rise due to the increased population, urbanization, and climate change.

DENV can be distinguished into four serotypes such as DENV-1,2,3, and 4 which can be related both genetically and antigenically. The DENV typically causes a flu-like illness that can affect all ages, but also more-severe disease manifestations are also common, including plasma leakage. The diagnostic test for the DENV depends upon the type of test, cost, and time of detection during the progression of the disease. There are several methods such as serological tests and nucleic acid amplification (PCR/RT-PCR) are developed to detect the virus in early stages. However, the RT-PCR is the gold standard method for the diagnosis of dengue infection at the preliminary stages. However, the detection of capsid protein of DENV, i.e. nonstructural protein 1 (NS1) protein has clinical significance for the detection of the virus in the early phase (0–14 days). Furthermore, based on the immune response of the human body, the measurement of IgG and IgM also proved to be an important parameter for the DENV diagnosis. However, most of the developed methods are not suitable to deploy in the resource-limited settings as DENV is prone in this region. Hence, in most of the cases, the samples are transferred to centralized laboratory settings for the confirmation of the results. With the advancement of nanotechnology, there are several other biosensor-based approaches have been developed and utilized for the detection of DENV. Therefore, there is unmet need for the development of highly specific POC diagnosis platforms fulfilling the WHO criteria for ASSURED.

There are several technologies such as electrochemical impedance spectroscopy (EIS)-based sensing, SPR, microarray-based sensing, paper-based lateral flow assays (LFA), and LOC are developed which can be employed in clinical settings. The application of nanotechnology also enhances the sensitivity, specificity, and other parameters of the sensor. Electrochemical sensing (ES) can measure changes in electrical properties (e.g. impedance, voltage, and current) from the biomolecular interactions for the detection of various biomolecules [65, 66]. EIS is also a type of sensor which can measure the vast array of frequencies, chemical reactions, and the conductance of the electrolyte from the biological interaction. Jin et al. developed an impedimetric biosensor to detect the RNA of DENV via primer hybridization using different oligonucleotide sequences [67]. The biosensor was designed by a composite of 3-Aminopropyltriethoxysilane (APTES) functionalized graphene oxide (APTES-GO) wrapped on SiO_2 particles (SiO_2@APTES-GO) was prepared via self-assembly and successful characterization was performed using TEM and ATR-FTIR. The results demonstrated that the SiO_2@APTES-GO electrode material led to enhanced dengue RNA detection sensitivity with LOD of 1 femto-Molar. Navakul et al. [68] developed a graphene–polymer-based biosensor based on the EIS and gold nanoparticle conjugated with graphene. The charge transfer resistance (Rct) of a gold electrode coated with graphene oxide was measured and the LOD was found to be 0.12 pfu ml^{-1} [68]. Carbon nanotube (CNTs) have also been incorporated into EIS technology to detect the DENV through the NS1 antigen and whole virus. Wasik et al. [69] developed a label-free chemiresistive immunosensor for the detection of NS1 antigen protein of

DENV by employing CNTs and monoclonal antibodies of NS1. The sensor capable to detect the NS1 was spiked artificial human saliva over the range of clinically relevant concentrations with high sensitivity and selectivity [69]. The same research group also developed single-walled CNTs network-based chemoresistive transducers functionalized with heparin for the label-free detection of whole DENV [70]. Herein, heparan sulfate proteoglycan was employed which acts as the receptors for the DENV in place of traditional antibodies. The biosensor demonstrated a sensitivity of DENV detection within a clinical range and has the potential of miniaturization POC device in resource-limited settings. Another ES method includes voltammetric sensing, i.e. either cyclic voltammetry (CV) or DPV. CV is used to analyze the redox potential and the electrochemical rates by measuring the resulting current change with respect to applied varying potential [71]. Palomar et al. 2018 developed an CNTs-based immunosensor using impedance spectroscopy and CV. The resulting impedimetric dengue biosensor was tested in bovine blood plasma and displays linearity in a wide concentration range ($10^{-13}/10^{-5}$ gm L^{-1}). Surface-enhanced Raman spectroscopy (SERS) is a sensing that enhances Raman spectroscopy by improving the electron cloud around metallic nanostructures, which is done through either electromagnetic or chemical enhancement. The benefits of SERS involve chemical fingerprint on an object to ensure higher specificity, requiring a less complex sample preparation mechanism, free from interference of signals capability, high reproductivity, and potential to be implemented in POC settings. SERS can be used for various areas such as various viral pathogen diagnostics, including HIV, Influenza virus, Respiratory viruses, SARS-CoV-2, and flaviviruses. Sanchez-Purra et al. demonstrated a multiplexed assay that can distinguish between Zika and dengue NS1 biomarkers. Herein, SERS-encoded gold nanostars were conjugated to specific antibodies for both diseases and used in a dipstick immunoassay, which exhibited 15-fold and 7-fold lower detection limits for Zika NS1 and dengue NS1, respectively [72, 73] developed a SERS assay of DENV gene was proposed via a cascade signal amplification strategy of localized catalytic hairpin assembly (LCHA) and hybridization chain reaction (HCR). The assay includes the operation of cascade signal amplification strategy and the following SERS measurements by transferring the products on SERS-active AgNRs arrays. The assay displays the linear calibration curve from 1 fM to 10 nM with the LOD of 0.49 fM and has good specificity, uniformity, and recovery, which indicates that the highly sensitive SERS assay provides an attractive tool for reliable, early diagnosis of DENV gene and is worth to be popularized in a wide detection of other viruses [73]. The SERS-based detection strategies hold the potential promises for the future but lack the robustness and reproducibility as compared to the currently used gold standard methods of DENV [74].The cost of Raman reader might not be available in the primary health care settings and rural areas which deters their infield application for the detection of DENV.

Microfluidic-based sensor consists of several types of modules such as LOC, lab-on-a-disc (LODc), LFA, and microfluidic paper-based devices (μPADs). The microfluidic sensor consists of several steps such as reactions, transportation, product detection, and sample/analyte concentrators in one single platform. The sensing strategies include paper-based or polymer platforms generally prepared using photolithography. Iswardy et al. incorporating the concept of dielectrophoretic (DEP), a microfluidic chip was prepared to rapidly detect the DENV using fluorescence immune-sensing [75]. The mechanism of the microfluidic sensor includes capture of virus particles using mouse anti-flavivirus monoclonal antibody-coated beads in the chip and subsequent fluorescent detection. The signal was obtained using fluorescence microscopy and quantified using image J software. The advantages of microfluidic chip is to enhance the chemical reaction, i.e. five minutes and detect the DENV as low as 104 PFU ml^{-1}. Due to the application of microfluidic chip, the sample volume is dramatically reduced (~50 to ~15 μl). Furthermore, the chip is reusable and (>50x) and portable as compared to the conventional methods. This proof of concept with regard to a microfluidic dielectrophoresis chip thus shows the potential of immunofluorescence-based-assay applications to meet diagnostic needs.

11.3.7 Biosensors for Tuberculosis

TB is one of the leading infectious diseases caused by Mycobacterium TB (M. tuberculosis) [76]. TB diagnostics were performed using conventional methods such as smear microscopy with Ziehl-Neelsen staining, immunological assays, and ELISA. The rapid diagnosis of TB in resource-limited settings is a concerning issue over the globe due to the unavailability of sophisticated laboratories. Hence, there is a need for alternate biosensing methods for the inexpensive detection of TB [77].

Thongchai Kaewphinit et al. combined loop-mediated isothermal amplification (LAMP) method with a lateral flow dipstick (LFD) for biosensor development. The FITC-labeled probe was added to the LAMP reaction for the hybridization. Furthermore, the hybridized product was added to an assay buffer solution, and the detection process was performed using the LFD [78]. Navjot Kaur et al. fabricated a POC inexpensive fluorescent isothermal paper-and-plastic nucleic acid amplification tests (FLIPP-NAATs) using LAMP. The sample was loaded on the FLIPP-NAAT, and the whole assembly was placed in the incubator for the reaction. Later, DNA intercalating dye was added before the detection. UV light source was used for excitation, and the results were captured for further analysis. Results showed that increasing the number of target molecules increases the green fluorescence intensity [79]. Figure 11.4 represents the whole process for a POC TB detection device.

11.3.8 Future Scope

The rapid innovations in biosensors result in POC approaches for disease detection and its real-time monitoring. Integrating internet-of-things-(IOT) based technologies enable biosensors to transmit and transfer data from the conducted tests. These tests can generate a vast amount of sensitive data. This generated data can be analyzed using proper data analytics tools soon [80, 81]. The conventional techniques are time-consuming, sophisticated, and require a highly skilled workforce. On the contrary, the sensitive biosensors could detect infectious diseases in a few minutes. This may show a ray of hope for early detection of the diseases. This will undoubtedly contribute to

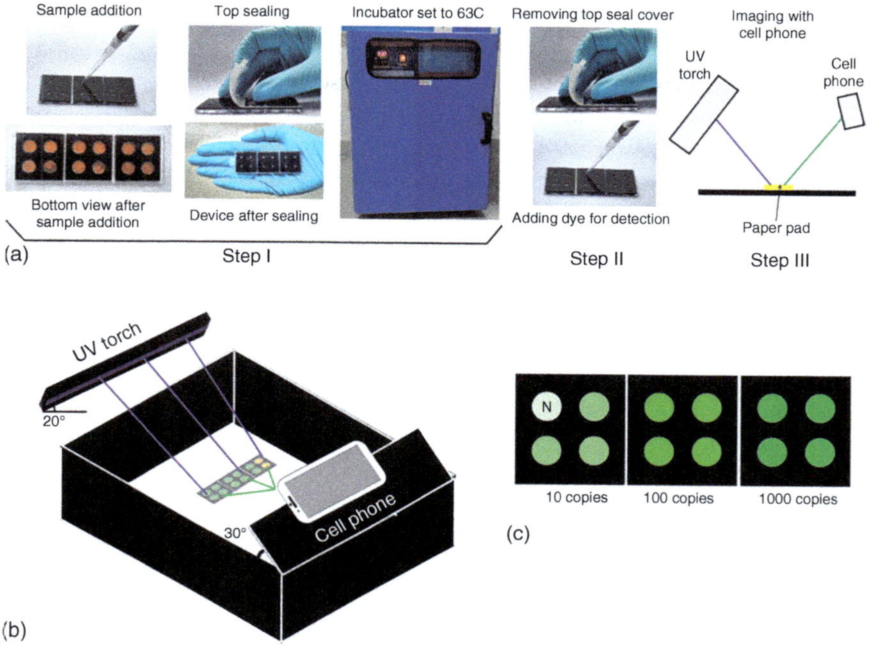

Figure 11.4 User experience of conducting a NAAT in FLIPP-NAAT. (a) Steps involved in operating FLIPPNAAT. Step I: Sample addition to FLIPP-NAAT, sealing the device and placing it in an incubator set to the reaction temperature. Step II: Removing FLIPP-NAAT from the incubator and adding the DNA intercalating dye. Step III: A UV torch is used for excitation and results are captured using a cell phone camera. (b) Schematic of an inexpensive imaging setup for end-point fluorescence detection. (c) Schematic of expected TB test results from FLIPP-NAAT. In one implementation, green fluorescence intensity increases with increasing number of target molecules. Source: The images are reproduced from Kaur et al. [79], Springer Nature, CC BY 4.0.

increasing their life expectancy and reducing disease-related mortality. Future trends will proceed with the further developments of the biochips capable of simultaneous detection and treatment. Advances in miniaturization technologies, microfluidics, nanotechnology, genomics, proteomics, and smartphone and internet technology developments lead to the generation of efficient, affordable POC diagnostics devices [82].

11.4 Conclusion

LOC devices are more than merely being the miniaturized versions of conventional analytical and diagnostic techniques. The advances in fabrication techniques, material sciences, surface modification strategies, and their integration with microfluidics and chemical and biological-based detection mechanisms have enormously enhanced the capabilities of these devices. The chapter describes about the fundamentals and applications of biosensors for the infectious diseases such as dengue and TB. Various material of construction was also discussed to provide information to the readers about the current trending materials. Components of biosensors such as transducers and bioreceptors were discussed in detail. The minuscule amount of sample and reagent requirements, faster transport phenomena, and ease of integration with various signal readout mechanisms make these platforms apt for use in resource-limited settings, especially in developing and underdeveloped parts of the world.

Acknowledgment

The authors are thankful to the management of Indian Institute of Technology, Ropar and SRM Institute of Science and Technology for their technical facilities and support throughout the work.

References

1 Naresh, V. and Lee, N. (2021). A review on biosensors and recent development of nanostructured materials-enabled biosensors. *Sensors (Switzerland)* 21: 1109. https://doi.org/10.3390/s21041109.
2 Kalyani, N., Goel, S., and Jaiswal, S. (2021). On-site sensing of pesticides using point-of-care biosensors: a review. *Environmental Chemistry Letters* 19: 345–354. https://doi.org/10.1007/s10311-020-01070-1.
3 Konwar, A.N. and Borse, V. (2020). Current status of point-of-care diagnostic devices in the Indian healthcare system with an update on COVID-19 pandemic. *Sensors International* 1: 100015. https://doi.org/10.1016/j.sintl.2020.100015.
4 Patel, H., Rawtani, D., and Agrawal, Y.K. (2019). A newly emerging trend of chitosan-based sensing platform for the organophosphate pesticide detection using Acetylcholinesterase – a review. *Trends in Food Science and Technology* 85: 78–91. https://doi.org/10.1016/j.tifs.2019.01.007.
5 Su, L., Jia, W., Hou, C., and Lei, Y. (2011). Microbial biosensors: a review. *Biosensors and Bioelectronics* 26: 1788–1799. https://doi.org/10.1016/j.bios.2010.09.005.
6 Hu, J., Cui, X., Gong, Y. et al. (2016). Portable microfluidic and smartphone-based devices for monitoring of cardiovascular diseases at the point of care. *Biotechnology Advances* 34: 305–320. https://doi.org/10.1016/j.biotechadv.2016.02.008.
7 Tripoliti, E.E., Fotiadis, D.I., Ioannidou, P. et al. (2020). Point-of-care testing devices for heart failure analyzing blood and saliva samples. *IEEE Reviews in Biomedical Engineering* 13: 17–31. https://doi.org/10.1109/RBME.2019.2905730.
8 Christodouleas, D.C., Kaur, B., and Chorti, P. (2018). From point-of-care testing to eHealth diagnostic devices (eDiagnostics). *ACS Central Science* 4: 1600–1616. https://doi.org/10.1021/acscentsci.8b00625.
9 Rajendran, R. and Rayman, G. (2014). Point-of-care blood glucose testing for diabetes care in hospitalized patients: an evidence-based review. *Journal of Diabetes Science and Technology* 8: 1081–1090. https://doi.org/10.1177/1932296814538940.

10 Oeschger, T., McCloskey, D., Kopparthy, V. et al. (2019). Point of care technologies for sepsis diagnosis and treatment. *Lab on a Chip* 19: 728–737. https://doi.org/10.1039/c8lc01102h.

11 Park, J., Han, D.H., and Park, J.K. (2020). Towards practical sample preparation in point-of-care testing: user-friendly microfluidic devices. *Lab on a Chip* 20: 1191–1203. https://doi.org/10.1039/d0lc00047g.

12 Cesewski, E. and Johnson, B.N. (2020). Electrochemical biosensors for pathogen detection. *Biosensors & Bioelectronics* 159: 112214. https://doi.org/10.1016/j.bios.2020.112214.

13 Cui, F., Zhou, Z., and Zhou, H.S. (2020). Review—measurement and analysis of cancer biomarkers based on electrochemical biosensors. *Journal of the Electrochemical Society* 167: 037525. https://doi.org/10.1149/2.0252003jes.

14 Jebelli, A., Oroojalian, F., Fathi, F. et al. (2020). Recent advances in surface plasmon resonance biosensors for microRNAs detection. *Biosensors & Bioelectronics* 169: 112599. https://doi.org/10.1016/j.bios.2020.112599.

15 Zeni, L., Perri, C., Cennamo, N. et al. (2020). A portable optical-fibre-based surface plasmon resonance biosensor for the detection of therapeutic antibodies in human serum. *Scientific Reports* 10: 1–9. https://doi.org/10.1038/s41598-020-68050-x.

16 Angelopoulou, M., Tzialla, K., Voulgari, A. et al. (2021). Rapid detection of *Salmonella typhimurium* in drinking water by a white light reflectance spectroscopy immunosensor. *Sensors* 21: 2683. https://doi.org/10.3390/s21082683.

17 Stavra, E., Petrou, P.S., Koukouvinos, G. et al. (2020). Fast, sensitive and selective determination of herbicide glyphosate in water samples with a White Light Reflectance Spectroscopy immunosensor. *Talanta* 214: 120854. https://doi.org/10.1016/j.talanta.2020.120854.

18 Stavra, E., Petrou, P.S., Koukouvinos, G. et al. (2018). Simultaneous determination of paraquat and atrazine in water samples with a white light reflectance spectroscopy biosensor. *Journal of Hazardous Materials* 359: 67–75. https://doi.org/10.1016/j.jhazmat.2018.07.029.

19 Rasmi, Y., Li, X., Khan, J. et al. (2021). Emerging point-of-care biosensors for rapid diagnosis of COVID-19: current progress, challenges, and future prospects. *Analytical and Bioanalytical Chemistry* 413: 4137–4159. https://doi.org/10.1007/s00216-021-03377-6/Published.

20 Thiha, A., Ibrahim, F., Muniandy, S. et al. (2018). All-carbon suspended nanowire sensors as a rapid highly-sensitive label-free chemiresistive biosensing platform. *Biosensors and Bioelectronics* 107: 145–152. https://doi.org/10.1016/j.bios.2018.02.024.

21 Tymm, C., Zhou, J., Tadimety, A. et al. (2020). Scalable COVID-19 detection enabled by lab-on-chip biosensors. *Cellular and Molecular Bioengineering* 13: 313–329. https://doi.org/10.1007/s12195-020-00642-z.

22 Bhalla, N., Jolly, P., Formisano, N., and Estrela, P. (2016). Introduction to biosensors. *Essays in Biochemistry* 60: 1–8. https://doi.org/10.1042/EBC20150001.

23 Bollella, P. and Katz, E. (2020). Biosensors—recent advances and future challenges. *Sensors (Switzerland).* https://doi.org/10.3390/s20226645.

24 Newman, J.D. and Setford, S.J. (2006). Enzymatic Biosensors. *Molecular Biotechnology* 32: 249–268.

25 Scheller, F.W., Schubert, F., Neumann, B. et al. (1991). Second generation biosensors. *Biosensors and Bioelectronics* 6: 245–253.

26 Wang, J. (2006). Electrochemical biosensors: towards point-of-care cancer diagnostics. *Biosensors and Bioelectronics* 21: 1887–1892. https://doi.org/10.1016/j.bios.2005.10.027.

27 Gribi, S., du Bois de Dunilac, S., Ghezzi, D., and Lacour, S.P. (2018). A microfabricated nerve-on-a-chip platform for rapid assessment of neural conduction in explanted peripheral nerve fibers. *Nature Communications* 9: 1–10. https://doi.org/10.1038/s41467-018-06895-7.

28 Berkenbrock, J.A., Grecco-Machado, R., and Achenbach, S. (2020). Microfluidic devices for the detection of viruses: aspects of emergency fabrication during the COVID-19 pandemic and other outbreaks. *Proceedings of the Royal Society A* 476: 20200398. https://doi.org/10.1098/rspa.2020.0398.

29 Dong, X., Liu, L., Tu, Y. et al. (2021). Rapid PCR powered by microfluidics: a quick review under the background of COVID-19 pandemic. *TrAC Trends in Analytical Chemistry* 143: 116377. https://doi.org/10.1016/j.trac.2021.116377.

30 Jiang, K., Jokhun, D.S., and Lim, C.T. (2021). Microfluidic detection of human diseases: from liquid biopsy to COVID-19 diagnosis. *Journal of Biomechanics* 117: 110235. https://doi.org/10.1016/j.jbiomech.2021.110235.

31 Zhou, X., Liu, D., Zhong, R. et al. (2004). Determination of SARS-coronavirus by a microfluidic chip system. *Electrophoresis* 25: 3032–3039. https://doi.org/10.1002/elps.200305966.

32 Chadwick, E., Le, K., Pei, Z. et al. (2021). Technical note: understanding the effect of COVID-19 on particle pollution using a low-cost sensor network. *Journal of Aerosol Science* 155: https://doi.org/10.1016/j.jaerosci.2021.105766.

33 Kenarkoohi, A., Noorimotlagh, Z., Falahi, S. et al. (2020). Hospital indoor air quality monitoring for the detection of SARS-CoV-2 (COVID-19) virus. *Sci. Total Environ.* 748: 141324. https://doi.org/10.1016/j.scitotenv.2020.141324.

34 Miller, T.C., Morgera, S.D., Saddow, S.E. et al. (2021). Electronic nose with detection method for alcohol, acetone, and carbon monoxide in coronavirus disease 2019 breath simulation model. *IEEE Sensors Journal* 21: 15935–15943. https://doi.org/10.1109/JSEN.2021.3076102.

35 Xiang, W., Lv, Q., Shi, H. et al. (2020). Aptamer-based biosensor for detecting carcinoembryonic antigen. *Talanta* 214: 120716. https://doi.org/10.1016/j.talanta.2020.120716.

36 Crivianu-Gaita, V. and Thompson, M. (2016). Aptamers, antibody sc Fv, and antibody Fab' fragments: an overview and comparison of three of the most versatile biosensor biorecognition elements. *Biosensors & Bioelectronics* 85: 32–45. https://doi.org/10.1016/j.bios.2016.04.091.

37 Villalonga, A., Pérez-Calabuig, A.M., and Villalonga, R. (2020). Electrochemical biosensors based on nucleic acid aptamers. *Analytical and Bioanalytical Chemistry* 412: 55–72. https://doi.org/10.1007/s00216-019-02226-x.

38 Du, Z., Li, H., and Gu, T. (2007). A state of the art review on microbial fuel cells: a promising technology for wastewater treatment and bioenergy. *Biotechnology Advances* 25: 464–482. https://doi.org/10.1016/j.biotechadv.2007.05.004.

39 Sassolas, A., Blum, L.J., and Leca-Bouvier, B.D. (2012). Immobilization strategies to develop enzymatic biosensors. *Biotechnology Advances* 30: 489–511. https://doi.org/10.1016/j.biotechadv.2011.09.003.

40 Lim, S.A. and Ahmed, M.U. (2016). Electrochemical immunosensors and their recent nanomaterial-based signal amplification strategies: a review. *RSC Advances* 6: 24995–25014. https://doi.org/10.1039/c6ra00333h.

41 Bancone, G., Gornsawun, G., Chu, C.S. et al. (2018). Validation of the quantitative point-of-care care start biosensor for assessment of G6PD activity in venous blood. *PLoS One* 13: e0196716. https://doi.org/10.1371/journal.pone.0196716.

42 Bhardwaj, H., Sumana, G., and Marquette, C.A. (2021). Gold nanobipyramids integrated ultrasensitive optical and electrochemical biosensor for Aflatoxin B_1 detection. *Talanta* 222: 121578. https://doi.org/10.1016/j.talanta.2020.121578.

43 Kulabhusan, P.K., Hussain, B., and Yüce, M. (2020). Current perspectives on aptamers as diagnostic tools and therapeutic agents. *Pharmaceutics* 12: 1–23. https://doi.org/10.3390/pharmaceutics12070646.

44 Grabowska, I., Sharma, N., Vasilescu, A. et al. (2018). Electrochemical aptamer-based biosensors for the detection of cardiac biomarkers. *ACS Omega* 3: 12010–12018. https://doi.org/10.1021/acsomega.8b01558.

45 Balasubramanian, S., Sorokulova, I.B., Vodyanoy, V.J., and Simonian, A.L. (2007). Lytic phage as a specific and selective probe for detection of *Staphylococcus aureus*-A surface plasmon resonance spectroscopic study. *Biosensors & Bioelectronics* 22: 948–955. https://doi.org/10.1016/j.bios.2006.04.003.

46 Lakshmanan, R.S., Guntupalli, R., Hu, J. et al. (2007). Detection of *Salmonella typhimurium* in fat free milk using a phage immobilized magnetoelastic sensor. *Sensors and Actuators B: Chemical* 126: 544–550. https://doi.org/10.1016/j.snb.2007.04.003.

47 Singh, A., Glass, N., Tolba, M. et al. (2009). Immobilization of bacteriophages on gold surfaces for the specific capture of pathogens. *Biosensors and Bioelectronics* 24: 3645–3651. https://doi.org/10.1016/j.bios.2009.05.028.

48 Cademartiri, R., Anany, H., Gross, I. et al. (2010). Immobilization of bacteriophages on modified silica particles. *Biomaterials* 31: 1904–1910. https://doi.org/10.1016/j.biomaterials.2009.11.029.

49 Kulabhusan, P.K., Rajwade, J.M., Sahul Hameed, A.S., and Paknikar, K.M. (2017). Lateral flow assay for rapid detection of white spot syndrome virus (WSSV) using a phage-displayed peptide as bio-recognition probe. *Applied Microbiology and Biotechnology* 101: 4459–4469. https://doi.org/10.1007/s00253-017-8232-6.

50 Xu, P., Ghosh, S., Gul, A.R. et al. (2021). Screening of specific binding peptides using phage-display techniques and their biosensing applications. *TrAC Trends in Analytical Chemistry* 137: 116229. https://doi.org/10.1016/j.trac.2021.116229.

51 Grieshaber, D., MacKenzie, R., Vörös, J., and Reimhult, E. (2008). Electrochemical biosensors - sensor principles and architectures. *Sensors* 8: 1400–1458. https://doi.org/10.3390/s8031400.

52 Ding, J. and Qin, W. (2020). Recent advances in potentiometric biosensors. *TrAC Trends in Analytical Chemistry* 124: 115803. https://doi.org/10.1016/j.trac.2019.115803.

53 Alaejos, M.S. and Montelongo, F.J.G. (2004). Application of amperometric biosensors to the determination of vitamins and α-amino acids. *Chemical Reviews* 104: 3239–3265. https://doi.org/10.1021/cr0304471.

54 Lazcka, O., Del Campo, F.J., and Muñoz, F.X. (2007). Pathogen detection: a perspective of traditional methods and biosensors. *Biosensors & Bioelectronics* 22: 1205–1217. https://doi.org/10.1016/j.bios.2006.06.036.

55 Martinkova, P., Kostelnik, A., Valek, T., and Pohanka, M. (2017). Main streams in the construction of biosensors and their applications. *International Journal of Electrochemical Science* 12: 7386–7403. https://doi.org/10.20964/2017.08.02.

56 Su, L., Zou, L., Fong, C.C. et al. (2013). Detection of cancer biomarkers by piezoelectric biosensor using PZT ceramic resonator as the transducer. *Biosensors and Bioelectronics* 46: 155–161. https://doi.org/10.1016/j.bios.2013.01.074.

57 Lim, J.Y. and Lee, S.S. (2020). Sensitive detection of microRNA using QCM biosensors: sandwich hybridization and signal amplification by TiO_2 nanoparticles. *Analytical Methods* 12: 5103–5109. https://doi.org/10.1039/d0ay01481h.

58 Länge, K., Rapp, B.E., and Rapp, M. (2008). Surface acoustic wave biosensors: a review. *Analytical and Bioanalytical Chemistry* 391: 1509–1519. https://doi.org/10.1007/s00216-008-1911-5.

59 Arduini, F., Micheli, L., Moscone, D. et al. (2016). Electrochemical biosensors based on nanomodified screen-printed electrodes: recent applications in clinical analysis. *TrAC Trends in Analytical Chemistry* 79: 114–126. https://doi.org/10.1016/j.trac.2016.01.032.

60 Camargo, J.R., Silva, T.A., Rivas, G.A., and Janegitz, B.C. (2022). Novel eco-friendly water-based conductive ink for the preparation of disposable screen-printed electrodes for sensing and biosensing applications. *Electrochimica Acta* 409: 139968. https://doi.org/10.1016/j.electacta.2022.139968.

61 Vu, Q.K., Tran, Q.H., Vu, N.P. et al. (2021). A label-free electrochemical biosensor based on screen-printed electrodes modified with gold nanoparticles for quick detection of bacterial pathogens. *Materials Today Communications* 26: 101726. https://doi.org/10.1016/j.mtcomm.2020.101726.

62 Zhang, S., Wright, G., Yang, Y. et al. (2016). Electrochemical immunosensors and their recent nanomaterial-based signal amplification strategies: a review. *Electrochimica Acta* 8: 24995–25014. https://doi.org/10.3389/fmats.2020.583739.

63 Bhatt, S., Gething, P.W., Brady, O.J. et al. (2013). The global distribution and burden of dengue. *Nature* 496: 504–507. https://doi.org/10.1038/nature12060.

64 Hasan, S., Jamdar, S.F., Alalowi, M., and Al Beaiji, A.A.S.M. (2016). Dengue virus: a global human threat: review of literature. *Journal of International Society of Preventive & Community Dentistry* 6: 1. https://doi.org/10.4103/2231-0762.175416.

65 Ilkhani, H. and Farhad, S. (2018). A novel electrochemical DNA biosensor for Ebola virus detection. *Analytical Biochemistry* 557: 151–155. https://doi.org/10.1016/j.ab.2018.06.010.

66 Tancharoen, C., Sukjee, W., Thepparit, C. et al. (2019). Electrochemical biosensor based on surface imprinting for zika virus detection in serum. *ACS Sensors* 4: 69–75. https://doi.org/10.1021/acssensors.8b00885.
67 Jin, S.A., Poudyal, S., Marinero, E.E. et al. (2019). Impedimetric dengue biosensor based on functionalized graphene oxide wrapped silica particles. *ACS Sensors* 194: 422–430. https://doi.org/10.1016/j.electacta.2016.02.116.
68 Navakul, K., Warakulwit, C., Yenchitsomanus, P.t. et al. (2017). A novel method for dengue virus detection and antibody screening using a graphene-polymer based electrochemical biosensor. *Nanomedicine: Nanotechnology, Biology and Medicine* 13: 549–557. https://doi.org/10.1016/j.nano.2016.08.009.
69 Wasik, D., Mulchandani, A., and Yates, M. (2018). Salivary detection of dengue virus NS1 protein with a label-free immunosensor for early dengue diagnosis. *Sensors* 18: 2641. https://doi.org/10.3390/s18082641.
70 Wasik, D., Mulchandani, A., and Yates, M.V. (2017). A heparin-functionalized carbon nanotube-based affinity biosensor for dengue virus. *Biosensors and Bioelectronics* 91: 811–816. https://doi.org/10.1016/j.bios.2017.01.017.
71 Anusha, J.R., Kim, B.C., Yu, K.H., and Raj, C.J. (2019). Electrochemical biosensing of mosquito-borne viral disease, dengue: a review. *Biosensors & Bioelectronics* 142: 111511. https://doi.org/10.1016/j.bios.2019.111511.
72 Sánchez-Purrà, M., Carré-Camps, M., De Puig, H. et al. (2017). Surface-enhanced Raman spectroscopy-based sandwich immunoassays for multiplexed detection of zika and dengue viral biomarkers. *ACS Infectious Diseases* 3: 767–776. https://doi.org/10.1021/acsinfecdis.7b00110.
73 Song, C., Zhang, J., Liu, Y. et al. (2020). Highly sensitive SERS assay of DENV gene via a cascade signal amplification strategy of localized catalytic hairpin assembly and hybridization chain reaction. *Sensors and Actuators B: Chemical* 325: 128970. https://doi.org/10.1016/j.snb.2020.128970.
74 Kabir, M.A., Zilouchian, H., Younas, M.A., and Asghar, W. (2021). Dengue detection: advances in diagnostic tools from conventional technology to point of care. *Biosensors* 11: 206. https://doi.org/10.3390/bios11070206.
75 Iswardy, E., Tsai, T.C., Cheng, I.F. et al. (2017). A bead-based immunofluorescence-assay on a microfluidic dielectrophoresis platform for rapid dengue virus detection. *Biosensors & Bioelectronics* 95: 174–180. https://doi.org/10.1016/j.bios.2017.04.011.
76 He, X., Zhou, L., He, D. et al. (2011). Biosensing technologies for mycobacterium tuberculosis detection: status and new developments. *Clinical & Developmental Immunology* 2011: https://doi.org/10.1155/2011/193963.
77 Srivastava, S.K., Van Rijn, C.J.M., and Jongsma, M.A. (2016). Biosensor-based detection of tuberculosis. *RSC Advances* 6: 17759–17771. https://doi.org/10.1039/c5ra15269k.
78 Kaewphinit, T., Santiwatanakul, S., Jaratsing, P. et al. (2012). Detection of Mycobacterium tuberculosis by using loop-mediated isothermal amplification combined with a lateral flow dipstick biosensor. In: *The 4th 2011 Biomedical Engineering International Conference*, 86–88. Chiang Mai: IEEE.

79 Kaur, N., Michael, J.S., and Toley, B.J. (2019). A modular paper-and-plastic device for tuberculosis nucleic acid amplification testing in limited-resource settings. *Scientific Reports* 9: 1–12. https://doi.org/10.1038/s41598-019-51873-8.

80 Ibrahim, A.U., Al-Turjman, F., Sa'id, Z., and Ozsoz, M. (2020). Futuristic CRISPR-based biosensing in the cloud and internet of things era: an overview. *Multimedia Tools and Applications* 81: 35143–35171. https://doi.org/10.1007/s11042-020-09010-5.

81 Vashistha, R., Dangi, A.K., Kumar, A. et al. (2018). Futuristic biosensors for cardiac health care: an artificial intelligence approach. *3 Biotech* 8: 1–11. https://doi.org/10.1007/s13205-018-1368-y.

82 Zarei, M. (2017). Portable biosensing devices for point-of-care diagnostics: recent developments and applications. *TrAC Trends in Analytical Chemistry* 91: 26–41. https://doi.org/10.1016/j.trac.2017.04.001.

12

Lab-on-a-Chip Devices for Point-of-Care Infectious Diseases Diagnostics

Snehal Jani[1], Vishakha Dave[2], Medha Pandya[3], Ranjeet Brajpuriya[4], and Sushma Dave[5]

[1]*Amity University Madhya Pradesh, Amity School of Pure & Applied Sciences, Department of Physics, Maharajpura Dang, Gwalior, Madhya Pradesh, 474005, India*
[2]*Maharaja Krishnakumarsinhji Bhavnagar University, Department of Physics, Bhavnagar, Gujarat, 364002, India*
[3]*Maharaja Krishnakumarsinhji Bhavnagar University, Department of Life Sciences, Bhavnagar, Gujarat, 364002, India*
[4]*University of Petroleum & Energy Studies, Department of Applied Physics, Dehradun, India*
[5]*JIET, Department of Applied Sciences, Jodhpur, India*

12.1 Introduction

An unhealthy lifestyle is dominant form of present modern lifestyle which leads a human community toward the biggest challenges of infectious diseases in the coming time. In the history of infectious disease outbreaks and vaccines, we have seen a long path of revolutions in medical sciences. It has taken decades for vaccines to be developed after disease outbreaks. The advancement of medical science has accelerated the detection and treatment of disease. In this regard, World Health Organization (WHO) has given utmost priority to the research and development of infectious diseases due to their potential for epidemics or even a pandemic [1]. Some deadly pathogens are known to be zoonosis like the Severe Acute Respiratory Syndrome, Middle East Respiratory Syndrome, Nipah virus, New Delhi metallo-ß-lactamase-1 Enterobacteriaceae, Rift Valley Fever virus, and Crimean-Congo Haemorrhagic Fever virus [1–3]. In addition to the mentioned pathogens environmental pathogens (bacteria, viruses, parasites, and toxins) every so often occurs in food and water [2, 4]. To save the human species from the attack of these deadly pathogens with successful and sustainable clinical approaches, a proactive approach of, "*Prevention is better than cure*" is required.

Point-of-Care Biosensors for Infectious Diseases, First Edition.
Edited by Sushma Dave and Jayashankar Das.

In the wake of the recent coronavirus (COVID-19) pandemic, it has been highlighted that point-of-care (POC) diagnostics should be used in place of traditional lab-based benchtop diagnoses [5–8] with a correspondingly long turnaround to get result [9, 10]. In addition, such a conventional laboratory setups are limited due to their high costs and unaffordability for horde of country like China and India. Lab-on-a-Chip (LOC) is a miniaturized device that integrates several laboratory analyses onto a single chip: analyses such as DNA sequencing and biochemical detection [11–13]. Research on LOC focuses on many applications of science and technology, including therapeutics. So as a first step toward prevention, crucial efforts are focused on POC tests. Ideally POCdiagnosis would be an independent and reliable which allow a non-trained operator to complete a primary diagnostic procedure at the patient site only [14].

LOC is a device that integrates several analytical functions on a single chip only a few square millimeters up to centimeters in size. LOC devices can perform the same functions as their full-scale counterparts. Processes, normally carried out in a laboratory, are miniaturized on a single chip to enhance efficiency and mobility as well as reduce sample and reagent volumes [15]. LOC is predicated to have several advantages over current test methods. There are two most important one first diagnostics at the POC and small amounts of materials/samples necessary for conducting test. Moreover, it is an invasive technique where detection of biomarkers is in extracted body fluids, such as metabolites, nucleic acids, proteins, and cells. Thus, LOC applications can provide benefits without compromising the quality of health care. Since a general practitioner can perform tests immediately, or even patients themselves can perform the test, LOC applications will contribute to the current trend of more self-reliance in healthcare (Figure 12.1). Providing healthcare professionals with the tools they need to develop new devices will increase their use of LOC applications.

S.C. Terry of Stanford University invented the first LOC analysis device, a gas chromatograph, in 1979 [16, 17]. LOC combines one or more laboratory tasks on a single integrated circuit (often referred to as a "chip") that is only a few square centimeters in size, allowing for automation and high-throughput screening [18] LOCs are capable of handling fluid quantities as tiny as picolitres.

Basically, a LOC is a platform where many advanced technologies like nanotechnology, microfluidics, biotechnology, and computer science technology work together to accomplish a single task of diagnosis with a great precision. Various nanomaterials are also being incorporated for signal amplification and to improve the total assay performance. A detailed overview of the primary LOC technologies in the field of POC medical diagnostics has been given by authors [14]. These include paper-based assays and microfluidics, magnetic-assisted detection, centrifugal microfluidics, and smartphone-based detection. Among all, the microfluidics-based LOC systems play a significant role due to precision of handling of the complex fluid over different microchannel geometries and identification of targets within minutes with a single-cell sensitivity level [13].

Microfluidics and molecular biology are two key technologies that have propelled the LOC sector forward. Microfluidics is the technique of manipulating and controlling fluids and particles at micron and submicron dimensions, as well as the

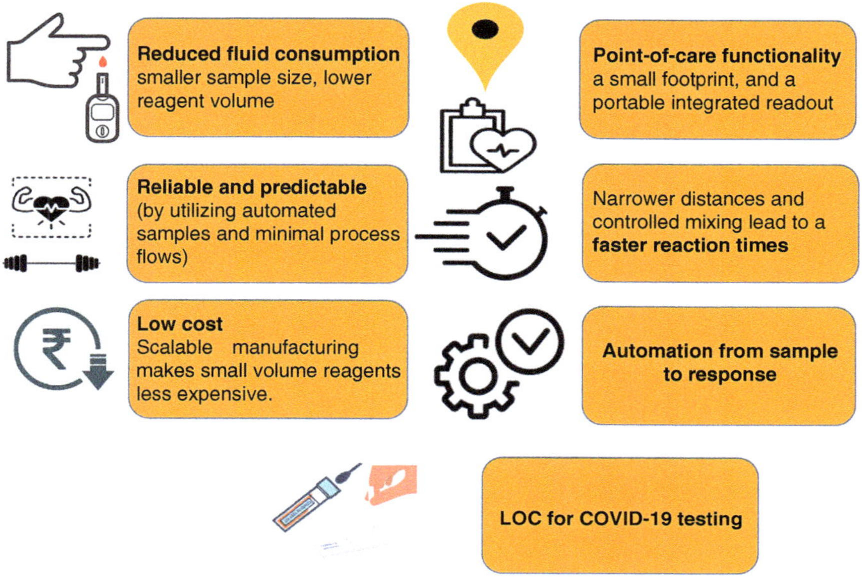

Figure 12.1 COVID-19 testing benefits from Lab-on-a-Chip devices. Benefits including smaller sample size, reliable and predictable, faster reaction times, low cost, point-of-care functionality and automation from sample to response.

technology connected with the creation of methods and systems to perform such operations [19]. Capillary-driven microfluidic chips are currently in use for POC diagnostics and advancements in 3D printing technology will likely positively intervene soon, at least with respect to plastic-based microfluidics-based LOC technology.

In this chapter, we provide a general overview of microfluidics-based LOC devices for a medical diagnostic as the analysis of a biological sample involves various processing steps, such as sample preparation, analyte enrichment, labeling, signal amplification, and signal detection, that are performed on the chip. Further, we will discuss role of LOC in diagnosis of infectious diseases and discuss the main challenges in the field and conclude by offering a future-oriented perspective.

12.2 Design of Lab-on-a-Chip Devices

An integration of several complex processes starting from sampling to data analysis on a single chip is the postulated concept named as micro total analysis systems (**μTAS**) by Menz et al. in early 1990 [20, 21]. In the modern era, the concept is emerged as state-of-the-art **LOC** device with potential advantages of cost reduction and increased sensitivity which proved best suitable for POC diagnosis. To complete the process flow, fabrication of several components (Figure 12.2) on a single piece (chip) is require where each component performs set of operations to carry out precise diagnosis.

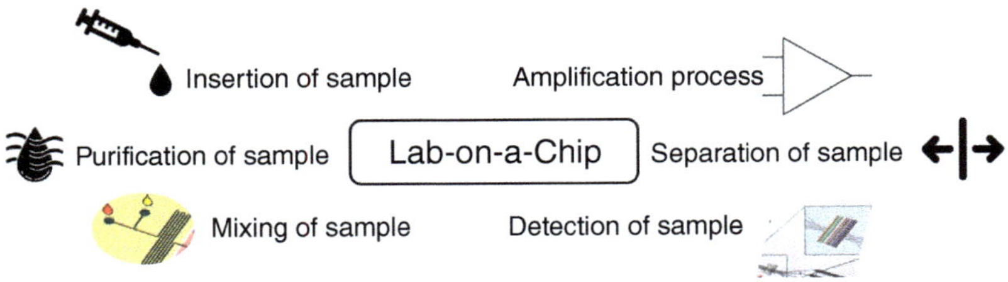

Figure 12.2 The Lab-on-a-Chip system is made up of individual components.

In a broad way if we define the unit components of the LOC then they are gone like sample purification, sample injection, sample mixing, sample separation, sample detection, and sample amplification. Each process has a different mechanism, choice of material, and fabrication techniques as per its need and application. Here in this section, we have provided an overview of two primary technologies, "Paper-Based LOC and Chip-based LOC" in the field of POC medical diagnostics. Further, this section discusses the involved processes and materials of LOC devices, the essential principles governing the design of these systems, and different detection techniques.

12.2.1 Microfluidic Paper-Based Analytical Devices (μPADs)

Paper-based diagnostic platforms have traditionally been positioned as a lifeline for developing countries, providing low-cost POC diagnosis. Microfluidic devices based on a paper can operate in a wide range of chemicals and are compatible with a wide range of chemical agents. It is also capable of transporting liquids by capillary action alone without having to use external aid. In addition to being portable, easily dispensable, and flexible, these analytical devices are easy to integrate. As paper chromatography was invented in the early twentieth century, diagnostic devices based on paper began to emerge [22, 23].

In 1956, Comer [24] created the first bioactive paper for glucose analysis in urine. Lateral flow assays were the next major breakthrough. LFAs first appeared in the 1970s for molecular detection, focusing on proteins. Lateral flow assays are commonly used in POC diagnostics. LFAs are being used in medical, food, environmental, and biodefense applications [25]. LFAs commonly capture tagged antibodies and identify biomarkers qualitatively and/or semiquantitatively by seeing a colored band [26]. A liquid sample containing the target analyte travels (through capillary forces) through various zones of polymeric strips on which capture probes capable of interacting with the analyte are immobilized in these tests (see Figure 12.3). One of the most popular lateral flow tests is the commercial pregnancy test, which uses a sandwich-based immunoassay to detect human chorionic gonadotropin in urine. The target protein is detected by a change in color that can be seen with the bare eyes [27]. The major benefits of lateral flow tests are their affordability, longer shelf life, simplicity, and convenience of use.

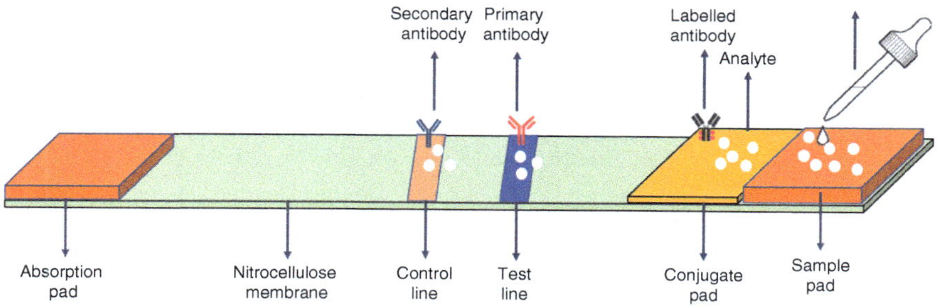

Figure 12.3 Components of an LFA test strip. An LFA strip consists of five components: samples, conjugate pads, test lines, control lines, and absorption pads.

When the Whitesides group published the first μPAD employing photolithography as the patterning approach, paper-based microfluidics gained popularity as a study field [28]. Because they utilize chemical printing and/or cutting to form flow channels, μPADs vary from LFAs and litmus paper. Consequently, multiplexed analysis utilizing tiny sample quantities became possible (Figure 12.4). Any kind of porous membrane with the ideal proportions of thickness, pore distribution, cost, and absorption rate may be used to create microfluidic devices. To pick the optimal membrane substrate for the microfluidic device, it is important to carefully analyze the paper's qualities. By designing the hydrophilic membrane so that isolated areas can flow through, hydrophobic barriers are produced. Hydrophobic barriers have been established using patterning techniques such as photolithography, printing, cutting, and chemical vapor deposition [29]. Once more, choosing the right barrier is crucial, and compatibility should be ensured by carefully considering the type of solutions to be used.

12.2.1.1 Fabrication of Paper-Based Microfluidic Devices

At production level, choice of efficient fabrication techniques and suitable materials is very important for cost-effective and bulk paper devices. There are number of

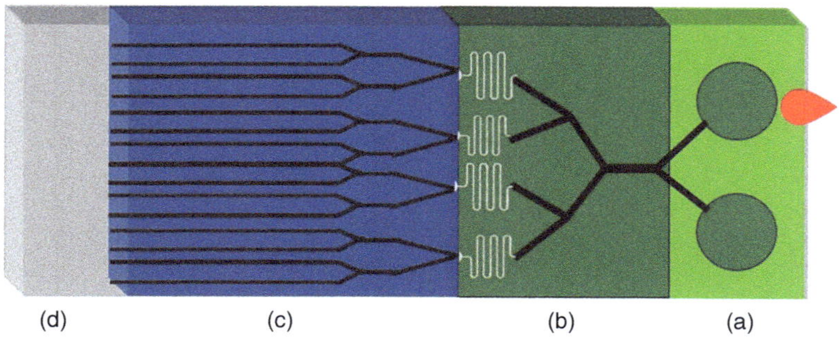

Figure 12.4 A paper-based diagnostic gadget with four essential operations is depicted. (a). Sample preparation and transport to the fluidic channels, (b). Microfluidic flow, further separation, and mixing stages, (c). The analyte's reaction and capture, (d). Collection of samples to guarantee that the whole volume goes through the reaction zone.

chemical and physical procedures and techniques available to pretreat the paper to make it further suitable for microfluidic devices or for indirect application use [30].

Few reported techniques are wax printing, photolithography, plasma treatment, screen printing, and laser treatment [29] which were selected accordingly type of paper material and need of modification. Very imperative aspect of this area is to focus on how to restore liquid on the paper as a separate segment likewise in chip-based microfluidic devices. In the same framework, we are going to discuss some important aspects of paper-based microfluidic devices.

A lithographic technique has been first introduced by Martinez et al. in 2007 to pave a fluidic way using a hydrophobic photoresist, i.e. SU-8 polymer (Figure 12.5). Photolithography is a very efficient method for formation of hydrophilic and hydrophobic areas on the paper. It has initially been using in fabrication of μPADs. Photolithography is based on transferring a geometric structure from a photomask, which is expensive and time-consuming to produce, to a light-sensitive photoresist impregnated in the paper. Through the hydrophobic walls, the liquid penetration pathway was decided. After the inception of liquid to the hydrophilic path, it further channelized in the paper matrix by capillary flow action. It is in demand for its simple and relatively cheap alternative to existing POC diagnosis. An advance approach in paper-based microfluidic devices is using paper origami by Asano and Shiraishi [31]. Using a 3D printer hydrophilic and hydrophobic zones can be created on the paper without using photomask. In this method, chromatography paper was cut to the desired dimensions and dipped in an octadecyltrichlorosilane *n*-hexane solution for five minutes. Next, the modified paper was exposed to UV light through the photomask to yield hydrophilic zones. Paper origami fabrication method offers features such as lightweight, simple and cost-effective synthesis instrumentation.

Wax printing is also one of the printing methods to form a microfluidic channel as shown by Carrilho et al. [32]. Solid wax patterns topologies can be reproduced on the surface of paper using a heat gun to melt the wax (Figure 12.6). The wax printing technique is an efficient process for prototyping a device in fast manner [32]. In comparison to photolithography, wax printing is very convenient technique for the deposition of sensing elements for POC devices, as very fewer steps are involved to

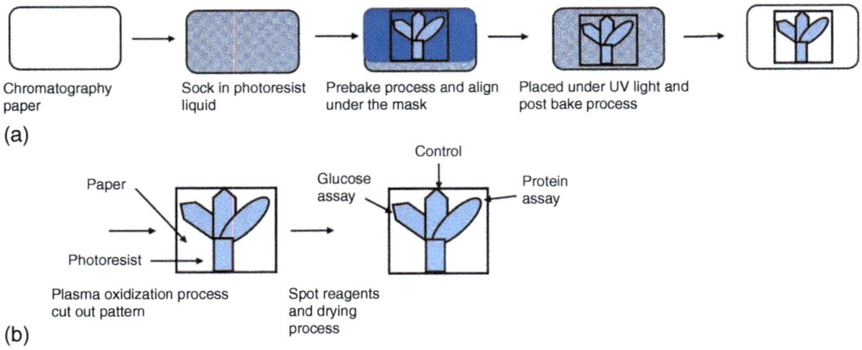

Figure 12.5 (a) Photolithography is used to create paper with millimeter-sized channels. (b) Paper spotting for glucose and protein tests.

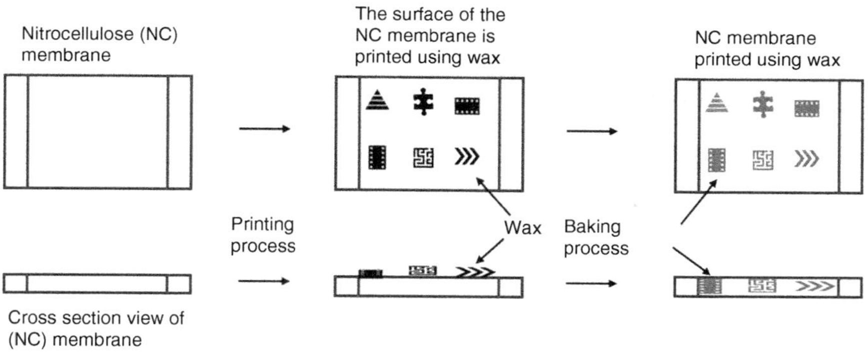

Figure 12.6 Printing wax onto a nitrocellulose membrane is seen schematically. The initial process is to design the wax onto the membrane with a wax printer, and the membrane is then placed in an oven at the desired temperature for the required time to melt the wax.

create hydrophobic barrier. Providing external heat, the wax can penetrate two dimensionally within the paper matrix. The applied wax can effectively confine the solution flow to plan region, if dispersion of the wax will be vertically through the paper. However due to horizontal nature of the paper fiber, wax dispersed faster in a horizontal direction, rather than vertical direction, causing a wider line compared to the designed width [32]. Therefore, the reproducibility of the fabrication method is highly dependent on the width of the wax line and the heating temperature.

To manage sample and reagent flow for glucose and protein detection, three-dimensional (3D) origami-based "μPADs" were also created [33]. Finally, new detection techniques including electrochemical detection, fluorescence, and chemiluminescence were introduced [29].

12.2.2 Chip-Based Microfluidic LOCs

The primary challenge of LOC is to manipulate and handle fluids on microscale and for that best performance of microfluidics is required. The component *Microfluidic unit* is two-dimensional (2D) device typically of 100 nm and 1 mm and manufactured in multilayer structure to accommodate number of assays. The *Microfluidic unit* performs operations like pumping, valving, mixing, separation, reagent storage, and sample preparation (see Figure 12.7) [34]. These fluidic microchips are contained domains interconnected by micrometer or nanometer of channels through which sample and other reagents are transferred. The fluid volumes are in the range of in the nanoliter to microliter range. In a multiplex, LOC or microarray contains immobilized reagents on the surface of a substrate made up of silicon, glass, plastic, or polymer and other materials like calcium alginate, cross-linked gelatine, or hydrogels [35, 36]. With this microfluidic unit, there is a built-in electronics component which has an interface with signal-generating molecules (antibodies and antigens) and other materials, together amplified signal from biomolecule to the detecting device [37]. Further, the chip is mounted on the signal processing unit within the chip connected to display control unit.

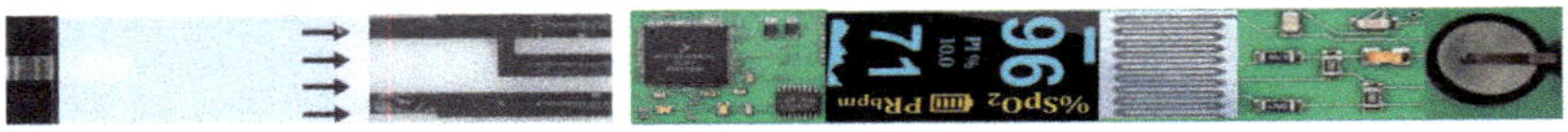

Figure 12.7 The perfect point-of-care diagnostic device. This device quantitatively measures multiple analytes within using appropriate sensitivity and transmits encrypted results to a healthcare record.

Microfabrication of the multilayer micro-assay unit is very crucial synthesis step of the LOC. Some of the best common methods of fabrication on different types of substrate are ultrasonic machining, electro-discharge machining, micro-electrochemical machining, laser ablation, electron beam machining, focused ion beam (FIB) machining, traditional microelectromechanical systems (MEMS) processes, as well as mold fabrication approaches for curved surfaces [38].

12.2.3 Chip-Based Microfluidic Device Substrate Materials

Initially, glass and silicon are found to be natural choice for LOC devices as manufacturing is uncomplicated due to already established fabrication techniques for MEMS/NEMS technologies. Glass is transparent, chemically inert, and has a successful chemical surface treatments and compatible with bioelectronics processes [39]. The first LOC was designed and built of silicon due to great accuracy of silicon microfabrication, silicon industrialization, and the ability to incorporate any type of microelectrode on the same chip. But from the point of view of POC diagnosis, cost affordable, disposable, and mass production of LOCs are required. But in terms of research, creating glass/silicon LOC necessitates clean rooms and fabrication experts together to raise final cost of the product. Over the period polymer-based microfluidic devices were found more practical because of low-cost microfluidics and biocompatibility [40, 41]. Typical substrate materials include polystyrene, polycarbonate, polymethylmethacrylate, and cyclic olefin copolymers [42]. Polydimethylsiloxane (PDMS) devices have found widespread use in the academic domain, as they can be manufactured by a simple casting process, and the surface can be functionalized for immunoassay requirements [43, 44].

12.2.4 Fundamentals of Flow of Liquid in Microchannels

Laws of Science is completely different at micro and nano level comparative to macro world. There is huge difference between macroscale, micro/nanoscale analytical systems [45–47] and these differences affect number of analyte molecules being sampled. At microlevel surface tension dominates over the cohesive forces among the molecules in aqueous biological fluids such as blood, urine, and sputum which ultimately increases viscosity of the fluid. By using hydrophilic surfaces, challenge of increased surface tension can be corrected. Corona discharge, plasma treatment, cleaning priming coating and etching are the methods to increase wettability of substrates [48]. In some special cases, surfactant can also be added to the sample or flow matrix to increase wettability of hydrophobic surfaces. Whereas in high surface-to-volume media including porous substrates (e.g. nitrocellulose-based lateral flow devices) and enclosed microstructured surfaces (e.g. microchip-based immunodiagnostics), capillary flow exploits surface tension effects [49]. Via electroosmotic flow, further manipulation of the fluids, i.e. *pumping and actuation* through the microfluidic channels is possible but does require charged surfaces and application of electromagnetic field [49]. In other cases, by application of pressure or many other techniques, this is possible.

12.2.5 Sampling

In the case of infectious diseases analytes are typically in a very low concentration range (micro or nano molar) in early stage of diagnosis. Further reduction of sample size in loc device, puts fundamental restriction on presence of analyte molecules for diagnosis. The authors have tabulated information on number of analyte molecules containing 1 nmol of analyte per liter. For example, if the sample has a concentration of 1 nmol, there is a reasonable chance that not one molecule of analyte will be in the sample. So accurate scaling of sample volume in a microscale scale device is necessary so that adequate amount of antigen and antibody brought into contact [50] to carry out successful diagnosis.

12.2.6 Diagnostics Material/Biomarkers in Microfluidic Devices

Infectious disease tests mainly cover *CD4+ T-lymphocyte* counting for HIV monitoring, dengue fever, influenza A, hepatitis C, malaria, and tuberculosis [50, 51]. For different diseases, different target materials are used, for example, nucleic acids, proteins, metabolites, cells, and pathogens. Brief overview regarding these biomarkers has been given below:

Pathogens: Bacteria, viruses, and parasites, particularly those that cause infectious illnesses, are essential analytical targets. Accurate detection of the organism responsible for an infection can decrease treatment costs, alleviate suffering, aid systems in preventing disease transmission, and save lives. Pathogens are frequently identified utilizing nucleic acid identification since species and strain identification are necessary. In rare circumstances, immunoassays are used to diagnose infections using antibodies found in infected hosts.

Cells: The identification and counting of individual (human) cells in blood and other samples is a fast-growing subject in POC diagnostics. In addition to basic blood cell counting, it is commonly acknowledged that POC cell assay-based devices might be used for diagnostic and prognostic testing for infectious diseases.

Proteins: Current LOC devices make use of immunoassay technologies, such as antigen–antibody binding. These tests look for disease-specific protein indicators such as bacterial and viral infection biomarkers of HIV, influenza, chlamydia, and hepatitis, D-dimer for thrombosis and troponin I or T for cardiac damage, prostate-specific antigen for prostate cancer, glycated hemoglobin (HbA1c) for diabetes and

C-reactive protein (CRP) for inflammation, pregnancy hormone human chorionic gonadotropin.

Nucleic acids: Nucleic acid diagnostics, also known as molecular diagnostics, evaluate DNA or other forms of RNA to screen targeted genomic or genetic information of a patient or to assay nucleic acid sequences specific to invading pathogens. Such techniques frequently use polymerase chain reaction (PCR) and various other ways of amplifying preselected nucleic acid sequences. Because of the additional steps required for sample preparation (e.g. cell sorting, isolation, lysis, and nucleic acid extraction), signal amplification, and target contamination and instability, these tests are among the most difficult to create.

12.2.7 Signal Generation and Detection

As discussed earlier in Section 12.2.3, due to very low concentration of analyte molecules present in immunoassays, it is essential that very high specific activity label is used. Together with achievement of very high surface-to-volume ratio further signal amplification can be achieved via labeling. Labels are typically chemical-linked or conjugated to the desired antibody or antigen. The silver and gold-coated labels are being in use as detection is optimum with high sensitivity using simple scanners including simultaneous benefits for POC diagnostics [52]. The semiconductor particles, such as quantum dots, are perfectly suited for multiplexed detection and have already been applied to the simultaneous microchip-based detection of HIV, HBV, and HCV [53–55]. Optical detection through nanotechnology-aided probes is as well in current focus for enhancing the signal [56, 57]. The electrochemical and electromechanical detection methods are nonoptical detection approaches, where very high surface-to-volume ratios of the sensing surfaces of these miniature devices would be exploited. Further magnetic bead-based immunoassay is also being used for insulin and interleukin-6 [58].

12.2.7.1 Electrochemical Method

Microfluidic channels, electrodes made of graphite ink, and electrical wires printed with silver ink made up the electrochemical "μPADs." On the dry μPAD that was inserted into the glucometer, a drop of the sample was applied. On the glucometer LCD, the electrochemical readout was then shown. The study served as the initial instance of integrating novel "μPAD" methodologies with already-in-use portable potentiostats [29]. Amperometric, potentiometric, and impedimetric measurements are the three categories into which electrochemical detection methods may be categorized. Amperometric biosensors are the most widely utilized. They are typically employed in glucose tests and provide current proportional to the concentration of the analyte being measured potentiometric detection analyses the potential difference between two reference electrodes that are separated by a selective permeable membrane. Impedimetric biosensors measure the change in impedance brought on by changes in resistance at the sensor to function [59]. The contemporary i-STAT® system may employ any of the three detection techniques, depending on the target analyte [60].

12.2.7.2 Magnetic Particle Labeling

Magnetic particles can be used to concentrate and localize analytes. Moreover, they can be used as labeling technology for detection without the requirements of fluorescent dyes. Stimulated by advances in memory devices, magnetic particle detection technology has evolved rapidly, the most promising and sensitive methods now using the giant magnetoresistance (GMR) effect, with detectors based on so-called spin valves or magnetic tunnel junction methods [61]. Philips Research (the Netherlands) reported the development of a compact biosensor platform to detect biomolecules with superparamagnetic particle labels using GMR sensors. The silicon detection chip is packaged in a disposable cartridge that integrates electrical connections for readout and fluidic subsystem. Recently, sensitive detection of amplified DNA on this system was reported using a miniaturized detection platform suitable for POC application [62] (Figure 12.8).

12.2.7.3 Optical Detection

Conventional optical detection methods, including absorbance, fluorescence, and chemiluminiscence, have all been applied in loc devices. Miniaturizing devices that use optical detection is generally difficult because of the expensive hardware it requires. Furthermore, due to the shorter optical paths through the sample, sensitivity is reduced and increased noise from nonspecific adsorption to the walls of the chamber can be caused by a lower surface-to-volume ratio [57]. To address these issues, many integrated optical systems are being explored in which new techniques are integrated into the microfluidic device to reduce costs and increase sensitivity.

Photonic biosensors are well-established technologies for the sensitive monitoring of molecular interactions. They could afford the requirements for the "on-chip" detection in loc platforms due to their outstanding characteristics of sensitivity, label-free, and real-time detection. The detection principle of most optical biosensors is based on the evanescent field detection. In the evanescent wave mechanism, a bioreceptor layer is immobilized onto the surface of a waveguide; the exposure to

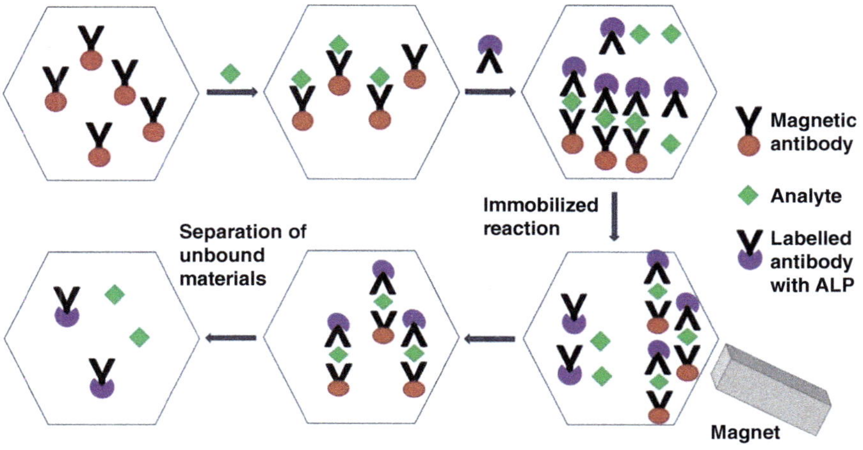

Figure 12.8 Design of a sandwich-type immunoassay based on magnetic beads.

the partner analyte produces a biomolecular interaction affecting the guiding properties of the waveguide (specifically, a variation of the refractive index) via the modification through the evanescent field [63]. The variation of the refractive index can be evaluated by any of the waveguiding optical properties (intensity, phase, resonant momentum, polarization, . . .) and this variation can be correlated with the concentration of the analyte and with the affinity constant of the interaction, resulting in a quantitative value of the interaction [64].

12.3 LOC for Diagnosis of Infectious Diseases

The detection of pathogenic microbes, such as bacteria, fungi, viruses, parasites, prions, and protozoa for clinical diagnostic applications, requires rapid and more accurate diagnostic technologies/tools. Rapid pathogen detection is vital in the early phases of disease therapy when choosing the right treatment might be critical. Despite the widespread use of traditional cell culture-based procedures in therapeutic settings, recently emerging microfluidic chips are becoming alluring benefits of cutting-edge procedures over traditional ones. To diagnose microorganisms, the miniature channels utilized in microfluidic systems analyze interactions between biological moieties and mimic interactions between cells and chemicals in microchannel structures. A new generation of traditional detection techniques, known as microfluidic systems, also called LOC devices, relies on phases including specimen preparation, reagent manipulation, bioreaction, and detection that may all be combined on a single platform.

12.3.1 LOC for Virus Detection

Only a century ago, the 1918 influenza pandemic claimed the lives of 50 million or more people, making it the deadliest catastrophe in recorded human history. The HIV/AIDS pandemic, first identified in 1981, has killed at least 37 million people. And the last decade has seen extraordinary pandemic outbreaks, including H1N1 "swine" influenza [65]. COVID-19, detected in late 2019, is just the most recent example of an unexpected, unique, and deadly pandemic disease. This recent experience suggests that we have entered a pandemic period. People who have been exposed to severe acute respiratory syndrome coronavirus 2 (SARS-CoV-2) may experience symptoms such as coughing, fever, pneumonia, and other complications that, in extreme situations, could be fatal. Additionally, some SARS-CoV-2-infected individuals may not exhibit any symptoms. Reverse transcription–polymerase chain reaction (RT–PCR), which evaluates patient samples including nasopharyngeal fluid, is currently the main diagnostic technique for COVID-19.

Point-of-care testing (POCT) is advantageous in the current pandemic situation due to its simplicity of use, improved approachability on the user's end, more prompt detection, and equivalent sensitivity and accuracy, which would lighten the testing burden on central hospitals. In POCT, Consequently, it is helpful for everyday pandemic control, early identification, and treatment.

An RNA-based POC device for COVID-19 that combines paper-based POC diagnostic tools and loop-mediated isothermal amplification (LAMP) assay technology has been described by Yang et al. [66]. This tool can be connected to a smartphone app to offer a quick, accurate, and user-friendly COVID-19 diagnostic tool. Individuals can independently collect nasal swab samples, carry out LAMP analysis, examine visible colorimetric test results, record the information, and share it with medical specialists or via the internet.

A half-strip lateral flow assay method for POCT SARS-CoV-2 antigen detection was created by Benjamin D. Grant et al. using commercially available reagents [67]. The test has a limit of detection of 0.65 ng ml^{-1} recombinant antigen. IgG and IgM antibodies against SARS-CoV-2 were precisely and repeatedly identified after Tian Wen and Chao Huang et al. improved the pH, antigen concentration, and a number of other factors for the lateral flow immunoassay strip. The team of Jilie Kong created a POC microfluidic platform for SARS-CoV-2 IgM, IgG, and antigen detection The diagnostic microchips have to be utilized in findings of a fluorescence detection analyzer are available in 15 minutes. There is a loading chamber, a waste reservoir, and a storage area. On the microchip, there is a fluid channel for fluorescence immunoassay. A capture channel is present in the fluorescence immunoassay fluid channel region as well as a test region. This platform provided a quick and simple SARS-CoV-2 detection technique [68].

An integrated microfluidic system with a microfluidic chip equipped with a hemagglutinin (HA)/neuraminidase (NA) array has been created by Shen et al. [69].

This technology enables one-step RT–PCR, fluorescent signal detection, virus lysis, and automatic virus purification. For viral isolation and extraction, magnetic microbeads are employed, and pneumatically driven microvalves are used for on-chip liquid control. With an analytical speed much faster than other techniques, up to 12 subtypes of the influenza A virus (InfA) can be automatically and simultaneously detected on this microfluidic platform in less than 100 minutes.

A microfluidic platform for ZIKV detection was created by Song et al. [70]. The procedure includes gathering and lysing saliva samples, nucleic acid extraction using a microfluidic cassette's isolation membrane, RT-LAMP amplification, and colorimetric DNA detection. Reagents and wash buffers must still be manually added even though the operation can be finished on the same microfluidic chip device. To identify many respiratory viruses, Wang et al. created a microfluidic chip technology with LAMP integration. The entire detecting procedure can be finished in one hour. This system has a high degree of specificity and sensitivity for detecting human adenoviruses, influenza B virus, and subtypes of the influenza A virus (H1N1, H3N2, H5N1, and H7N9, 69).

12.3.2 LOC for Detection of Bacteria

One of the most crucial applications in medicine is the detection of bacteria since, in many situations, the selection of an antibiotic depends on determining the source of the illness. Microfluidics-based devices have been used to identify a wide range of bacteria; here, we'll talk about the detection of the pathogens most frequently linked to human pathology, including *Escherichia coli*, *Mycobacterium tuberculosis* (TB), Salmonella spp., and *Vibrio cholerae*. Stokes et al. [71] reported a multilayer detection platform in 2000, in which *E. coli* cells were immobilized on a substrate and allowed to bind to Cy5 tagged antibodies; binding was detected using an array of photodiodes, amplifiers, discriminators, and logic circuits.

Gau and colleagues [72] described a MEMS biosensor with a "Self-Assembled Monolayer" (SAM) surface. On a Si wafer, several electrodes were formed, and a monolayer of streptavidin was fixed on the Au electrode surface. The biotinylated single-stranded DNA (ssDNA) probes were subsequently thiol-attached to the Au to collect rRNA from *E. coli*. Following that, a second set of fluorescein-coated ssDNA probes were added, followed by an anti-fluorescein antibody linked to peroxidase. The method detected 1000 *E. coli* cells using enzymatic amplification instead of PCR.

For the detection of Salmonella, novel biosensing approaches from the microfluidic systems category and impedimetric methods have been used. Kim et al. [73] described an antibody-sandwich-based microfluidic chip. Another smartphone-based device was utilized in conjunction with paper-based microfluidic chips as a portable, easy-to-use, extremely sensitive, and accurate approach for *Salmonella typhimurium* detection [74].

Several methods for detecting *V. cholerae* have been tried. Jarvius et al. described a microfluidic device that uses rolling circle amplification (RCA) to detect single molecules [70]. In another study, Batalla and colleagues used an electrochemical microfluidic chip to separate and trace D-amino acid (D-AAs) such as D-Methionine (D-Met) and D-Leucine (D-Leu), molecules that act as biomarkers for *Vibrio Cholerae*-related disorders [75] (Table 12.1).

12.3.2.1 Future Perspectives and Conclusion

This emerging field of research is now confronting a variety of obstacles, but the initiatives taken to overcome these obstacles offer promising potential. The development of a wearable LoC-based system for continuous glucose monitoring may offer an effective tool for the glucose control in patients with type 2 diabetes. For instance, the majority of current LoC systems need to demonstrate their competitive advantages over conventional systems in cost efficiency and test accuracy. With their improved capacity for sample processing and analyte enrichment, LoC-based sensors have the potential to resolve this problem. By using a basic photodiode, new reactions, such as the improved chemiluminescence reaction of hydrogen peroxide with creatinine, might increase detection accuracy while also lowering the cost of the analyzer.

Generally speaking, the LoC-based approaches have distinct qualities that make them useful alternatives or complimentary tests to the standard methods rather than necessarily replacing them. In addition, enhanced and trustworthy test results as well as significant new benefits in terms of test operation, data processing, result reporting, and management are required for clinical acceptance of novel diagnostic technologies. Few LoC systems available today, though, can fulfill all of these demands. A lot of assays still need the use of extra tools for sample preparation, signal detection, and signal processing. It is necessary to use portable devices for test signal reading, processing, and transmission along with highly integrated, affordable, and automated LoC systems.

Table 12.1 Lab-on-a-Chip devices for detection of pathogens.

Type of LOC	Name of pathogen	Target	References
Microfluidic device	SARS-CoV-2	Blood serum	[76]
Paper-based CRISPR based	SARS-CoV-2	polymerase amplification (RT-RPA) and CRISPR–Cas12a detection	[77]
Paper-based POC	SARS-CoV-2		[66]
Microfluidic chip	influenza A virus (		[69]
Lateral flow immunoassay strip	SARS-CoV-2	IgG and IgM antibodies	[78]
Microfluidic platform	SARS-CoV-2	IgG and IgM antibodies	[68]
Microfluidic System	HIV	DNA probe attaching to HIV	[79]
mChip design	HIV		[80]
Microfluidic System	HBV	Targeting of HBV DNA using micro-beads and streptavidin conjugation system	[81]
Microfluidic Chip	HBV	HBV DNA with complete genome	[81]
Microfluidic device	HIV	Whole blood	[82]
Microfluidic System	Zika virus		[70]
Smartphone signals	Zika virus		[70]
Microfluidic	*E. coli*	Cy5 labeled antibody	[71]
Electromechanical	*E. coli*	DNA a	[83]
Microfluidic system	Salmonella	Anti-Salmonella polyclonal antibodies/QDots	[37, 84]
optical detection	*Vibrio cholerae*		[85]
Microfluidic technology coupled with nuclear magnetic resonance (NMR)	*Mycobacterium tuberculosis*	TB surface receptor	[86]
Microfluidic Chip	*M. tuberculosis*	Quantitation of rpoB gene from *M. tuberculosis*	[87]
Microfluidic and NMR microcoil combination	*M. tuberculosis*	Barcoding of nucleic acid segments of TB	[88]

References

1 Watkins, K. (2018). Emerging infectious diseases: a review. *Current Emergency and Hospital Medicine Reports* 6 (3): 86–93.

2 Tan, J.J.L., Capozzoli, M., Sato, M. et al. (2014). An integrated lab-on-chip for rapid identification and simultaneous differentiation of tropical pathogens. *PLoS Neglected Tropical Diseases* 8 (7): e3043.

3 Zhu, H., Fohlerová, Z., Pekárek, J. et al. (2020). *Recent Advances in Lab-on-a-Chip Technologies for Viral Diagnosis*, Biosensors and Bioelectronics, vol. 153 (ed. P.R. Brown and E. Grushka). Elsevier Ltd.

4 Daw, R. and Finkelstein, J. (2006). Insight: lab on a chip. *Nature* 442 (7101): 367–418.

5 Panwar, R., Churi, H., Dave, S. (2022). Point-of-care electrochemical biosensors using CRISPR/Cas for RNA analysis. In: *Biosensors for Emerging and Re-Emerging Infectious Diseases*, pp. 317–333. Academic Press.

6 Dave, S., Dave, A., Radhakrishnan, S., Das, J., Dave, S., (2022). Biosensors for healthcare: an artificial intelligence approach. *Biosensors for Emerging and Re-Emerging Infectious Diseases:* 365–383.

7 Das, J., Dave, S., Radhakrishnan, S., Mohanty, P. (2022). Biosensors for Emerging and Re-Emerging Infectious Diseases. In: *Elsevier Science & Technology.*

8 Dave, S. and Jone Kirubavathy, S., (2022). Biosensors based on metal-organic framework (MOF): Paving the way to point-of-care diagnosis. In: *Electrochemical Applications of Metal-Organic Frameworks*, pp. 255–267. Elsevier.

9 Tymm, C., Zhou, J., Tadimety, A. et al. (2020). Scalable COVID-19 detection enabled by lab-on-chip biosensors. *Cellular and Molecular Bioengineering* 13 (4): 313–329.

10 Berkenbrock, J.A., Grecco-Machado, R., and Achenbach, S. (2020). Microfluidic devices for the detection of viruses: aspects of emergency fabrication during the COVID-19 pandemic and other outbreaks. *Proceedings of the Royal Society A: Mathematical, Physical and Engineering Sciences* 476 (2243): 20200398.

11 Azizipour, N., Avazpour, R., Rosenzweig, D.H. et al. (2020). Evolution of biochip technology: a review from lab-on-a-chip to organ-on-a-chip. *Micromachines (Basel)* 1 (6): 1–15.

12 Wu, J., Dong, M., Rigatto, C. et al. (2018). Lab-on-chip technology for chronic disease diagnosis. *npj Digital Medicine* 1 (1).

13 Dutse, S.W. and Yusof, N.A. (2011). Microfluidics-based lab-on-chip systems in DNA-based biosensing: an overview. *Sensors* 11 (6): 5754–5768.

14 Sofia, A.-G. and Segal, E. *Lab-on-a-Chip Devices for Point-of-Care Medical Diagnostics*, 1–19. Berlin, Heidelberg: Springer Berlin Heidelberg https://doi.org/10.1007/10_2020_127.

15 Chin, C.D., Linder, V., and Sia, S.K. (2012). Commercialization of microfluidic point-of-care diagnostic devices. *Lab on a Chip* 12 (12): 2118.

16 Angell, J.B., Terry, S.C., and Barth, P.W. (1983). Silicon micromechanical devices. *Scientific American* 248 (4): 44–55.

17 Terry, S.C., Jerman, J.H., and Angell, J.B. (1979). A gas chromatographic air analyzer fabricated on a silicon wafer. *IEEE Transactions on Electron Devices* 26 (12): 1880–1886.

18 Volpatti, L.R. and Yetisen, A.K. (2014). Commercialization of microfluidic devices. *Trends in Biotechnology* 32 (7): 347–350.

19 Whitesides, G.M. (2006). The origins and the future of microfluidics. *Nature* 442 (7101): 368–373.

20 Manz, A., Graber, N., and Widmer, H.M. (1990). Miniaturized total chemical analysis systems: a novel concept for chemical sensing. *Sensors and Actuators B: Chemical* 1 (1–6): 244–248.

21 Manz, A., Harrison, D.J., Verpoorte, E., and Widmer, H.M. (2021). Planar chips technology for Miniaturization of separation systems: a developing perspective in chemical monitoring. In: *Advances in Chromatography* (eds. N. Grinberg, P. W. Carr), 1–66. CRC Press.

22 Yu, J., Ge, L., Huang, J. et al. (2011). Microfluidic paper-based chemiluminescence biosensor for simultaneous determination of glucose and uric acid. *Lab on a Chip* 11 (7): 1286.

23 Toennies, G. and Kolb, J.J. (1951). Techniques and reagents for paper chromatography. *Analytical Chemistry* 23 (6): 823–826.

24 Comer, J.P. (1956). Semiquantitative specific test paper for glucose in urine. *Analytical Chemistry* 28 (11): 1748–1750.

25 Nie, Z., Deiss, F., Liu, X. et al. (2010). Integration of paper-based microfluidic devices with commercial electrochemical readers. *Lab on a Chip* 10 (22): 3163.

26 Abe, K., Suzuki, K., and Citterio, D. (2008). Inkjet-printed microfluidic multianalyte chemical sensing paper. *Analytical Chemistry* 80 (18): 6928–6934.

27 Hemmig, E., Temiz, Y., Gökçe, O. et al. (2020). Transposing lateral flow immunoassays to capillary-driven microfluidics using self-coalescence modules and capillary-assembled receptor carriers. *Analytical Chemistry* 92 (1): 940–946.

28 Martinez, A.W., Phillips, S.T., Butte, M.J., and Whitesides, G.M. (2007). Patterned paper as a platform for inexpensive, low-volume, portable bioassays. *Angewandte Chemie International Edition* 46 (8): 1318–1320.

29 Ozer, T., McMahon, C., and Henry, C.S. (2020). Advances in paper-based analytical devices. *Annual Review of Analytical Chemistry* 13 (1): 85–109.

30 Li, X., Ballerini, D.R., and Shen, W. (2012). A perspective on paper-based microfluidics: current status and future trends. *Biomicrofluidics* 6 (1): 011301.

31 Asano, H. and Shiraishi, Y. (2015). Development of paper-based microfluidic analytical device for iron assay using photomask printed with 3D printer for fabrication of hydrophilic and hydrophobic zones on paper by photolithography. *Analytica Chimica Acta* 883: 55–60.

32 Carrilho, E., Martinez, A.W., and Whitesides, G.M. (2009). Understanding wax printing: a simple micropatterning process for paper-based microfluidics. *Analytical Chemistry* 81 (16): 7091–7095.

33 McNeill, L., Pearson, C., Megson, D. et al. (2021). Origami chips: development and validation of a paper-based lab-on-a-chip device for the rapid and cost-effective detection of 4-methylmethcathinone (mephedrone) and its metabolite, 4-methylephedrine in urine. *Forensic Chemistry* 22: 100293.
34 Haeberle, S. and Zengerle, R. (2007). Microfluidic platforms for lab-on-a-chip applications. *Lab on a Chip* 7 (9): 1094.
35 Schena, M. (2000). *Microarray Biochip Technology* (ed. M. Schena). Natick, MA: Eaton Publishing.
36 Schena, M. (1999). DNA microarrays: a practical approach. *Practical Approach Series.*
37 Cho, I.H., Kim, D.H., and Park, S. (2020). Electrochemical biosensors: perspective on functional nanomaterials for on-site analysis. *Biomaterials Research* 24 (1): 6.
38 Scott, S. and Ali, Z. (2021). Fabrication methods for microfluidic devices: an overview. *Micromachines (Basel)* 12 (3): 319.
39 Madou, M.J. (2011). *Fundamentals of Microfabrication and Nanotechnology, Three-Volume Set*, 3rde. CRC Press.
40 Becker, H. (2002). Polymer microfluidic devices. *Talanta* 56 (2): 267–287.
41 Liu, C. (2007). Recent developments in polymer MEMS. *Advanced Materials* 19 (22): 3783–3790.
42 Nunes, P.S., Ohlsson, P.D., Ordeig, O., and Kutter, J.P. (2010). Cyclic olefin polymers: emerging materials for lab-on-a-chip applications. *Microfluidics and Nanofluidics* 9 (2–3): 145–161.
43 McDonald, J.C., Duffy, D.C., Anderson, J.R. et al. (2000). Fabrication of microfluidic systems in poly(dimethylsiloxane). *Electrophoresis* 21 (1): 27–40.
44 Sia, S.K. and Kricka, L.J. (2008). Microfluidics and point-of-care testing. *Lab on a Chip* 8 (12): 1982.
45 Janasek, D., Franzke, J., and Manz, A. (2006). Scaling and the design of miniaturized chemical-analysis systems. *Nature* [Internet] 442 (7101): 374–380. http://www.nature.com/articles/nature05059.
46 Gad-el-Hak, M. (ed.) (2005). *MEMS*. CRC Press.
47 Kirby, B.J. (2012). *Micro- and Nanoscale Fluid Mechanics*. Cambridge University Press.
48 Ebnesajjad, S. (2011). Material surface preparation techniques. In: *Handbook of Adhesives and Surface Preparation*, 49–81. Elsevier.
49 Rattle, S., Hofmann, O., Price, C.P. et al. (2013). Lab-on-a-chip, micro- and nanoscale immunoassay systems, and microarrays. In: *The Immunoassay Handbook*, 175–202. Elsevier.
50 (2007). Scaling revisited. *Lab on a Chip* 7 (12): 1630.
51 Yager, P., Edwards, T., Fu, E. et al. (2006). Microfluidic diagnostic technologies for global public health. *Nature* 442 (7101): 412–418.
52 Nam, J.M., Thaxton, C.S., and Mirkin, C.A. (1979). Nanoparticle-based bio-bar codes for the ultrasensitive detection of proteins. *Science* 301 (5641): 1884–1886.
53 Klostranec, J.M., Xiang, Q., Farcas, G.A. et al. (2007). Convergence of quantum dot barcodes with microfluidics and signal processing for multiplexed high-throughput infectious disease diagnostics. *Nano Letters* 7 (9): 2812–2818.

54 Chin, C.D., Laksanasopin, T., Cheung, Y.K. et al. (2011). Microfluidics-based diagnostics of infectious diseases in the developing world. *Nature Medicine* 17 (8): 1015–1019.

55 Yan, Z., Zhou, L., Zhao, Y. et al. (2006). Rapid quantitative detection of Yersinia pestis by lateral-flow immunoassay and up-converting phosphor technology-based biosensor. *Sensors and Actuators B: Chemical* 119 (2): 656–663.

56 Myers, F.B. and Lee, L.P. (2008). Innovations in optical microfluidic technologies for point-of-care diagnostics. *Lab on a Chip* 8 (12): 2015.

57 Azzazy, H.M., Mansour, M.M., and Kazmierczak, S.C. (2006). Nanodiagnostics: a new frontier for clinical laboratory medicine. *Clinical Chemistry* 52 (7): 1238–1246.

58 Sista, R.S., Eckhardt, A.E., Srinivasan, V. et al. (2008). Heterogeneous immunoassays using magnetic beads on a digital microfluidic platform. *Lab on a Chip* 8 (12): 2188.

59 Mohammed, M.I. and Desmulliez, M.P.Y. (2011). Lab-on-a-chip based immunosensor principles and technologies for the detection of cardiac biomarkers: a review. *Lab on a Chip* 11 (4): 569–595.

60 Chin, C.D., Linder, V., and Sia, S.K. (2007). Lab-on-a-chip devices for global health: past studies and future opportunities. *Lab on a Chip* 7 (1): 41–57.

61 Gubala, V., Harris, L.F., Ricco, A.J. et al. (2012). Point of care diagnostics: status and future. *Analytical Chemistry* 84 (2): 487–515.

62 Foudeh, A.M., Fatanat Didar, T., Veres, T., and Tabrizian, M. (2012). Microfluidic designs and techniques using lab-on-a-chip devices for pathogen detection for point-of-care diagnostics. *Lab on a Chip* 12 (18): 3249.

63 Estevez, M.C., Alvarez, M., and Lechuga, L.M. (2012). Integrated optical devices for lab-on-a-chip biosensing applications. *Laser & Photonics Reviews* 6 (4): 463–487.

64 Morens, D.M. and Fauci, A.S. (2020). Emerging pandemic diseases: how we got to COVID-19. *Cell* 182 (5): 1077–1092.

65 Yang, T., Wang, Y.C., Shen, C.F., and Cheng, C.M. (2020). Point-of-care RNA-based diagnostic device for COVID-19. *Diagnostics* 10 (3): 165.

66 Grant, B.D., Anderson, C.E., Williford, J.R. et al. (2020). SARS-CoV-2 coronavirus nucleocapsid antigen-detecting half-strip lateral flow assay toward the development of point of care tests using commercially available reagents. *Analytical Chemistry* 92 (16): 11305–11309.

67 Lin, Q., Wen, D., Wu, J. et al. (2020). Microfluidic immunoassays for sensitive and simultaneous detection of IgG/IgM/antigen of SARS-CoV-2 within 15 min. *Analytical Chemistry* 92 (14): 9454–9458.

68 Shen, K.M., Sabbavarapu, N.M., Fu, C.Y. et al. (2019). An integrated microfluidic system for rapid detection and multiple subtyping of influenza a viruses by using glycan-coated magnetic beads and RT-PCR. *Lab on a Chip* 19 (7): 1277–1286.

69 Song, J., Mauk, M.G., Hackett, B.A. et al. (2016). Instrument-free point-of-care molecular detection of Zika virus. *Analytical Chemistry* 88 (14): 7289–7294.

70 Wang, R., Zhao, R., Li, Y. et al. (2018). Rapid detection of multiple respiratory viruses based on microfluidic isothermal amplification and a real-time colorimetric method. *Lab on a Chip* 18 (22): 3507–3515.

71 Gau, J.J., Lan, E.H., Dunn, B. et al. (2001). A MEMS based amperometric detector for *E. Coli* bacteria using self-assembled monolayers. *Biosensors and Bioelectronics* 16 (9–12): 745–755.

72 Kim, S.H., Lee, Y.J., Lee, B.H. et al. (2006). Characteristics of nanostructured Fe-33 at.%Si alloy powders produced by high-energy ball milling. *Journal of Alloys and Compounds* 424 (1–2): 204–208.

73 Park, T.S., Li, W., McCracken, K.E., and Yoon, J.Y. (2013). Smartphone quantifies salmonella from paper microfluidics. *Lab on a Chip* 13 (24): 4832.

74 Batalla, P., Martín, A., López, M.Á. et al. (2015). Enzyme-based microfluidic chip coupled to graphene electrodes for the detection of D-amino acid enantiomer-biomarkers. *Analytical Chemistry* 87 (10): 5074–5078.

75 Rodriguez-Moncayo, R., Cedillo-Alcantar, D.F., Guevara-Pantoja, P.E. et al. (2021). A high-throughput multiplexed microfluidic device for COVID-19 serology assays. *Lab on a Chip* 21 (1): 93–104.

76 Yin, K., Ding, X., Li, Z. et al. (2021). Autonomous lab-on-paper for multiplexed, CRISPR-based diagnostics of SARS-CoV-2. *Lab on a Chip* 21 (14): 2730–2737.

77 Wen, T., Huang, C., Shi, F.J. et al. (2020). Development of a lateral flow immunoassay strip for rapid detection of IgG antibody against SARS-CoV-2 virus. *Analyst* 145 (15): 5345–5352.

78 Eid, J., Mougel, M., and Socol, M. (2020). Advances in continuous microfluidics-based technologies for the study of HIV infection. *Viruses* 12 (9): 982.

79 Ackerman, C.M., Myhrvold, C., Thakku, S.G. et al. (2020). Massively multiplexed nucleic acid detection with Cas13. *Nature* 582 (7811): 277–282.

80 Zhi, X., Deng, M., Yang, H. et al. (2014). A novel HBV genotypes detecting system combined with microfluidic chip, loop-mediated isothermal amplification and GMR sensors. *Biosensors and Bioelectronics* 54: 372–377.

81 Phillips, E.A., Moehling, T.J., Ejendal, K.F.K. et al. (2019). Microfluidic rapid and autonomous analytical device (microRAAD) to detect HIV from whole blood samples. *Lab on a Chip* 19 (20): 3375–3386.

82 Stokes, D.L., Griffin, G.D., and Vo-Dinh, T. (2001). Detection of *E. coli* using a microfluidics-based antibody biochip detection system. *Fresenius' Journal of Analytical Chemistry* 369 (3–4): 295–301.

83 Kim, G., Moon, J.H., Moh, C.Y. et al. (2015). A microfluidic nano-biosensor for the detection of pathogenic *Salmonella*. *Biosensors and Bioelectronics* 67: 243–247.

84 Jarvius, J., Melin, J., Göransson, J. et al. (2006). Digital quantification using amplified single-molecule detection. *Nature Methods* 3 (9): 725–727.

85 Mbano, I.M., Mandizvo, T., Rogich, J. et al. (2020). Light forge: a microfluidic DNA melting-based tuberculosis test. *The Journal of Applied Laboratory Medicine* 5 (3): 440–453.

86 Nasseri, B., Soleimani, N., Rabiee, N. et al. (2018). Point-of-care microfluidic devices for pathogen detection. *Biosensors and Bioelectronics* 117: 112–128.

87 Liong, M., Hoang, A.N., Chung, J. et al. (2013). Magnetic barcode assay for genetic detection of pathogens. *Nature Communications* 4 (1): 1752.

88 Srivastava, S.K., Van Rijn, C.J.M., and Jongsma, M.A. (2016). Biosensor-based detection of tuberculosis. *RSC Advances* 6 (22): 17759–17771.

Index

Point-of-Care Biosensors for Infectious Diseases, First Edition.
Edited by Sushma Dave and Jayashankar Das.